Multivariate Probability

Multivariate Probability

John H. McColl
Department of Statistics,
University of Glasgow

John Wiley & Sons, Ltd

First published in Great Britain in 2004 by
Arnold, a member of the Hodder Headline Group,
338 Euston Road, London NW1 3BH

John Wiley & Sons Ltd, The Atrium, Southern Gate, Chichester, West
Sussex, PO19 8SQ, United Kingdom

For details of our global editorial offices, for customer services and
for information about how to apply for permission to reuse the copyright
material in this book please see our website at www.wiley.com.

Library of Congress Cataloging-in-Publication Data
A catalog record for this book is available from the Library of Congress

ISBN: 978-0-470-68926-4

1 2 3 4 5 6 7 8 9 10

Typeset in 10/12 Times Ten Roman by Charon Tec Pvt. Ltd, Chennai, India

Contents

Series preface

Arnold Texts in Statistics is a series that is designed to provide an introductory account of key subject areas in statistics. Each book focuses on a particular area, and subjects covered include statistical modelling, time series, applied stochastic modelling, multivariate probability and randomized controlled clinical trials. Texts in this series combine theoretical development with practical examples. Indeed, a distinguishing feature of the texts is that they are copiously illustrated with examples drawn from range of applications. These illustrations take full account of the widespread availability of statistical packages and other software for data analysis. The theoretical content of the texts is sufficient for an appreciation of the techniques being presented, but mathematical detail is only included when necessary. The texts are designed to be accessible to undergraduate and postgraduate students of statistics. In addition, they will enable statisticians and quantitative scientists working in research institutes, industry, government organizations, market research agencies, financial institutions, and so on, to update their knowledge in those areas of statistics that are of direct relevance to them.

David Collett
Series Editor

Preface

This book is intended to provide the reader with the basic tools required to handle collections of random variables. Many of the scientists, social scientists and engineers who could make use of multivariate probability distributions in their work seem reluctant to do so and instead attempt to get by using only univariate results. It is hoped that this book will appeal to these groups of potential users as well as specialists in the mathematical sciences.

Multivariate probability is a large and potentially complex subject, but an attempt has been made to keep this book readily accessible to anyone with a mathematics background up to second year university level. It is definitely not intended to delve into the measure-theoretic properties of probability distributions, whether univariate or multivariate, so attention is largely restricted to absolutely continuous and discrete random variables and their multivariate analogues.

The theory of random variables is built up slowly, from consideration of a single variable to two variables and only then to the general case. Substantial parts of the book deal with functions of random variables, and the link with simulation is exploited in several places. A variety of measures of association are introduced, and the point is made that correlation is not always the best indicator of the strength of a relationship between two random variables. A contrast is drawn between the conditional correlation and the partial correlation (as usually defined in practice) between two variables in the presence of a third (or more), since these are not generally equivalent measures. Sampling and sampling distributions are introduced towards the end of the book as important applications of the earlier theory.

This book does not deal with the theory of stochastic processes, which is the subject of many good, specialist textbooks.

I have based this book on lecture courses I have given in the Department of Statistics, University of Glasgow over several years. I would like to record my thanks to the colleagues and students who have discussed those courses with me. I am particularly indebted to Dr. Jim Kay (Statistics) and Dr. Ian Murphy (Mathematics) for their encouragement to complete this book.

I also wish to thank successive staff at Arnold who have encouraged me to finish this project and who have borne with me very patiently over the time I have been writing it. I am grateful to Mrs. Kathleen Mosson, who typed a substantial part of the manuscript. Finally, I acknowledge the very great contribution of my family, especially Isabel, Stephanie and Richard; this book could never have been written without their support and patience.

John H. McColl
Glasgow
February 2004

1
Fundamentals of probability

This first chapter is intended to give a brief overview of some basic ideas in probability theory. These concepts are usually covered in a first course in probability, so most readers will already know much of this material. This chapter is primarily intended to act as a reminder and to establish consistent terminology, so the pace is fairly brisk. Any reader who is unfamiliar with the topics covered in this chapter, and who finds it particularly difficult as a result, is directed to a first book on probability such as McColl (1995).

1.1 Modelling uncertainty

Probability is the mathematical study of situations in which the outcome is uncertain. There are many important areas of life in which this is usually the case. Here are some examples.

- Financial institutions which manage portfolios of stocks and shares accept that, at least in the short term, some of their investments will decline in value while others will increase in value. They do not know with certainty what will happen to a particular stock when they make a decision to invest in it.

- Electronic and mechanical engineers are concerned with the reliability of the systems they design. The components they use might fail if they are overloaded and will all eventually wear out anyway. Under apparently identical operating conditions, apparently identical components continue to operate for variable lengths of time, so an engineer cannot tell with certainty how long a particular system will function.

- Epidemiologists chart and predict the course of epidemics of diseases such as measles and AIDS. One individual with an infectious disease might infect many other people, while an apparently equivalent individual infects no one else. So the future course of a particular epidemic is very uncertain.

In probability, an *experiment* is any process by which information is obtained. Here are four examples.

Example 1.1

(a) Put an individual suffering from clinical depression on an established drug treatment. Record whether or not this individual experiences side effects within 3 months of beginning to use the drug.

(b) Vaccine for human use should be microbiologically sterile. In order to establish that this is so in a particular case, UK government regulations require 20 ampoules from a batch of 5,000 ampoules of a vaccine to be sampled and tested. The number of tested ampoules that are not sterile is recorded. Even when some of the 5,000 ampoules are not sterile, it is possible that 20 sterile ampoules are sampled for test.

(c) Count and record the number of cars that pass a fixed point on an urban motorway in a one-minute interval during the morning 'rush hour'. This will fluctuate from minute to minute, even on the same day at the same spot.

(d) Ask a subject to draw a line of length 20 cm (without the aid of a measuring instrument). Give the subject training in length estimation and then repeat the exercise. Record the lengths, x and y (say), of the two lines. These lengths will vary from person to person, and even from attempt to attempt by the same person.

A single performance of an experiment is known as a *replicate* (or, sometimes, as a *trial*). The information recorded as the result of one replicate of an experiment is called the *outcome*. All the experiments described above are *random* (or *stochastic*) *experiments*. This means that, even when the experiment is repeated under identical experimental conditions, there is more than one possible outcome and it is not certain beforehand which of these outcomes will occur when the experiment is next carried out. Probability is the branch of mathematics that is concerned with building models of stochastic experiments and with providing mathematical tools to draw logical conclusions from these models.

A *sample space*, S, of a stochastic experiment is a set that contains all the possible outcomes of the experiment.

Example 1.1 – continued

The following sets are possible sample spaces for the experiments described above:

(a) $S_1 = \{$'no side effect', 'side effect'$\}$;

(b) $S_2 = \{0, 1, 2, \ldots, 20\}$;

(c) $S_3 = \{0, 1, 2, \ldots\}$;

(d) $S_4 = \{(x, y) : x > 0, y > 0\}$.

Of these, S_1 and S_2 are called *finite* sample spaces since they contain only a finite number of possible outcomes. S_3 and S_4 are both infinite sample spaces. S_3 is called *countably infinite*, since the outcomes in it can be mapped onto the set of natural numbers. On the other hand, S_4 is said to be *uncountably infinite*, meaning that the outcomes in S_4 are so rich that there is no one-to-one mapping from this set onto the set of natural numbers.

An *event* is a collection of outcomes. Every event can be represented by a subset of the sample space. Any event that consists of a single outcome is called a *simple event*. Other events are called *compound events*.

Example 1.1   continued

(a) In S_1, {'side effect'} is a simple event.

(b) In S_2, the compound event 'at least one of the tested ampoules is sterile' is represented by $\{1, 2, \ldots, 20\}$.

(c) In S_3, the simple event 'no car passes by' is represented by $\{0\}$. The compound event 'at least 20 cars pass by' is represented by $\{20, 21, \ldots\}$.

(d) In S_4, the compound event 'the length of the second line is closer to 20 cm than that of the first line' is represented by the subset $\{(x, y): x > 0, y > 0, |x - 20| > |y - 20|\}$.

S itself represents an event that is certain to occur. The empty set, $\emptyset$, represents an event that is impossible. If S is finite or countably infinite, then all possible subsets of S represent events. When S is an uncountably infinite sample space, typically an interval of the real line or a region of k-dimensional real space, then it is possible to construct subsets of S that we would not wish to treat as events. A general discussion of this issue is outwith the scope of this book and we generally ignore it in what follows. We let $\mathcal{E}$ denote the collection of all events in the sample space S. $\mathcal{E}$ is defined in such a way that complements, unions and intersections of events in $\mathcal{E}$ are also events in $\mathcal{E}$. In the case where S is the real line, for example, all subsets of the following forms are events: $\{a\}$, (a, b), $[a, b]$, $[a, b)$, $(-\infty, b)$, $(-\infty, b]$, (a, ∞). The interested reader is directed to a book on measure theory for further details, e.g. Shiryaev (1995).

Two events are called *disjoint* (or *mutually exclusive*) if they have no outcome in common. This means that they cannot both occur on the same replicate of the underlying experiment. So, E and F are disjoint if and only if $EF = \emptyset$ (where EF denotes the intersection of E and F). In any sample space, any two (different) simple events must be disjoint. A collection of events, $E_1, E_2, \ldots$ is said to be *pairwise disjoint* if every pair of them is disjoint, i.e. $E_i E_j = \emptyset$ (where $i \neq j$).

When $E \subseteq F$, i.e. E is a subset of F, then the set $F \backslash E$ consists of all the outcomes that are in F but not in E, so $F \backslash E$ represents the event that 'F occurs but E does not occur'. In general, this event is represented by FE' and the notation $F \backslash E$ is reserved for the special case where $E \subseteq F$.

Set-theoretic notation can be used to express relationships among events. The main notation is shown in Table 1.1.

The *probability* of an event quantifies the uncertainty of that event happening when the experiment is carried out. Probability is measured in the range $[0, 1]$, with S, the certain event, being assigned probability 1 and $\emptyset$, the impossible event, being assigned probability 0. Events that are almost certain to occur have probabilities near 1, while events that almost never occur have probabilities

Table 1.1 *Set notation and events*

The set ...	represents ...
S (the universal set)	an event that is certain
$\emptyset$ (the empty set)	an impossible event
E' (the complement of E)	the event E does not occur
$E \cup F$ (the union of E and F)	either E occurs or F occurs or both occur
EF (the intersection of E and F)	both E and F occur
The relationship ...	**means that** ...
$E \subseteq F$ (E is a subset of F)	if E occurs, then F is certain to occur also
$EF = \emptyset$ (E and F are disjoint)	if E occurs, then F cannot occur simultaneously (and vice versa)

near 0. For example, in Example 1.1(a), experience suggests that the probability of the event {'side effect'} is very small, perhaps about 0.05. It is usual to denote the probability of the event E by $P(E)$.

Over the centuries, there have been various attempts to formalize the concept of a probability. One that has survived from very early in the history of probability is the idea of *equally likely outcomes* in a finite sample space. In an equally likely outcomes model, there are k different outcomes of the experiment. Each is as likely to occur as any other, so each outcome has probability $1/k$. For example, suppose a standard die is rolled, and the number of dots on the uppermost face of the die recorded. In this experiment, $S = \{1, 2, 3, 4, 5, 6\}$. Each outcome in S might be assumed to be equally likely to occur, in which case the probability $\frac{1}{6}$ would be assigned to each of them. It follows that any event, E, has probability $n(E)/k$, where $n(E)$ is the number of outcomes in the event. For example, if E is 'the score on the die is even', then $E = \{2, 4, 6\}$ and $n(E) = 3$. So E has probability $\frac{3}{6}$ or 0.5.

This definition is all right in so far as it goes, but not every random experiment has equally likely outcomes. For example, if we were to roll two fair dice and take the outcome as the sum of the two scores, then we could get the values 2, 3, ..., 12. These are not equally likely. For example, a total of 2 can only be obtained from the ordered outcomes (1, 1) whereas a total of 7 can be obtained from either (1, 6) or (2, 5) or ... or (6, 1). So, the outcome 7 is much more likely to occur than the outcome 1.

The most popular definition of probability in current use is based on the idea of *relative frequency*. This views the probability of an event as the long-term proportion of times that the event would occur. More formally, suppose that the experiment is carried out, or replicated, n times under identical experimental conditions. If the event E occurs on n_E of those replicates, then the relative frequency of E is the proportion of times it occurs:

$$rf_n(E) = \frac{n_E}{n}.$$

The probability of E is defined to be the long-term, or limiting, value of this proportion:

$$P(E) = \lim_{n \to \infty} rf_n(E).$$

This is sometimes called the *frequentist* definition of probability. It is not guaranteed by the definition of relative frequency that the required limit always exists, nor that the same limit is always reached for every possible sequence of replicates under identical experimental conditions. This is an assumption, known as *statistical regularity*, that has been borne out by widespread experience of replicating stochastic experiments in the past.

Although the frequentist definition is designed to give an objective meaning to probability, it is tied conceptually to experiments which may be repeated endlessly under identical conditions. Under the frequentist scheme, it is not possible to make a statement such as 'The probability that I will pass my forthcoming Probability exam is 0.9', since I cannot sit the same exam more than once under identical conditions (thankfully!).

In recent times, the view has become popular that a probability is better treated as a statement of personal belief. This is sometimes known as *subjective probability*. With this approach, I might well express my view on the forthcoming Probability exam by saying that I have probability 0.9 of passing it. But, if you know me well, you might disagree with my assessment; you might think that my probability of passing is just 0.7. Neither of these probabilities can be 'right' or 'wrong'; since they are our subjective views they do not need to be in agreement with one another. In extending the range of probability to include 'one-off' events, the subjectivist school has also broken the (at least apparent) link between frequentist probability and objective reality.

Whichever of these views of the meaning of a probability they favour, almost all probabilists are willing to use the same mathematical methods for manipulating probabilities within a probability model. The next section will lay the foundations for all such calculations.

Exercises

1 For each of the following experiments, write down an appropriate sample space. State whether this sample space is finite, countably infinite or uncountably infinite.

(a) Count and record the number of votes won by the Labour candidate in a particular Parliamentary constituency at the next general election.

(b) In a secondary school class of 30 pupils, record the number of pupils who have a diagnosis of dyslexia.

(c) Record the pressure in a boiler.

(d) Record the pressure in a boiler at hourly intervals during a 10-hour shift.

(e) A student sits a multiple-choice examination consisting of 20 questions, each with four possible answers (A, B, C, D). Record the student's answers to all the questions.

(f) Measure and record a child's height (m) and weight (kg) on her first day at school.

2 Below are descriptions of some events in the sample spaces of Exercise 1. Write down the subset of S that represents each event.

(a) The candidate gets fewer than 10,000 votes.

(b) No more than one-fifth of the class have a diagnosis of dyslexia.

(c) The pressure is above a certain critical threshold, c (where $c > 0$).

(d) The pressure is never above c.

(e) The student answers D for the first three questions.

(f) The child's body-mass index is greater than 20, where the body-mass index is defined to be weight/(height)^2.

3 The finite sample space, S, consists of just four distinct outcomes, $S = \{s_1, s_2, s_3, s_4\}$. Write down all 16 distinct events in $\mathcal{E}$. Assuming that the outcomes are equally likely, write down the probability of each of these events in turn.

4 Explain why, for a finite sample space consisting of k distinct outcomes, $\mathcal{E}$ contains exactly 2^k possible events.

1.2 The axioms of probability

Probability theory is fundamentally concerned with building a mathematical model to describe the uncertainty in the outcome of a stochastic experiment. In mathematical terms, a probability model is a function, P, which associates with every event, $E \in \mathcal{E}$, a real value $P(E)$ in the range $0 \le P(E) \le 1$. We call this value the *probability* of E.

A crucial question is how to determine $P(E)$. The relative frequency definition of probability might suggest taking an empirical approach, replicating the experiment many times under identical conditions. Although this is a sensible strategy for estimating $P(E)$, and underpins the whole science of statistics, it can not yield the exact value of a probability. For example, if we determine $rf(E)$ after 10,000 replicates, $rf(E)$ will change on the 10,001st replicate. For this reason, probability models are often built up from theoretical considerations.

In order to give consistent results for all calculations, a probability model needs only to obey the four *axioms of probability*, usually attributed to Kolmogorov:

1. For any event $E \in \mathcal{E}, 0 \le P(E) \le 1$.

2. $P(S) = 1$.

3. If $E, F \in \mathcal{E}$ are disjoint events (i.e. $EF = \emptyset$), then $P(E \cup F) = P(E) + P(F)$.

4. If $E_1, E_2, \ldots \in \mathcal{E}$ are pairwise disjoint events, then $P(E_1 \cup E_2 \cup \cdots) = P(E_1) + P(E_2) + \cdots$.

Axiom 3 can be extended by mathematical induction only to cover finite sequences of events, so Axiom 4 is required to deal with infinite sequences of events that might be required in an infinite sample space. Axiom 3 may be recovered from the more general Axiom 4 by writing $E_1 = E, E_2 = F, E_3 = E_4 = \ldots = \emptyset$ (since $P(\emptyset) = 0$, as we shall shortly show). These two axioms may not be extended consistently to uncountable sequences of events. This is related to the reason why, in an uncountably infinite sample space, $\mathcal{E}$ does not contain all possible subsets of S. (For further details, the reader is referred to a book on measure theory.)

It can easily be shown that relative frequencies have properties equivalent to these axioms and that probabilities in an equally likely outcomes model obey the axioms (see Exercises 1 and 2 below).

Within a general probability model, we want to be able to calculate unknown probabilities from ones we do know. We could perform such calculations directly from the Axioms each time, but it is much easier to make use of standard results, some of which will be obtained now.

Proposition 1.1

For any event $E \in \mathcal{E}$,

$$P(E') = 1 - P(E).$$

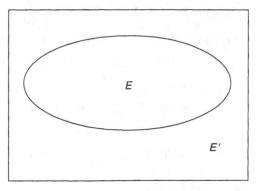

Proof.

$$E \cup E' = S.$$

Using Axiom 2,

$$P(E \cup E') = P(S) = 1.$$

E and E' are disjoint, so by Axiom 3,

$$P(E \cup E') = P(E) + P(E')$$
$$\therefore \quad P(E) + P(E') = 1$$
$$\text{i.e.} \quad P(E') = 1 - P(E)$$

Since $\emptyset = S'$, it follows from Axiom 2 and Proposition 1.1 that $P(\emptyset) = 0$. So, an event that is certain has probability 1 and an impossible event has probability 0, which is intuitively what we would expect, but the converses of these statements are not true. In other words, $P(E) = 1$ does not necessarily imply that $E = S$ and $P(E) = 0$ does not necessarily imply that $E = \emptyset$. Sometimes, we assign probability 1 to an event that is not certain to occur and probability 0 to an event that can occur, as we shall now show. ∎

Example 1.2

Consider the experiment of taking a random real number, X, between 0 and 1. We shall interpret this to mean that the probability that X lies in any interval $0 < a < X < b < 1$ is equal to the width of the interval, which is $b - a$. We will find the probability that there is a 7 in the (shortest possible) decimal expansion of X.

Let $F = $'there is a 7 in the decimal expansion of X' and $E_i = $'there is a 7 in the ith place of the decimal expansion of X, but not in the earlier places' $(i = 1, 2, \ldots)$. Then

$$P(E_1) = P(0.7 \leq X \leq 0.799\ldots) = \frac{1}{10}$$
$$P(E_2) = P(0.07 \leq X \leq 0.0799\ldots)$$
$$+ P(0.17 \leq X \leq 0.1799\ldots)$$
$$+ \cdots$$
$$+ P(0.67 \leq X \leq 0.6799\ldots)$$
$$+ P(0.87 \leq X \leq 0.8799\ldots)$$
$$+ P(0.97 \leq X \leq 0.9799\ldots)$$
$$= 9 \times \frac{1}{100}$$

In general,

$$P(E_i) = \frac{9^{i-1}}{10^i} \qquad (i = 1, 2, \dots).$$

But $E_1, E_2, \dots$ are pairwise disjoint events, and $F = \bigcup_i E_i$. So, by Axiom 4,

$$P(F) = \sum_{i=1}^{\infty} P(E_i) = \sum_{i=1}^{\infty} \frac{9^{i-1}}{10^i} = \frac{1}{10} \sum_{i=1}^{\infty} \left(\frac{9}{10}\right)^{i-1} = \frac{1}{10} \cdot \frac{1}{1 - 9/10} = 1.$$

By Proposition 1.1, then, $P(F') = 0$. Clearly, though, F' can occur, e.g. $0.65 \in F'$. The reason for this apparently paradoxical result is that the points in F' are discretely scattered along the real line, in the sense that any two points in F' are separated by an interval which does not lie in F'. (In measure theory, the set F' is said to have measure 0.)

Proposition 1.2

For any events $E, F \in \mathcal{E}$,

$$P(E \cup F) = P(E) + P(F) - P(EF).$$

Proof. $E \cup F$ is the union of the three subsets EF, EF' and $E'F$, which are pairwise disjoint (see diagram below). So, by Axiom 4,

$$P(E \cup F) = P(EF) + P(EF') + P(E'F).$$

Similarly,

$$P(E) = P(EF) + P(EF'),$$
$$P(F) = P(EF) + P(E'F).$$

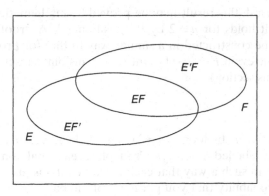

Therefore,

$$P(E) + P(F) = [P(EF) + P(EF')] + [P(EF) + P(E'F)]$$
$$= [P(EF) + P(EF') + P(E'F)] + P(EF)$$
$$= P(E \cup F) + P(EF),$$

i.e.

$$P(E \cup F) = P(E) + P(F) - P(EF). \qquad \blacksquare$$

Proposition 1.2 extends Axiom 3, which holds only for disjoint events. Notice that if E and F are disjoint, then we can recover the axiom as follows:

$$P(E \cup F) = P(E) + P(F) - P(EF) = P(E) + P(F) - P(\emptyset) = P(E) + P(F),$$

since $P(\emptyset) = 0$.

Proposition 1.3

For any events $E_1, E_2, \ldots, E_n \in \mathcal{E}$ $(n \geq 2)$,

$$P(E_1 \cup E_2 \cup \cdots \cup E_n)$$

$$= \sum_{i=1}^{n} P(E_i) - \sum_{i=1}^{n-1} \sum_{j=i+1}^{n} P(E_i E_j) + \sum_{i=1}^{n-2} \sum_{j=i+1}^{n-1} \sum_{k=j+1}^{n} P(E_i E_j E_k) - \cdots$$

$$+ (-1)^{n-1} P(E_1 E_2 \ldots E_n).$$

In particular, when $n = 3$,

$$P(E_1 \cup E_2 \cup E_3) = P(E_1) + P(E_2) + P(E_3) - P(E_1 E_2) - P(E_1 E_3)$$
$$- P(E_2 E_3) + P(E_1 E_2 E_3).$$

Proof. In general, this result may be proved by mathematical induction on n, noting that it holds for $n = 2$ by Proposition 1.2. A proof for the special case $n = 3$ can be constructed in a similar way to that for Proposition 1.2, by decomposing the event $E_1 \cup E_2 \cup E_3$ into seven disjoint subsets (see Exercise 7 at the end of this section). $\qquad \blacksquare$

Example 1.3

Suppose you have n balloons $(n \geq 2)$, labelled $1, 2, \ldots, n$ respectively, and n baskets, also labelled $1, 2, \ldots, n$. You place each balloon in a randomly selected basket in such a way that each basket contains just one balloon. We will find the probability that you put none of the balloons in the basket with the corresponding number.

Let $F =$ 'none of the balloons is correctly matched with the corresponding basket', and $E_i =$ 'balloon i is correctly matched with basket i' $(i = 1, 2, \ldots, n)$. Then

$$F' = E_1 \cup E_2 \cup \cdots \cup E_n.$$

By Proposition 1.1,

$$P(F) = 1 - P(E_1 \cup E_2 \cup \cdots \cup E_n).$$

We can use Proposition 1.3 to obtain $P(E_1 \cup E_2 \cup \cdots \cup E_n)$.

There are n available baskets for balloon 1; $n - 1$ available baskets for balloon 2, once balloon 1 has been placed; $n - 2$ available baskets for balloon 3, once balloons 1 and 2 have been placed; .... This means that the total number of ways of placing the balloons in the baskets (in other words, the total number of distinct permutations of the balloons among the baskets) is:

$$n \times (n - 1) \times (n - 2) \times \cdots \times 1 = n!$$

Assuming that the process of placing balloons is carried out randomly, all of these arrangements are equally likely, so each possible permutation has probability $1/n!$.

Suppose now that balloon i is correctly placed in basket i. Then the other $n - 1$ balloons may be placed in any of the remaining $n - 1$ baskets. This may be done in $(n - 1)!$ different ways. So,

$$P(E_i) = \frac{(n - 1)!}{n!} = \frac{1}{n}, \qquad i = 1, 2, \ldots, n.$$

In a similar way, suppose that balloons i and j are correctly placed in baskets i and j respectively $(i \neq j)$. Then the other $n - 2$ balloons may be placed in any of the remaining $n - 2$ baskets. This may be done in $(n - 2)!$ different ways. So,

$$P(E_i \cap E_j) = \frac{(n - 2)!}{n!}, \qquad i \neq j.$$

In general, the intersection of any m of the n events has probability

$$\frac{(n - m)!}{n!}, \qquad m = 1, 2, \ldots, n.$$

Using Proposition 1.3, then,

$$P(E_1 \cup E_2 \cup \cdots \cup E_n) = \sum_{i=1}^{n} \frac{(n - 1)!}{n!} - \sum_{i=1}^{n-1} \sum_{j=i+1}^{n} \frac{(n - 2)!}{n!}$$

$$+ \sum_{i=1}^{n-2} \sum_{j=i+1}^{n-1} \sum_{k=j+1}^{n} \frac{(n - 3)!}{n!} - \cdots + (-1)^{n-1} \frac{1}{n!}.$$

Now, each sum in this series is taken over all possible combinations of m different events from the list $E_1, E_2, \ldots, E_n$. There are $\binom{n}{m}$, or ^nC_m, such combinations, and so:

$$P(E_1 \cup E_2 \cup \cdots \cup E_n)$$

$$= n\frac{(n-1)!}{n!} - \binom{n}{2}\frac{(n-2)!}{n!} + \binom{n}{3}\frac{(n-3)!}{n!} - \cdots + (-1)^{n-1}\frac{1}{n!}$$

$$= 1 - \frac{1}{2!} + \frac{1}{3!} - \cdots + (-1)^{n-1}\frac{1}{n!}.$$

So,

$$P(F) = \frac{1}{2!} - \frac{1}{3!} + \cdots + (-1)^n\frac{1}{n!}.$$

This is a surprising result. We might expect $P(F)$ to decline monotonically as n increases, i.e. as we add more and more balloons, but this is not so. $P(F)$ increases at the even-numbered terms, then decreases again at the odd-numbered terms (suggesting that there are slightly more problems matching even numbers of balloons and baskets). In the limit, as $n \to \infty$,

$$P(F) \to \sum_{m=2}^{\infty} \frac{(-1)^m}{m!} = \sum_{m=0}^{\infty} \frac{(-1)^m}{m!} = e^{-1} = 0.3679.$$

Due to the rapid convergence of this series, $P(F)$ is within 1% of its limiting value whenever $n \geq 5$, and within 0.01% of its limiting value whenever $n \geq 7$. So, remarkably, the chance of matching at least one balloon with the correct basket hardly depends at all on the number of balloons and baskets for $n \geq 7$.

The events $E_1, E_2, \ldots \in \mathcal{E}$ are said to form an *increasing sequence* if

$$E_1 \subseteq E_2 \subseteq E_3 \subseteq \cdots$$

and a *decreasing sequence* if

$$E_1 \supseteq E_2 \supseteq E_3 \supseteq \cdots.$$

For an increasing sequence of events, we define

$$\lim_{i \to \infty} E_i = E_1 \cup E_2 \cup E_3 \cup \cdots$$

For a decreasing sequence of events, we define

$$\lim_{i \to \infty} E_i = E_1 E_2 E_3 \cdots.$$

Proposition 1.4

If the events $E_1, E_2, \ldots \in \mathcal{E}$ form an increasing or decreasing sequence, then

$$\lim_{i \to \infty} P(E_i) = P(\lim_{i \to \infty} E_i).$$

Proof. We shall prove the result for an increasing sequence of events; the case of a decreasing sequence is left as an exercise.

Let $F_1 = E_1$. For $i = 2, 3, \ldots$, define the event $F_i = E_i \backslash E_{i-1}$. Since the events $F_1, F_2, \ldots$ are disjoint,

$$P(F_1 \cup F_2 \cup \cdots) = P(F_1) + P(F_2) + \cdots.$$

Now

$$E_1 \cup E_2 \cup \cdots = F_1 \cup F_2 \cup \cdots,$$

so

$$P\left(\lim_{i \to \infty} E_i\right) = P(F_1) + P(F_2) + \cdots.$$

But

$$E_i = F_1 \cup F_2 \cup \cdots \cup F_i$$

so

$$P(E_i) = P(F_1) + P(F_2) + \cdots + P(F_i)$$

and

$$\lim_{i \to \infty} P(E_i) = P\left(\lim_{i \to \infty} E_i\right). \qquad \blacksquare$$

Exercises

1 Show that relative frequencies must satisfy Axioms 1 to 4.

2 Show that an equally likely outcomes model satisfies Axioms 1 to 3. Why is Axiom 4 irrelevant for a finite sample space like this?

3 Consider the finite sample space, $S = \{s_1, s_2, \ldots, s_n\}$. For some m $(1 \le m \le n)$, suppose that the event E consist of any m distinct outcomes in S, $E = \{s_{i_1}, s_{i_2}, \ldots, s_{i_m}\}$. Prove that E has probability $\sum_{k=1}^{m} P(s_{i_k})$.

4 Suppose that E and F are events in a sample space S, such that $E \subseteq F$. Prove that $P(E) \le P(F)$.

5 If E is an event with probability $P(E)$, then the *odds* on E are defined by $P(E)/P(E')$ or $P(E)/1 - P(E)$. Find the odds on E when (a) $P(E) = 0.2$, (b) $P(E) = 0.5$, (c) $P(E) = 0.99$.

6 Suppose that E and F are events in the sample space S, such that $P(E)$, $P(F)$ and $P(EF)$ are known. Write down an expression for the probability that E occurs but F does not occur. Hence show that the probability that exactly one of the events E and F occurs is

$$P(E) + P(F) - 2 \times P(EF).$$

Find the probability that exactly k of the events E and F occur, for $k = 0, 1, 2$.

7 Suppose that E and F are events in the sample space S. Prove that

$$P(EF) \leq P(E) \leq P(E \cup F) \leq P(E) + P(F).$$

8 Find an expression for $P(E_1 \cup E_2 \cup E_3)$ by decomposing the event $E_1 \cup E_2 \cup E_3$ as a union of seven disjoint events and applying Axiom 4.

9 Prove Proposition 1.3 for general n, using mathematical induction.

10 Prove that, for any events $E_1, E_2, \ldots, E_n \in \mathcal{E}$,

$$P(E_1 \cup E_2 \cup \cdots \cup E_n) = P(E_1) + P(E_1' E_2) + P(E_1' E_2' E_3)$$
$$+ \cdots + P(E_1' \ldots E_{n-1}' E_n).$$

11 Prove Boole's inequality, which states that, for any events $E_1, E_2, \ldots, E_n$

$$P\left(\bigcup_{i=1}^{n} E_i\right) \leq \sum_{i=1}^{n} P(E_i).$$

12 Prove Proposition 1.4 for a decreasing sequence of events.

1.3 Conditional probability

Suppose that E and F are events in the sample space S and that $P(E) > 0$. Then the *conditional probability of F given E* is defined to be

$$P(F|E) = \frac{P(FE)}{P(E)}.$$

We cannot 'prove' a definition, but a relative frequency argument can help us to see why this definition of conditional probability produces reasonable answers. Suppose that the events E, F and FE occur on (respectively) n_E, n_F and n_{FE} out of the first n replicates of an experiment. So, $rf_n(F) = n_F/n$, for example. Now F must occur on exactly n_{FE} of the n_E replicates on which E occurred. In other words, the conditional relative frequency of F given that E occurs is n_{FE}/n_E. In the limit,

$$\lim_{n \to \infty} \frac{n_{FE}}{n_E} = \lim_{n \to \infty} \frac{n_{FE}/n}{n_E/n} = \frac{\lim\limits_{n \to \infty} n_{FE}/n}{\lim\limits_{n \to \infty} n_E/n}.$$

The limit on the left-hand side is what we might reasonably consider to be the conditional probability $P(F|E)$, while the limits on the right-hand side have already been defined as the unconditional probabilities $P(FE)$ and $P(E)$.

For any given event $E \in \mathcal{E}$ such that $P(E) > 0$, the conditional probabilities $P(\cdot|E)$ satisfy the axioms of probability as before, and the rules of probability that are derived from them. For example, it is obvious that

$$P(E|E) = 1.$$

It is easily proved that, for any events $F, G \in \mathcal{E}$,

$$0 \le P(F|E) \le 1,$$
$$P(F'|E) = 1 - P(F|E),$$
$$P(F \cup G|E) = P(F|E) + P(G|E) - P(FG|E).$$

From the definition of conditional probability, it follows that, for any events E and F such that $P(E) > 0$,

$$P(FE) = P(F|E)P(E).$$

This general result is sometimes known as the *multiplication theorem* of probability. We can use a tree diagram to represent it. The one below shows the event E occurring logically before F. The branches of the diagram are annotated with the *unconditional* probability of the first event, E, and the *conditional* probabilities of later events given E. The probability of E and F both occurring, $P(FE)$, is found by multiplying the probabilities on the branch that leads to F, following a path from left to right across the diagram.

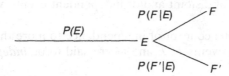

The multiplication theorem is useful for calculating probabilities when an experiment can be split logically into two sub-experiments that occur one after the other.

The multiplication theorem can be extended to n events, as follows. Suppose that $E_1, E_2, \ldots, E_{n-1}$ and F are events such that $P(E_{n-1} \ldots E_2 E_1) > 0$. Then

$$P(FE_{n-1} \ldots E_1) = P(F|E_{n-1} \ldots E_1)P(E_{n-1}|E_{n-2} \ldots E_1) \ldots P(E_2|E_1)P(E_1).$$

This result is proved in one of the exercises at the end of this section.

One of the most critical concepts in probability is that of the independence of two events. Informally, we say that E and F are independent events if the occurrence or non-occurrence of E does not change the probability of F (and vice versa). Formally, the events E and F are defined to be *independent* if

$$P(FE) = P(F)P(E).$$

Whenever $P(E) > 0$, so that $P(F|E)$ is sensibly defined,

$$P(FE) = P(E)P(F)$$

$$\Leftrightarrow \frac{P(FE)}{P(E)} = P(F)$$

$$\Leftrightarrow P(F|E) = P(F).$$

In other words, when $P(E) > 0$, then E and F are independent if and only if the conditional probability of F given E is the same as the unconditional probability of F. Similarly, if $P(F) > 0$, then E and F are independent if and only if the conditional probability of E given F is the same as the unconditional probability of E. Intuitively, these are precisely the relationships we want the formal definition of independence to express.

In practice, it is very rare to discover that two events are independent by working out separately the probabilities of E, F and EF and showing that they are related by the above formula. Almost invariably, as in the following examples, independence is a model assumption justified by our knowledge of experimental conditions. When we can be sure that events are independent, even apparently complex probabilities become relatively easy to calculate.

If two disjoint events, E and F, both have non-zero probability, then they cannot be independent. This is intuitively obvious, since knowing that E has occurred tells us that F cannot also occur. Formally, $P(FE) = 0$ whereas $P(E) > 0$ and $P(F) > 0$ and so $P(FE) \neq P(E)P(F)$. Table 1.2 summarizes our results about pairs of disjoint and/or independent events, when $P(E) > 0$ and $P(F) > 0$.

We can extend the concept of independence to more than two events. For example, the three events E, F and G are said to be *independent* if they are

Table 1.2 *Some probability results when P(E) > 0 and P(F) > 0*

| | **E, F not disjoint** | |
E, F disjoint	**E, F independent**	**E, F not independent**
$P(EF) = 0$	$P(EF) = P(E)P(F)$	
$P(E \cup F) = P(E) + P(F)$	$P(E \cup F) = P(E) + P(F) - P(E)P(F)$	$P(E \cup F) = P(E) + P(F) - P(EF)$
$P(F\|E) = 0$	$P(F\|E) = P(F)$	$P(F\|E) = \dfrac{P(FE)}{P(E)}$

pairwise independent (i.e. E and F are independent, E and G are independent, F and G are independent) *and* $P(EFG) = P(E)P(F)P(G)$. Notice that both these conditions are necessary, since neither necessarily implies the other (see the exercises at the end of this section).

In general, for $n \geq 2$, the events $E_1, E_2, \ldots, E_n$ are said to be *independent* if, for any integer $k, 2 \leq k \leq n$, and any k of the events $E_{i_1}, E_{i_2}, \ldots, E_{i_k}$,

$$P(E_{i_1} E_{i_2} \ldots E_{i_k}) = P(E_{i_1})P(E_{i_2}) \ldots P(E_{i_k}).$$

Example 1.4

Suppose you have a bag that contains n beads, which are all identical apart from colour. r of the beads are red, the remaining $b = n - r$ are black. This collection of beads is your population, from which you intend to obtain a small random sample of size m $(1 \leq m \leq \min(r, b))$. (This problem is identical conceptually to many real-life problems, such as opinion polling where the beads represent a population of voters with the red beads holding one opinion on a political question and the black beads holding the opposing view.) We shall find the probability of obtaining a sample that contains exactly x red beads, where $0 \leq x \leq m$.

First of all, consider *random sampling with replacement*. This means that, in order to obtain a bead for the sample, we mix the beads thoroughly in the bag, pick one out without looking at the beads, note its colour and return it to the bag. We do this m times, to obtain a sample of m beads (not necessarily all different). Each time, the probability of obtaining a red bead must be r/n, since there are always n beads in the bag, of which r are red, and each bead is equally likely to be picked out.

If E_i is the event 'the ith bead picked out is red', then the events $E_1, E_2, \ldots,$ E_m are independent. This is intuitively obvious, since the outcome of the ith draw from the bag is unaffected by what happened at previous draws.

In order to obtain exactly x red beads (out of m), we need a sequence of outcomes like

$$\underbrace{RR\ldots R}_{x} \quad \underbrace{BB\ldots B}_{m-x}$$

Since the E_i are independent, this sequence of outcomes has probability

$$P(E_1 \ldots E_x E'_{x+1} \ldots E'_n) = P(E_1) \ldots P(E_x)P(E'_{x+1}) \ldots P(E'_m)$$

$$= \frac{r}{n} \cdots \frac{r}{n} \left(1 - \frac{r}{n}\right) \cdots \left(1 - \frac{r}{n}\right)$$

$$= \left(\frac{r}{n}\right)^x \left(1 - \frac{r}{n}\right)^{m-x}.$$

Any sequence of outcomes consisting of exactly x red beads and exactly $m - x$ black beads has this same probability. There are $\binom{m}{x}$ different combinations of

m outcomes of which exactly x are red. So, the probability of obtaining exactly x red beads, when sampling randomly with replacement, is

$$\binom{m}{x}\left(\frac{r}{n}\right)^{x}\left(1-\frac{r}{n}\right)^{m-x}, \qquad x = 0,1,\ldots,m.$$

Now consider *random sampling without replacement*. This means that a bead may appear at most once in the sample. So, whenever a bead is drawn from the bag, its colour is noted but the bead is not replaced in the bag. This means that the events $E_1, E_2, \ldots, E_m$ are not independent. For example,

$$P(E_2|E_1) = \frac{r-1}{n-1},$$

while

$$P(E_2|E_1') = \frac{r}{n-1},$$

since after picking the first bead there are only $n-1$ beads left in the bag, of which $r-1$ are red if E_1 occurred (i.e. the first bead drawn out of the bag was red) but of which r are red if E_1 did not occur (i.e. the first bead drawn out of the bag was black).

Again consider a particular sequence that yields a sample containing exactly x red beads:

$$\underbrace{RR\ldots R}_{x} \quad \underbrace{BB\ldots B}_{m-x}$$

Using the multiplication theorem, the probability of this sequence of events is

$$P(E_1)P(E_2|E_1)\cdots P(E_x|E_{x-1}\ldots E_1)P(E_{x+1}'|E_x\ldots E_1)\cdots P(E_m'|E_{m-1}'\ldots E_1)$$

$$= \frac{r}{n}\frac{r-1}{n-1}\cdots\frac{r-(x-1)}{n-(x-1)}\frac{b}{n-x}\frac{b-1}{n-(x+1)}\cdots\frac{b-(m-x-1)}{n-(m-1)}$$

$$= \frac{\{r(r-1)\cdots(r-(x-1))\}\{b(b-1)\cdots(b-((m-x)-1))\}}{\{n(n-1)\cdots(n-(m-1))\}}$$

$$= \frac{\dfrac{r!}{(r-x)!}\dfrac{b!}{(b-(m-x))!}}{\dfrac{n!}{(n-m)!}}.$$

It can be shown that all $\binom{m}{x}$ sequences of outcomes that yield exactly r red beads in the sample have the same probability. (The terms in r and b, in the numerator of the above expression, are permuted for each possible sequence.)

So, the probability of obtaining exactly x red beads ($x = 0, 1, \ldots, m$), when sampling randomly without replacement, is

$$\binom{m}{x} \frac{\dfrac{r!}{(r-x)!} \dfrac{b!}{(b-(m-x))!}}{\dfrac{n!}{(n-m)!}} = \frac{m!}{x!(m-x)!} \frac{\dfrac{r!}{(r-x)!} \dfrac{b!}{(b-(m-x))!}}{\dfrac{n!}{(n-m)!}}$$

$$= \frac{\dfrac{r!}{x!(r-x)!} \dfrac{b!}{(m-x)!(b-(m-x))!}}{\dfrac{n!}{m!(n-m)!}}$$

$$= \frac{\binom{r}{x}\binom{b}{m-x}}{\binom{n}{m}}.$$

This result may be obtained more easily by noting that all $\binom{n}{m}$ possible samples (ignoring ordering) are equally likely. There are $\binom{r}{x}$ (unordered) different ways of choosing x red beads, and $\binom{b}{m-x}$ (unordered) different ways of choosing $m - x$ black beads. Hence there are $\binom{r}{x}\binom{b}{m-x}$ different samples containing exactly x red beads (and, hence, exactly $m - x$ black beads), and the result follows.

Probabilities of the above form, obtained respectively by random sampling with and without replacement, are called binomial and hypergeometric distributions, as we shall see in Chapter 2.

We have already seen that, for any event E in a sample space S,

$$EE' = \emptyset,$$
$$E \cup E' = S.$$

These results arise because E and E' are disjoint, and every outcome in S appears in either E or E'. In general, a collection of events $E_1, E_2, \ldots$ is said to form a *partition* of the sample space S if every outcome appears in one and only one of the events.

More formally, $E_1, E_2, \ldots$ is a partition of S if (see diagram overleaf)

1. $\bigcup_i E_i = S$,

2. $E_i E_j = \emptyset$ when $i \neq j$,

3. $P(E_i) > 0$ for each i.

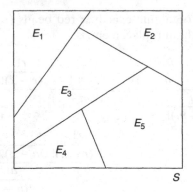

In particular, then, E and E' partition S whenever $E \in \mathcal{E}$ and $0 < P(E) < 1$.

Proposition 1.5 – The law of total probability

Suppose that the events $E_1, E_2, \ldots$ partition the sample space S. Let F be any event in S. Then

$$P(F) = \sum_i P(F|E_i) \times P(E_i).$$

Proof. It follows from the distributive laws of set theory that

$$F = FS = F \left\{ \bigcup_i E_i \right\} = \bigcup_i \{FE_i\}$$

Since the E_i are disjoint, the FE_i are disjoint too. So,

$$P(F) = \sum_i P(FE_i) \quad \text{[by Axiom 4]}$$

$$= \sum_i P(F|E_i) \times P(E_i) \quad \text{[by definition of conditional probability].} \quad \blacksquare$$

Drawing a *tree diagram* helps us keep track of the probabilities in situations like this one, as in the example below.

Example 1.5

According to Matthews (1996), the short-term forecasting record of the UK Meteorological Office is impressively accurate. For example, on 83% of rainy days, rain has been forecast, while on 83% of dry days, no rain has been forecast. Assuming that it rains on 40% of days, let us find the probability that rain is forecast.

Define the events $R =$ 'it rains' and $F =$ 'rain is forecast'. We are told that

$$P(R) = 0.4, \qquad \therefore P(R') = 1 - P(R) = 0.6,$$

where, clearly, R and R' partition the sample space. Also,

$$P(F|R) = 0.83, \qquad \therefore P(F'|R) = 1 - P(F|R) = 0.17$$

$$P(F'|R') = 0.83, \qquad \therefore P(F|R') = 1 - P(F'|R') = 0.17.$$

A tree diagram looks like this:

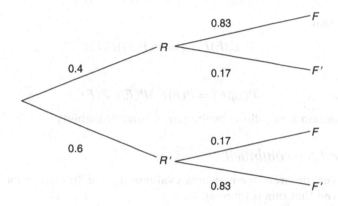

In an extension to what we did when considering the multiplication theorem, Proposition 1.5 tells us that we can find $P(F)$ by multiplying the probabilities on every branch of this tree that leads to the event F, and then adding them together. So,

$$P(F) = P(F|R) \times P(R) + P(F|R') \times P(R')$$

$$= (0.83 \times 0.4) + (0.17 \times 0.6)$$

$$= 0.434$$

i.e. rain is forecast on 43.4% days.

In the above example, we might be more interested in a different conditional probability, the conditional probability that it rains *given* that rain is forecast, $P(R|F)$. To evaluate such a probability, we require another important result, known as Bayes' theorem.

Proposition 1.6 – Bayes' theorem

Suppose that the events $E_1, E_2, \ldots$ partition the sample space S. Let F be any event in S, with $P(F) > 0$. Then

$$P(E_j|F) = \frac{P(F|E_j)P(E_j)}{\sum_i P(F|E_i)P(E_i)} \qquad j = 1, 2, \ldots.$$

Proof. By the multiplication theorem,

$$P(FE_j) = P(F|E_j)P(E_j)$$

and

$$P(FE_j) = P(E_jF) = P(E_j|F)P(F)$$

It follows that

$$P(E_j|F)P(F) = P(F|E_j)P(E_j)$$

i.e.

$$P(E_j|F) = P(F|E_j)P(E_j)/P(F)$$

Bayes' theorem now follows by the law of total probability. ∎

Example 1.5 – continued

Using Bayes' theorem, we may now evaluate the conditional probability that it rains given that rain is forecast:

$$P(R|F) = \frac{P(F|R)P(R)}{P(F|R)P(R) + P(F|R')P(R')} = \frac{0.83 \times 0.4}{(0.83 \times 0.4) + (0.17 \times 0.6)} = 0.765.$$

So, rain occurs about three-quarters of the time it is forecast. Not hearing a forecast, one might assume that there is probability 40% of rain. Hearing that rain has been forecast increases this probability to 76.5%. The latter figure is called the *posterior probability* of rain, and is contrasted with the original or *prior probability*.

The posterior probability depends not just on the conditional probabilities $P(F|R)$ and $P(F|R')$, but also on the *incidence* or *base rate*, $P(R)$. In order to see this, consider a different time of year when $P(R) = 0.2$. Then

$$P(R|F) = \frac{0.83 \times 0.2}{(0.83 \times 0.2) + (0.17 \times 0.8)} = 0.550.$$

At this time of year, then, it rains on just over half the days on which rain is forecast. This is still a lot higher than the prior probability of rain, which is just 0.2, but a good bit lower than we might expect from the accuracy rates of 0.83. The effect of the base rate is discussed again, in a different context, in the next section.

Exercises

1 Suppose that A and B are two events in the sample space S. Assuming that A and B are independent, show that (a) A and B' are independent; (b) A' and B are independent; (c) A' and B' are independent. Can you explain why this makes sense intuitively?

2 Under what conditions can the events E and F be both disjoint and independent?

3 (a) Show that S is independent of any event $A \subseteq S$ (including S itself).

(b) Show that $\emptyset$ is independent of any event $A \subseteq S$ (including $\emptyset$ itself).

(c) Show that, if the event $A \subseteq S$ is independent of itself, then $P(A) = 0$ or 1.

4 Suppose that E and F are two events in a sample space S, such that $P(F) > 0$.

(a) Show formally that, if $E \subseteq F$, then $P(E \mid F) = P(E)/P(F)$.

(b) Show formally that, if $F \subseteq E$, then $P(E \mid F) = 1$.

5 Suppose that E and F are two events in a sample space S, such that $0 < P(E) < 1$. Show that:

(a) $P(F \mid E) + P(F' \mid E) = 1$;

(b) $P(F \mid E) + P(F \mid E')$ need not equal 1;

(c) $P(F \mid E) + P(F' \mid E')$ need not equal 1.

6 Suppose that A and B are two events in the sample space S, such that $0 < P(A) < 1$ and $0 < P(B) < 1$. Assuming that $P(A \mid B) > P(A)$, show that:

(a) $P(B \mid A) > P(B)$;

(b) $P(A' \mid B) < P(A')$;

(c) $P(A \mid B') < P(A)$.

Can you explain why this makes sense intuitively?

7 Suppose that E, F and G are independent events in a sample space S such that $P(FG) > 0$. Show that

$$P(E) = P(E \mid F) = P(E \mid G) = P(E \mid FG).$$

8 (This example is attributed to Bernstein.) Toss two 'fair' coins. Let A be the event 'the first coin lands heads up', B the event 'the second coin lands heads up' and C the event 'exactly one of the coins lands heads up'. Show that A, B and C are pairwise independent but not independent.

9 Suppose that E, F and G are independent events in a sample space S. Show that

(a) E, F and G' are independent;

(b) E, F' and G' are independent;

(c) E', F' and G' are independent.

10 Suppose that E, F and G are independent events in a sample space S. Find, in terms of $P(E)$, $P(F)$ and $P(G)$, the probability that exactly k of these events occur ($k = 0, 1, 2, 3$).

11 Prove that, if $E_1, E_2, \ldots, E_{n-1}$ and F are events such that $P(E_{n-1} \ldots E_2 E_1) > 0$, then

$$P(FE_{n-1} \ldots E_1) = P(F|E_{n-1} \ldots E_1) \times P(E_{n-1}|E_{n-2} \ldots E_1) \times P(E_2|E_1) \times P(E_1).$$

12 Suppose that $E_1, E_2, \ldots, E_n$ are independent events in a sample space S. Show that

$$P(E_1 \cup E_2 \cup \cdots \cup E_n) = 1 - P(E_1')P(E_2') \ldots P(E_n').$$

13 Suppose that $E_1, E_2, \ldots, E_n$ are independent events in a sample space S, with respective probabilities $1 - \theta_i$. Show that the probability that none of them occurs is $\theta_1 \theta_2 \ldots \theta_n$.

14 In a certain electrical circuit, two identical components are placed *in series*. As shown in the diagram below, this means that current can only flow from point X to point Y as long as both components are working.

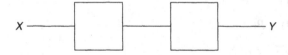

The probability that, in a normal operating period, a component of this type fails is θ. Assuming that the two components fail or not independently, find the probability that current fails to flow from point X to point Y in a normal operating period.

15 In a certain electrical circuit, two identical components are placed *in parallel*. As shown in the diagram below, this means that current can flow from point X to point Y as long as at least one of the two components is working.

The probability that, in a normal operating period, a component of this type fails is θ. Assuming that the two components fail or not independently, find the probability that current fails to flow from point X to point Y in a normal operating period.

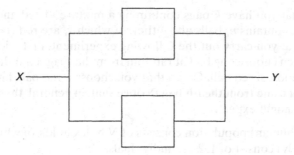

16 Try to extend Exercises 14 and 15 by obtaining expressions for the probability of failure of a circuit consisting of n (>2) identical components, when the components are placed (a) in series, (b) in parallel. Prove that the design with n components in parallel is less likely to fail than the alternative design with n components in series as long as $0 < \theta < 1$.

17 The diagram below shows part of an electric circuit. The points X and Y are joined by an assembly containing five switches.

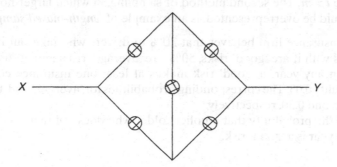

Current will pass from X to Y as long as the appropriate switches close when the assembly is turned on. Suppose that there is probability θ that a switch of this kind does not close when required, and that the five switches in this assembly fail to function independently of one another. Find the probability that current flows from X to Y.

18 (See Holmes (1994).) When tossing a coin repeatedly, it is plausible to believe that the outcome on one trial (i.e. whether the coin lands heads or tails up) is independent of the outcome of every other trial. A 'run' is a sequence of trials all with the same outcome (i.e. all heads or all tails). If θ is the probability that the coin lands heads up on a trial, find the probability of an even-numbered and an odd-numbered run. Show that, if $\theta \neq \frac{1}{2}$, then odd-numbered runs are more likely.

19 Suppose that you have k bags containing a mixture of red and black balls. The ith bag contains n_i balls altogether, of which x_i are red $(i = 1, 2, \ldots, k)$. Suppose that you carry out the following experiment: first, pick one bag at random; then choose one ball at random from that bag. Find the probability that you choose a red ball. Given that you choose a red ball, find the probability that it came from the ith bag. Notice that, in general, these are not the values you might expect.

20 A certain (human) population consists of N households, of which $n_1, n_2, \ldots$ (respectively) consist of $1, 2, \ldots$ individuals.

(a) If a *household* is sampled at random from this population, find the probability that it consists of exactly i individuals $(i = 1, 2, \ldots)$.

(b) On average, the households in this population contain m members, so that the population consists of exactly Nm individuals. If an *individual* is sampled at random from this population, find the probability that he or she belongs to a household containing exactly i individuals $(i = 1, 2, \ldots)$.

(c) These two sets of probabilities are not equal, which has implications for the way in which sample surveys are carried out. Show that the probability in (b) is greater than the probability in (a) only for household size $i > m$. The second method of sampling, in which larger households would be overrepresented, is an example of *length-biased sampling*.

21 A car insurance firm believes that 20% of drivers who take out insurance policies with it are 'good' risks, 50% are 'average' risks and 30% are 'bad' risks. In any year, a 'good' risk makes at least one insurance claim with probability 0.05. The corresponding probabilities for 'average' and 'bad' risks are 0.15 and 0.30, respectively.

Find the probability that a policyholder who does not make a claim in a certain year is a 'good' risk.

Summary

This chapter has laid the foundations of probability theory. After reviewing some situations in which probability is useful, it went on to discuss briefly the meaning of a probability, from both the frequentist and subjectivist points of view. Axioms of probability were then introduced, from which other useful results were obtained. Conditional probability was defined, leading to derivation of the multiplication theorem, the law of total probability and Bayes' theorem. These basic concepts and results will be used throughout the remainder of this book. The next chapter introduces the idea of a random variable, which is crucial for all that we do from now on.

2
Random variables

The previous chapter discussed how to calculate probabilities associated with general events in general sample spaces. Next, we will see how to adapt these general principles to meet the situation in which each outcome of an experiment may be represented adequately by a single numerical value. A function that associates a real number with each outcome in the sample space is called a random variable. It is often easier and more informative to work directly with a suitable random variable than with events in the original sample space. The remainder of this book deals with random variables and collections of random variables.

2.1 Basic definitions

Example 2.1

A set of ten small, brightly coloured lights is connected in series. When one of the light bulbs fuses, the whole set of lights goes out and the only way to find out which bulb is faulty is to replace each bulb in turn until the set lights up again. If we record the total number of lights that have to be replaced, then a suitable sample space is $S = \{1, 2, \ldots, 10\}$. The outcome of this stochastic experiment is intrinsically numerical.

Example 2.2

Subjects suffering from clinical depression may be prescribed a certain drug treatment. A record is kept of whether or not each subject who is prescribed this treatment suffers side effects (since these may be serious enough for the subject eventually to be taken off treatment). A suitable sample space for recording the status of a subject is $S = \{$'side effect absent', 'side effect present'$\}$. The outcomes in S are not numerical, but we could decide to associate a number with each of them, for example we might record a 0 for 'side effect absent' and a 1 for 'side effect present'.

Example 2.3

Suppose the operational lifetimes of standard batteries made by a certain manufacturer are to be assessed. This can be done by putting each of a sample of batteries into a torch, with the beam of the torch shining on a photo-electric cell connected to a timer. When the battery fails, the photo-electric cell switches off and the timer records the elapsed time, x minutes. A suitable sample space for this experiment is $S = \{x: x \geq 0\}$. Again, this experiment has intrinsically numerical outcomes.

Example 2.4

Suppose that an X-ray is taken of the liver of a subject who is suffering from liver cancer. The outcome of this experiment is a complicated, black-and-white image that potentially provides a great deal of information. The sample space consists of all possible black-and-white images of the given size. There will be particular interest, however, in the length of the tumour, which can usually be determined using software supplied with the X-ray equipment. This is a non-negative real value that may be determined from the outcome in the original sample space.

A *random variable*, X, is a *function* that associates a real number, $X(s)$, with each elementary outcome, s, in the sample space, S. In symbols,

$$X: S \rightarrow \mathbb{R}.$$

When the outcomes in S are intrinsically numerical, the identity function often defines the random variable of particular interest.

Every time the underlying stochastic experiment is conducted, a particular value of the random variable is associated with the outcome that is recorded. This is known as a *realization* of the random variable. It is usual to denote the random variable itself by a capital letter (such as X) and a general realization of it by the corresponding lower-case letter (such as x). The *range space*, $R_X \subseteq \mathbb{R}$, of the random variable X is the set of all possible realizations of X.

Example 2.1 – continued

Determine the order in which the bulbs will be tested, begin testing and continue to test bulbs until the defective one is identified. Let the random variable X be the total number of bulbs that have to be tested until the defective one is identified. Then X is defined by the identity function, $X(s) = s$. The range space of X is $R_X = \{1, 2, \ldots, 10\} = S$.

Example 2.2 – continued

Define the random variable X by

$$X(\text{'no side effect'}) = 0, \qquad X(\text{'side effect'}) = 1.$$

Then X has range space $R_X = \{0, 1\}$.

Any two distinct numerical values would define a different, but equally valid, random variable. For example, suppose the additional cost of treatment is £200 for a patient who experiences side effects. Then the additional cost of treatment (in £) due to side effects is a random variable, Y, defined by

$$Y(\text{'no side effect'}) = 0, \qquad Y(\text{'side effect'}) = 200.$$

Example 2.3 – continued

Let the random variable X be the time to failure (hours) of the battery. Then X is defined by the identity function, $X(s) = s$, and has range space $R_X = \{x: x > 0\}$.

Example 2.4 – continued

Let the random variable X be the length of the liver tumour (mm). Then X has the range space $R_X = \{x: x \geq 0\}$.

A probability may be associated with every possible realization of X. In general, the probability that X lies in some interval $(a, b) \subseteq \mathbb{R}$ is induced from the probabilities associated with the original sample space, $P(a < X < b) = P(\{s \in S: a < X(s) < b\})$.

We now need to find a convenient way in which to summarize the probability information about a random variable. There are a number of different functions that can perform this role. We will begin by considering one function that is well defined for any random variable. The *distribution function* (d.f.) of a random variable, X, is the function

$$F_X(x) = P(X \leq x), \qquad x \in \mathbb{R}.$$

Example 2.1 – continued

We may assume that exactly one of the ten bulbs is defective, and that the defect is equally likely to have occurred in each of the bulbs. Once we decide the order in which we shall test the bulbs, then:

$$P(X \leq 1) = P(\text{the first bulb for test is the faulty one}) = \frac{1}{10} = 0.1,$$

$$P(X \leq 2) = P(\text{one of the first two bulbs for test is the faulty one})$$

$$= \frac{2}{10} = 0.2,$$

etc.

In this example, $F(x)$ must be a step function. Its value can only change at the points $x = 1, 2, \ldots, 10$, which represent increases in the (whole) number of bulbs tested. And so:

$$F_X(x) = \begin{cases} 0, & x < 1, \\ 0.1, & 1 \leq x < 2, \\ 0.2, & 2 \leq x < 3, \\ \ldots \\ 1, & 10 \leq x. \end{cases}$$

A plot of this distribution function is shown in Figure 2.1(a).

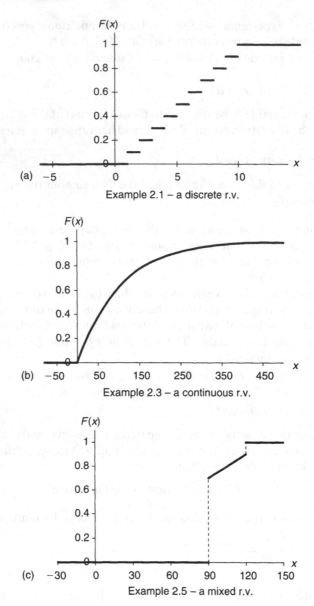

(a) Example 2.1 – a discrete r.v.

(b) Example 2.3 – a continuous r.v.

(c) Example 2.5 – a mixed r.v.

Figure 2.1 *Distribution functions of three random variables*

Suppose that, as in Example 2.1, the random variable X can take only a finite or countably infinite number of different possible values, which we may write as $R_X = \{x_1, x_2, \dots\}$, and that it has a distribution function that is a step function. Then we say that X is a *discrete random variable*. It is mathematically possible to construct random variables that are defined on discrete sets of values but whose distribution functions are continuous (Shiryaev, 1995). These random variables are usually distinguished by being called *singular*. We do not discuss singular random variables further in this book.

Example 2.3 – continued

The range space in this example is an interval of the real line, $[0, \infty)$, so it is uncountably infinite. This implies that this random variable is not discrete.

Suppose that a battery of this kind fails only due to a 'shock' to the electrical system (e.g. a surge of electrical power). Suppose also that these 'shocks' occur randomly across time such that, in any time interval of length t hours, the probability that no shock occurs is $\exp(-\theta t)$ (where θ is a positive constant). Then, for any $x \geq 0$,

$$F_X(x) = P(X \leq x) = P(\text{at least one 'shock' in the interval } [0, x]) = 1 - e^{-\theta x}.$$

We will usually write just

$$F_X(x) = 1 - e^{-\theta x} \qquad x > 0.$$

More properly, though, $F_X(x)$ should be specified for all x as follows:

$$F_X(x) = \begin{cases} 0, & x \leq 0, \\ 1 - e^{-\theta x}, & x > 0. \end{cases}$$

This is an example of an *exponential distribution* for a continuous random variable, written $X \sim \text{Ex}(\theta)$. Its distribution function is plotted in Figure 2.1(b). Notice that it is continuous. In addition, letting

$$f_X(x) = \begin{cases} 0, & x \leq 0, \\ \theta e^{-\theta x}, & x > 0, \end{cases}$$

then we see that $F_X(x)$ can be written as the integral

$$F_X(x) = \int_{-\infty}^{x} f_X(t) \, dt \qquad (2.1)$$

If a random variable has a distribution function that is continuous and can be written in the form (2.1), then it is said to be a *continuous random variable*. (Properly, in the language of mathematical analysis, this is an *absolutely continuous random variable*.) A random variable whose distribution function is continuous but cannot be written in the form (2.1) is *singular* (Shiryaev, 1995).

Example 2.5

Here is an example of a random variable that is neither discrete nor continuous (and is not singular either). In some international football (or 'soccer') competitions, a match lasts for at least the usual 90 minutes. A team that has scored more goals than the opposition at the end of 90 minutes wins the match outright. If the teams are drawing after 90 minutes, then the match goes into extra playing time and the first team to score a goal wins the match immediately. If no goal is scored in 30 minutes of extra time, then play is stopped and the match is decided on a penalty shoot-out.

X, the playing time (minutes) of a match played under these rules, is a random variable. There is non-zero probability that play lasts exactly 90 minutes or exactly 120 minutes. On the other hand, every value in the range (90, 120) is also possible, since the match could stop instantaneously at any time between 90 and 120 minutes if a deciding goal is scored. A possible d.f. for this random variable is shown in Figure 2.1(c). The step to $F_X(90) = 0.7$ indicates that there is probability 0.7 of a match being won decisively within the 90 minutes of normal play. $F_X(x)$ then increases linearly on the range (90, 120), reaching the value 0.9 at $x = 120$. This suggests that there is probability $0.2 = 0.9 - 0.7$ of a game being decided by a goal scored at some point in extra time. Finally, $F_X(x)$ steps up to the value 1.0 at $x = 120$, since play must stop then. This suggests that there is probability $0.1 = 1.0 - 0.9$ of a game not being decided by the end of 120 minutes play (and so requiring a penalty shoot-out).

This distribution function is given by the following formula:

$$F_X(x) = \begin{cases} 0, & x < 90, \\ 0.7 + \dfrac{0.2(x-90)}{(120-90)}, & 90 \le x < 120, \\ 1, & 120 \le x. \end{cases}$$

Notice that $F_X(x)$ can be expressed as the weighted sum of a discrete distribution function, $F_d(x)$, and a continuous distribution function, $F_c(x)$, since

$$F_X(x) = 0.8F_d(x) + 0.2F_c(x)$$

where

$$F_d(x) = \begin{cases} 0, & x < 90, \\ \dfrac{7}{8}, & 90 \le x < 120, \\ 1, & 120 \le x, \end{cases} \qquad F_c(x) = \begin{cases} 0, & x < 90, \\ \dfrac{x-90}{30}, & 90 \le x < 120, \\ 1, & 120 \le x. \end{cases}$$

X in this example is a *mixed random variable*. In general, a random variable is called 'mixed' if its distribution function can be written in the form

$$F_X(x) = \alpha F_d(x) + (1 - \alpha)F_c(x),$$

where $0 < \alpha < 1$, $F_d(x)$ is a discrete distribution function and $F_c(x)$ is a continuous distribution function.

In this book, we shall concentrate on discrete and (absolutely) continuous random variables, since these are much the most important types in practice. This decision greatly simplifies the terminology we will need to use, but it is worth noting that many of the results derived in the chapters ahead do generalize to the case of a mixed random variable.

Any valid distribution function, $F_X(x)$, is defined for all $x \in \mathbb{R}$ and has the following properties:

(i) $F_X(-\infty) = 0$ and $F_X(\infty) = 1$;

(ii) whenever $a \leq b$, then $F_X(a) \leq F_X(b)$, [i.e. $F_X(x)$ is non-decreasing on $\mathbb{R}$]

(iii) for all $x \in \mathbb{R}$,

$$\lim_{h \to 0^+} F_X(x + h) = F_X(x)$$

[i.e. $F_X(x)$ is continuous on the right at all values of $x \in \mathbb{R}$]

The distribution function incorporates all the probability information about X. Other useful probabilities can be recovered from it, for example,

$$P(a < X \leq b) = P(X \leq b) - P(X \leq a) = F(b) - F(a).$$

In general, though, the distribution function is not the most informative way in which to summarize the probability information about a random variable. The general properties listed above mean that many random variables have very similar distribution functions, and this reduces the usefulness of the distribution function as a display.

For *discrete* random variables, the most basic probability information comes in the form

$$P(X = x) = p_X(x), \qquad x \in \mathbb{R}.$$

Clearly, $P(X = x) = 0$ whenever $x \notin R_X$. The list of values $\{(x, p_X(x)), x \in R_X\}$ is called the *probability mass function* of the discrete random variable X.

Example 2.2 – continued

Suppose that a psychiatrist prescribes this drug for n (≥ 1) depressed patients. Let the discrete random variable X be the number of them who suffer side effects. Then, X has range space $R_X = \{0, 1, \ldots, n\}$. Suppose that θ ($0 \leq \theta \leq 1$) is the probability that a randomly selected depressed individual suffers side effects when treated with this drug. It seems reasonable to assume that the n individuals respond independently of one another. So,

$$p_X(x) = \binom{n}{x} \theta^x (1 - \theta)^{n-x}, \qquad x = 0, 1, \ldots, n,$$

since there are $\binom{n}{x}$ possible outcomes in which exactly x of the patients suffer side effects, and each of these combinations has probability $\theta^x (1 - \theta)^{n-x}$.

A random variable with this form of probability mass function is said to follow a *binomial distribution* with parameters n and θ, written $X \sim \text{Bi}(n, \theta)$. This family of distributions gets its name from the well-known *binomial theorem* (Result A1.8 in Appendix A). Using this theorem, it follows that for any binomial distribution

$$\sum_{x=0}^{n} p_X(x) = \sum_{x=0}^{n} \binom{n}{x} \theta^x (1-\theta)^{n-x} = (\theta + 1 - \theta)^n = 1.$$

Alternatively, suppose that the psychiatrist is interested in a different random variable. Let Y be the number of patients treated with this drug until the first one is found to suffer side effects. Y must have the range space $R_Y = \{1, 2, \ldots\}$. Again assume that the subjects respond independently of one another, and that each has probability θ of suffering side effects. Then, denoting a patient who suffers side effects by S and a patient who does not suffer side effects by N,

$$p_Y(y) = P(\underbrace{NN\ldots N}_{y-1}S) = (1-\theta)^{y-1}\theta, \qquad y = 1, 2, \ldots.$$

This is known as a *geometric distribution*, written $Y \sim \text{Ge}(\theta)$. This name arises because the terms of $p_Y(y)$ form a geometric sequence (Results A1.3 and A1.4). Notice again that

$$\sum_{y=1}^{\infty} p_Y(y) = \sum_{y=1}^{\infty} (1-\theta)^{y-1}\theta = \theta \sum_{y=1}^{\infty} (1-\theta)^{y-1} = \theta \frac{1}{1-(1-\theta)} = 1.$$

The binomial and geometric distributions are two very important special discrete probability distributions. Table 2.1 lists the most common discrete and continuous distributions.

Table 2.1 *Some special probability distributions*

Name	Notation	Range space	Probability mass function	Restrictions
(a) Discrete distributions				
Binomial	$\text{Bi}(n, \theta)$	$\{0, 1, \ldots, n\}$	$\binom{n}{x}\theta^x(1-\theta)^{n-x}$	$n \in \mathbb{Z}^+$ $0 < \theta < 1$
Geometric	$\text{Ge}(\theta)$	$\{1, 2, \ldots\}$	$(1-\theta)^{x-1}\theta$	$0 < \theta < 1$
Hyper-geometric	Hyp (n, N, M)	$\{0, 1, \ldots, n\}$	$\dfrac{\binom{M}{x}\binom{N-M}{n-x}}{\binom{N}{n}}$	$n, N, M \in \mathbb{Z}^+$ $n \leq N$ $M \leq N$
Negative Binomial	$\text{NeBi}(k, \theta)$	$\{k, k+1, \ldots\}$	$\binom{x-1}{k-1}(1-\theta)^{x-k}\theta^k$	$k \in \mathbb{Z}^+$ $0 < \theta < 1$
Poisson	$\text{Po}(\theta)$	$\{0, 1, \ldots\}$	$\dfrac{e^{-\theta}\theta^x}{x!}$	$\theta > 0$

(*continued*)

Table 2.1 *Continued*

Name	Notation	Range space	Probability density function	Restrictions
(b) Continuous distributions				
Beta	$\mathrm{Be}(\alpha, \beta)$	$(0, 1)$	$\dfrac{x^{\alpha-1}(1-x)^{\beta-1}}{\mathbb{B}(\alpha, \beta)}$	$\alpha > 0$ $\beta > 0$
Chi-squared	χ^2_ν	$(0, \infty)$	$\dfrac{x^{\frac{1}{2}\nu-1}e^{-\frac{1}{2}x}}{2^{\frac{1}{2}\nu}\Gamma(\frac{1}{2}\nu)}$	$\nu > 0$
Exponential	$\mathrm{Ex}(\theta)$	$(0, \infty)$	$\theta e^{-\theta x}$	$\theta > 0$
F	F_{ν_1, ν_2}	$(0, \infty)$	$\dfrac{\nu_1^{\frac{1}{2}\nu_1} \nu_2^{\frac{1}{2}\nu_2}}{\mathbb{B}(\frac{1}{2}\nu_1, \frac{1}{2}\nu_2)} \times$ $\dfrac{x^{\frac{1}{2}\nu_1-1}}{(\nu_2 + \nu_1 x)^{\frac{1}{2}(\nu_1+\nu_2)}}$	$\nu_1 > 0$ $\nu_2 > 0$
Gamma	$\mathrm{Ga}(\alpha, \theta)$	$(0, \infty)$	$\dfrac{\theta^\alpha x^{\alpha-1} e^{-\theta x}}{\Gamma(\alpha)}$	$\theta > 0$
Normal	$N(\mu, \sigma^2)$	$(-\infty, \infty)$	$\dfrac{1}{\sigma\sqrt{2\pi}} \times$ $\exp\left\{-\dfrac{1}{2}\left(\dfrac{x-\mu}{\sigma}\right)^2\right\}$	$\sigma > 0$
Pareto	$\mathrm{Pa}(k, \theta)$	(k, ∞)	$\theta k^\theta / x^{\theta+1}$	$\theta > 0$ $k > 0$
Student's t	t_ν	$(-\infty, \infty)$	$\left[\sqrt{\nu}\mathbb{B}(\frac{1}{2}\nu, \frac{1}{2}) \times \left(1 + \dfrac{x^2}{\nu}\right)^{\frac{\nu+1}{2}}\right]^{-1}$	$\nu \in \mathbb{Z}^+$
Uniform	$\mathrm{Un}(a, b)$	(a, b)	$\dfrac{1}{b-a}$	$b > a$
Weibull	$\mathrm{We}(\alpha, \theta)$	$(0, \infty)$	$\alpha\theta x^{\alpha-1} \exp\{-\theta x^\alpha\}$	$0 < \alpha$ $0 < \theta$

There are two general properties that are common to the probability mass functions of all discrete random variables:

(i) $0 \le p_X(x) \le 1$, for all $x \in \mathbb{R}$ (this follows because $p_X(x)$ is a probability);

(ii) $\sum_{x \in R_X} p_X(x) = 1$ (this follows because the events $\{s: X(s) = x\}$, $x \in R_X$, partition the sample space).

When the random variable X is not discrete, the probability mass function is not a useful concept, as the following result shows.

Proposition 2.1

Let X be a continuous random variable. Then $P(X = x) = 0$ for all x.

Proof. This is a consequence of the continuity of $F_X(x)$. For $x \in \mathbb{R}$, define the decreasing sequence of events:

$$E_n = \left(x - \frac{1}{n}, x + \frac{1}{n} \right], \qquad n = 1, 2, 3, \ldots.$$

Then

$$
\begin{aligned}
\lim_{n \to \infty} P(E_n) &= \lim_{n \to \infty} \left\{ P\left(x - \frac{1}{n} < X \le x + \frac{1}{n} \right) \right\} \\
&= \lim_{n \to \infty} \left\{ F\left(x + \frac{1}{n} \right) - F\left(x - \frac{1}{n} \right) \right\} \\
&= \lim_{n \to \infty} \left\{ F\left(x + \frac{1}{n} \right) \right\} - \lim_{n \to \infty} \left\{ F\left(x - \frac{1}{n} \right) \right\} \\
&= F(x) - F(x) \\
&= 0,
\end{aligned}
$$

using the fact that F is continuous (i.e. left as well as right continuous). But

$$\lim_{n \to \infty} E_n = \bigcap_{i=1}^{\infty} E_i = \{x\}.$$

Using Proposition 1.4, then,

$$P\left\{ \lim_{n \to \infty} E_n \right\} = \lim_{n \to \infty} P(E_n),$$

i.e. $\qquad\qquad\qquad\qquad P(\{x\}) = 0.$ ∎

This means that, when X is continuous and a and b are real values such that $b \ge a$,

$$P(a < X < b) = P(a < X \le b) = P(a \le X < b) = P(a \le X \le b)$$
$$= F_X(b) - F_X(a).$$

For a continuous random variable, a function that does usefully summarize the available probability information is the *probability density function* (p.d.f.).

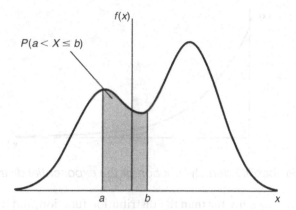

We have seen in the (absolutely) continuous case that there is a function $f_X(x)$ such that

$$F_X(x) = \int_{-\infty}^{x} f_X(t)\,dt, \qquad x \in \mathbb{R}.$$

$f_X(x)$ is the probability density function, which may therefore be defined by

$$f_X(x) = \frac{d}{dx}F_X(x).$$

Notice that $f_X(x) = 0$ for all x outside the range space of X.

The probability that a continuous random variable X lies in the interval (a, b) can be found by integrating its probability density function over the appropriate range:

$$P(a < X \le b) = P(X \le b) - P(X \le a)$$
$$= F_X(b) - F_X(a)$$
$$= \int_{-\infty}^{b} f(x)\,dx - \int_{-\infty}^{a} f(x)\,dx$$
$$= \int_{a}^{b} f(x)\,dx$$

Since $P(X = x) = 0$ for all x, it should be clear that the p.d.f. does not represent probabilities in the same way as the probability mass function of a discrete random variable. There is a connection, though, which can be seen from the following argument.

Let x be any real number and let δ be a small positive value. Then,

$$P\left(x - \frac{\delta}{2} < X < x + \frac{\delta}{2}\right) = \int_{x-\delta/2}^{x+\delta/2} f(t)\,dt \approx f(x) \cdot \delta.$$

This means that, although $f(x)$ is not the probability that X equals x, it is proportional to the probability that X lies in a small interval (of fixed width) centred on x. This means that the probability density function of a continuous random

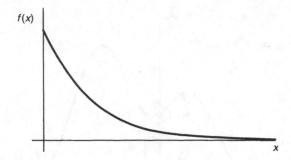

Figure 2.2 *Probability density function of the exponential distribution*

variable has more character than the distribution function, and it is for this reason that the probability density function is preferred as a way of summarizing the probabilities associated with X. The probability density function of several common continuous distributions are listed in Table 2.1(b) on page 35.

Example 2.3 – continued

Since

$$F_X(x) = \begin{cases} 0, & x \le 0, \\ 1 - e^{-\theta x}, & x > 0, \end{cases}$$

it follows that

$$f_X(x) = \begin{cases} 0, & x \le 0, \\ \theta e^{-\theta x}, & x > 0. \end{cases}$$

This function is plotted in Figure 2.2.

For a continuous random variable, $f_X(x)$ must always be 0 for $x \notin R_X$. This justifies simplifying the way $f_X(x)$ is written by restricting attention to its non-zero values; for example, in this case writing just

$$f_X(x) = \theta e^{-\theta x}, \qquad x > 0.$$

There are two general properties that are shared by the probability density functions of all continuous random variables:

(i) $0 \le f_X(x)$, for all $x \in \mathbb{R}$ (this follows because $f_X(x)$ is the derivative of a non-decreasing function, $F_X(x)$);

(ii) $\int_{-\infty}^{\infty} f_X(x)\mathrm{d}x = 1$ (this follows because $F_X(-\infty) = 0$ and $F_X(\infty) = 1$).

Example 2.6

A continuous random variable with range space $R_X = [0, 1]$ and probability density function

$$f_X(x) = \frac{x^{\alpha-1}(1-x)^{\beta-1}}{\mathbb{B}(\alpha, \beta)}, \qquad 0 < x < 1,$$

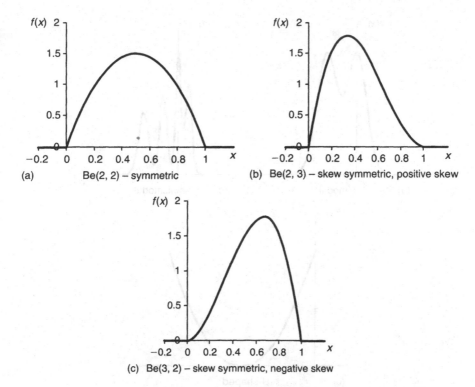

Figure 2.3 *The probability density functions of three beta distributions*

is said to follow a *beta distribution*, $X \sim \text{Be}(\alpha, \beta)$. This is a family of probability distributions with two parameters, α and β, both of which are positive constants. $\mathbb{B}(\alpha, \beta)$ denotes the usual *beta function* [Results A2.4–A2.6], which is defined by

$$\mathbb{B}(\alpha, \beta) = \int_0^1 x^{\alpha-1}(1-x)^{\beta-1} \, dx.$$

Three particular beta distributions have the following probability density functions:

(a) $\alpha = 2, \beta = 2$: $f_X(x) = 6x(1-x),$ $0 < x < 1;$

(b) $\alpha = 2, \beta = 3$: $f_X(x) = 12x(1-x)^2,$ $0 < x < 1;$

(c) $\alpha = 3, \beta = 2$: $f_X(x) = 12x^2(1-x),$ $0 < x < 1.$

These functions are plotted in Figure 2.3. All three of these distributions are *unimodal*, which means that their probability density functions have exactly one *mode* (or local maximum).

Not all distributions are unimodal; Figure 2.4 shows some other possible shapes.

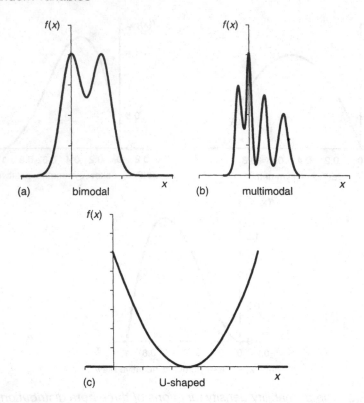

Figure 2.4 *Some possible shapes of probability density functions*

Example 2.6 – continued

The $\text{Be}(\alpha, \beta)$ distribution is unimodal whenever $\alpha > 1$ and $\beta > 1$. Otherwise, it is U-shaped. For example

(d) $\alpha = \beta = \frac{1}{2}$: $\quad f(x) = \dfrac{1}{\pi\sqrt{x(1-x)}}, \qquad 0 < x < 1.$

Distributions are also classified as *symmetric* or *skew-symmetric*. In general, a random variable X is said to have a symmetric distribution if there is some real constant c such that

$$F_X(x) = 1 - F_X(2c - x), \qquad \text{for all } x \in \mathbb{R}.$$

When X is continuous, this is equivalent to there being a real constant c such that

$$f_X(x) = f_X(2c - x), \qquad \text{for all } x \in \mathbb{R}.$$

Example 2.6 – continued

Of the three unimodal beta distributions discussed earlier, only (a) is symmetric, and it is symmetric around $c = \frac{1}{2}$. Distribution (b) is *positively skewed* or *skewed*

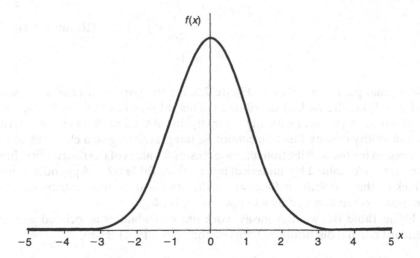

Figure 2.5 *The probability density function of the standard normal distribution*

to the right. Distribution (c) is *negatively skewed* or *skewed to the left*. The U-shaped distribution, (d), is also symmetric around $c = \frac{1}{2}$.

Example 2.7

Perhaps the most famous and most widely used probability model for a continuous random variable is the *normal distribution*. Normal distributions all have a symmetric, 'bell-shaped' probability density function. A particular member of the family is identified by two parameters, often denoted μ and σ^2, where μ is the centre of symmetry of the distribution and σ is a measure of the amount of variability in the distribution. We shall write $X \sim N(\mu, \sigma^2)$ when the continuous random variable X follows the normal distribution with p.d.f.

$$f_X(x) = \frac{1}{\sqrt{2\pi}\sigma} \exp\left\{-\frac{1}{2}\left(\frac{x-\mu}{\sigma}\right)^2\right\}, \qquad -\infty < x < \infty.$$

In order to prove that this is a well-defined probability density function, we need to integrate $f_X(x)$ over the real line, making the change of variable $u = (x-\mu)/\sigma$. Then

$$\int_{-\infty}^{\infty} \frac{1}{\sqrt{2\pi}\sigma} \exp\left\{-\frac{1}{2}\left(\frac{x-\mu}{\sigma}\right)^2\right\} dx = \int_{-\infty}^{\infty} \frac{1}{\sqrt{2\pi}} \exp\{-u^2/2\} \, du$$

$$= 2\int_{0}^{\infty} \frac{1}{\sqrt{2\pi}} \exp\{-u^2/2\} \, du$$

$$= \frac{1}{\sqrt{\pi}} \int_{0}^{\infty} e^{-v} v^{-\frac{1}{2}} \, dv$$

$$= \frac{1}{\sqrt{\pi}} \Gamma\left(\frac{1}{2}\right) \qquad \text{(Result A2.3)}$$

$$= 1$$

The normal p.d.f. is graphed in Figure 2.5, for the particular case when $\mu = 0$ and $\sigma^2 = 1$, i.e. the N(0, 1) distribution. This is known as the *standard normal distribution*. A problem with any normally distributed random variable is that its probability density function cannot be integrated to give a closed algebraic expression for the distribution function. Instead, values of the distribution function must be obtained by numerical integration. Table B2 in Appendix B gives values of the distribution function of the standard normal distribution; this function is often denoted by an upper-case phi, Φ.

Using Table B2, we may easily work out probabilities associated with the standard normal distribution. Suppose that $Z \sim N(0,1)$. Then,

$$P(Z \leq 1) = \Phi(1) = 0.8413 \qquad \text{(using the table directly)}$$

$$P(Z > 1) = 1 - P(Z \leq 1) = 1 - \Phi(1) = 1 - 0.8413 = 0.1587$$

$$P(Z \leq -1) = P(Z \geq 1) = 1 - \Phi(1) = 1 - 0.8413 = 0.1587 \quad \text{(symmetry)}$$

$$P(-1 < Z < 1) = P(Z < 1) - P(Z \leq -1) = \Phi(1) - \Phi(-1)$$

$$= 0.8413 - 0.1587 = 0.6826.$$

Exercises

1 In the 2×2 determinant,

$$D = \begin{vmatrix} x_{11} & x_{12} \\ x_{21} & x_{22} \end{vmatrix} = x_{11}x_{22} - x_{12}x_{21},$$

suppose that each x_{ij} independently takes the values 0 and 1 with probability $\frac{1}{2}$. Find the probability mass function of D. Hence find the distribution function of X, and plot it appropriately.

2 Suppose that the discrete random variable X follows the Bi$(4, \theta)$ distribution, where $0 < \theta < 1$. Write down $P(X = x)$ for all values of x in the range space of X. Plot the probability mass function of X for various values of θ (e.g. $\theta = 0.1, 0.3, 0.5, 0.7, 0.9$). Confirm that these distributions follow the general rule for binomials, which are symmetric only when $\theta = \frac{1}{2}$, positively skewed when $\theta < \frac{1}{2}$ and negatively skewed when $\theta > \frac{1}{2}$.

3 Suppose that $X \sim \text{Bi}(n, \theta)$. Confirm the following recursive relationship, which is often useful for evaluating binomial probabilities (e.g. on a spreadsheet or other computer package):

$$P(X = 0) = (1 - \theta)^n,$$

$$P(X = x) = \frac{n - (x-1)}{x} \cdot \frac{\theta}{1-\theta} \cdot P(X = x-1), \qquad x = 1, 2, \ldots, n.$$

At what value or values of x must $P(X = x)$ reach its maximum?

4 A discrete random variable X is said to follow a Poisson distribution with parameter $\theta > 0$ if its probability mass function is of the form:

$$P(X = x) = \frac{e^{-\theta}\theta^x}{x!}, \qquad x = 0, 1, 2, \ldots.$$

This family of distributions is used to model such random variables as the number of emissions from a radioactive source and the number of cases of a rare disease in a geographical region of a particular area. Plot $P(X = x)$ for various values of θ (e.g. $\theta = 0.5, 1, 2, 4, 8$). What is the effect on the shape of the distribution of increasing θ?

5 Suppose that $X \sim \text{Po}(\theta)$. Confirm the recursive relationship

$$P(X = 0) = e^{-\theta},$$

$$P(X = x) = \frac{\theta}{x}.P(X = x-1), \qquad x = 1, 2, \ldots, n.$$

At what value or values of x must $P(X = x)$ reach its maximum?

6 Suppose that the continuous random variable X has range space $R_X = (0, \theta_2)$ and the probability density function shown in the diagram below.

This is a two-parameter family of distributions, with parameters θ_1 and θ_2 (where $0 \le \theta_1 \le \theta_2$). Show that

$$f(\theta_1) = \frac{2}{\theta_2},$$

and hence write down $f(x)$ explicitly.
Find $P(X \le \theta_1)$.
Derive $F(x)$ and plot it appropriately.

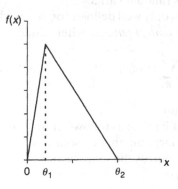

7 The two-parameter gamma distribution has the following probability density function:

$$f(x) = \frac{\theta^\alpha x^{\alpha-1} e^{-\theta x}}{\Gamma(\alpha)}, \qquad x > 0.$$

Here, the parameters α and θ are constrained to be positive. $\Gamma(\alpha)$ denotes the gamma function (see Appendix A2), which is a generalization of the factorial function. When $\alpha = n$, a positive integer, the form of the p.d.f. becomes

$$f(x) = \frac{\theta^n x^{n-1} e^{-\theta x}}{(n-1)!}, \qquad x > 0.$$

(a) Hold θ constant and plot $f(x)$ for a variety of values of n. Do this for a few different values of θ. Describe the effect of increasing n. n (or α) is often called the *shape* parameter of the gamma distribution.

(b) Now hold n constant and plot $f(x)$ for a variety of values of θ. θ is called the *scale* parameter of the gamma distribution.

2.2 Expected value and variance

In the previous section, we discussed the shape of the probability mass function or probability density function of a random variable. Now, we shall investigate the location (or centre) of a distribution and the spread. One natural measure of location is the *expected value* of X, defined by

$$\mathbb{E}(X) = \sum_{x \in R_X} x p(x)$$

when X is a discrete random variable, and

$$\mathbb{E}(X) = \int_{x \in R_X} x f(x) \, dx$$

when X is a continuous random variable.

The expected value is only well defined (or *finite*) when the appropriate sum or integral is *absolutely convergent*, i.e. when either

$$\sum_{x \in R_X} |x| p(x) \quad \text{or} \quad \int_{x \in R_X} |x| f(x) \, dx$$

converges to a finite limit.

The interpretation of $\mathbb{E}(X)$ is as follows. If the underlying stochastic experiment is conducted on a large number of occasions, and the value of X recorded each time, then in the limit, as the number of trials tends to infinity, the mean value recorded is $\mathbb{E}(X)$.

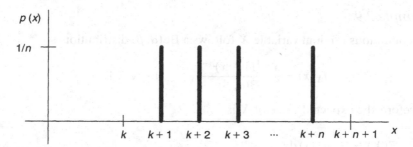

Figure 2.6 *Probability mass function for equally likely outcomes model*

Example 2.8

The equally likely outcomes model (mentioned in Chapter 1) gives rise to the uniform distribution for discrete random variables. X follows a *discrete uniform distribution* if its probability mass function is of the form

$$P(X = x) = \frac{1}{n}, \qquad x = k+1, \dots, k+n,$$

for some integer k and positive integer n (see Figure 2.6). For example, the score on a 'fair' die follows a discrete uniform distribution with $k=0$ and $n=6$.
In general,

$$\mathbb{E}(X) = \sum_{x=k+1}^{k+n} x.P(X = x)$$

$$= \frac{1}{n} \sum_{x=k+1}^{k+n} x$$

$$= \frac{1}{n} \sum_{y=1}^{n} (y + k) \qquad \text{(substituting } y = x - k\text{)}$$

$$= k + \frac{1}{n} \sum_{y=1}^{n} y$$

$$= k + \frac{1}{n}.\frac{1}{2}n(n+1)$$

$$= k + \frac{1}{2}(n+1).$$

This is an intuitively reasonable result. The probability mass function is symmetric about $x = k + \frac{1}{2}(n+1)$, so this value seems a reasonable measure of location. In the particular example of a 'fair' die $(k=0, n=6)$, $\mathbb{E}(X) = 3.5$.

For discrete random variables in general, as in the above example, $\mathbb{E}(X)$ need not lie in R_X.

Example 2.9

The continuous random variable X follows a $\text{Be}(\alpha, \beta)$ distribution, so

$$f_X(x) = \frac{x^{\alpha-1}(1-x)^{\beta-1}}{\mathbb{B}(\alpha, \beta)}, \qquad 0 < x < 1.$$

Therefore, the expected value of X is

$$\mathbb{E}(X) = \int_0^1 x f(x)\, dx$$

$$= \frac{1}{\mathbb{B}(\alpha, \beta)} \int_0^1 x^{\alpha}(1-x)^{\beta-1}\, dx$$

$$= \frac{1}{\mathbb{B}(\alpha, \beta)} \mathbb{B}(\alpha + 1, \beta)$$

$$= \frac{\Gamma(\alpha + \beta)}{\Gamma(\alpha)\Gamma(\beta)} \cdot \frac{\Gamma(\alpha + 1)\Gamma(\beta)}{\Gamma(\alpha + \beta + 1)} \qquad \text{(see Appendix A2)}$$

$$= \frac{\Gamma(\alpha + \beta)}{(\alpha + \beta)\Gamma(\alpha + \beta)} \cdot \frac{\alpha \cdot \Gamma(\alpha)}{\Gamma(\alpha)} \qquad \text{(since } \Gamma(v + 1) = v\Gamma(v)\text{)}$$

$$= \frac{\alpha}{\alpha + \beta}.$$

When $\alpha = \beta$, the probability density function is symmetric around $x = \frac{1}{2}$, and this is also the expected value of X.

In general, if X has a symmetric distribution then $\mathbb{E}(X)$ occurs at the centre of symmetry, as long as $\mathbb{E}(X)$ converges to a finite limit.

If $g(X)$ is a function of the random variable X, then it is itself a random variable. The expected value of $g(X)$ is defined as follows. When X is discrete,

$$\mathbb{E}\{g(X)\} = \sum_{x \in R_X} g(x) p_X(x).$$

When X is continuous,

$$\mathbb{E}\{g(X)\} = \int_{x \in R_X} g(x) f_X(x)\, dx.$$

This assumes that the sum or integral in question is absolutely convergent.

Proposition 2.2

Let X be a random variable with finite expected value $\mathbb{E}(X)$. Let a and b be real constants. Then $aX + b$ has finite expected value

$$\mathbb{E}(aX + b) = a\mathbb{E}(X) + b.$$

Proof. We prove this result only for the case when X is a discrete random variable. The proof is easily adapted for the continuous case.

$$\mathbb{E}\{aX + b\} = \sum_{x \in R_X} (ax + b)p_X(x)$$

$$= a \sum_{x \in R_X} xp_X(x) + b \sum_{x \in R_X} p_X(x)$$

$$= a\mathbb{E}(X) + b.$$ ∎

A measure of the spread of the distribution around its expected value is the *variance* of X:

$$\mathrm{var}(X) = \mathbb{E}\left\{[X - \mathbb{E}(X)]^2\right\}$$

$$= \begin{cases} \sum_{x \in R_X} (x - \mathbb{E}(X))^2 p_X(x), & X \text{ discrete,} \\ \int_{x \in R_X} (x - \mathbb{E}(X))^2 f_X(x)\, dx, & X \text{ continuous.} \end{cases}$$

The variance of X is only well defined if this sum or integral is absolutely convergent.

The variance is the average squared distance of X from its expected value. Since it is an expected value (or long-term average) of a non-negative quantity, then the variance itself must be non-negative, i.e. $\mathrm{var}(X) \geq 0$.

$\mathbb{E}\{X - \mathbb{E}(X)\}$ cannot be used as a measure of spread since this is always identically equal to 0. To prove this, put $a = 1, b = -\mathbb{E}(X)$ in Proposition 2.2. An alternative measure of spread that could be useful is $\mathbb{E}\{|X - \mathbb{E}(X)|\}$, the average absolute distance of X from its expected value. Historically, $\mathbb{E}\left\{[X - \mathbb{E}(X)]^2\right\}$ has been preferred as it is more amenable to mathematical manipulation.

Example 2.10

(a) A random variable that can take only two distinct values is said to be *binary*. Suppose that the binary random variable X has the following probability mass function (where $k \geq 0$):

x	$-k$	$+k$
$p(x)$	$\frac{1}{2}$	$\frac{1}{2}$

By symmetry, $\mathbb{E}(X) = 0$. So,

$$\mathrm{var}(X) = \mathbb{E}\left(X^2\right) = (-k)^2 \cdot \frac{1}{2} + (k)^2 \cdot \frac{1}{2} = k^2.$$

So, $\mathrm{var}(X)$ increases as k increases. The reason for this is that, as k increases, the two possible values of X become more separated from one another so the probability distribution of X becomes more spread out. When $k = 0$, the whole probability mass is concentrated at $x=0$, so there is no variability in X and $\text{var}(X)=0$.

(b) Suppose now that the binary random variable X has the following probability mass function (where $1 \geq \theta \geq 0$)

x	-1	0	$+1$
$p(x)$	$\frac{1}{2}\theta$	$1-\theta$	$\frac{1}{2}\theta$

By symmetry, $\mathbb{E}(X)=0$. So,

$$\text{var}(X) = \mathbb{E}(X^2) = (-1)^2 \cdot \frac{1}{2}\theta + (1)^2 \cdot \frac{1}{2}\theta = \theta.$$

In this example, $\text{var}(X)$ increases as θ increases even though the range space of X does not change. The reason for this is that the probability mass becomes more spread out as θ increases, since more of the probability is concentrated at the points -1 and $+1$ and less of the probability at the centre of the distribution $(x=0)$.

Proposition 2.3

Suppose that X is a random variable with finite expected value $\mathbb{E}(X)$ and finite variance $\text{var}(X)$. Let a and b be real constants. Then $aX + b$ has finite variance $a^2 \, \text{var}(X)$.

Proof.

$$\text{var}\,(aX + b) = \mathbb{E}\left\{[aX + b - \mathbb{E}\,(aX + b)]^2\right\}$$

$$= \mathbb{E}\left\{[aX + b - a\mathbb{E}(X) - b]^2\right\} \qquad \text{(using Proposition 2.2)}$$

$$= \mathbb{E}\left\{a^2\,[X - \mathbb{E}(X)]^2\right\}$$

$$= a^2\mathbb{E}\left\{[X - \mathbb{E}(X)]^2\right\} \qquad \text{(using Proposition 2.2)}$$

$$= a^2 \, \text{var}(X). \qquad\blacksquare$$

Proposition 2.4

$$\text{var}(X) = \mathbb{E}(X^2) - \{\mathbb{E}(X)\}^2$$

Proof. For the continuous case,

$$\text{var}(X) = \mathbb{E}\{[X - \mathbb{E}(X)]^2\}$$

$$= \int_{x \in R_X} \{x - \mathbb{E}(X)\}^2 f(x)\, dx$$

$$= \int_{x \in R_X} x^2 f(x) \, dx - 2\mathbb{E}(X) \int_{x \in R_X} x f(x) \, dx + \{\mathbb{E}(X)\}^2 \int_{x \in R_X} f(x) \, dx$$

$$= \mathbb{E}(X^2) - 2\mathbb{E}(X)\mathbb{E}(X) + \{\mathbb{E}(X)\}^2 . 1$$

$$= \mathbb{E}(X^2) - \{\mathbb{E}(X)\}^2 . \qquad \blacksquare$$

Example 2.8 – continued

Suppose that the discrete random variable X follows the discrete uniform distribution:

$$P(X = x) = \frac{1}{n}, \qquad x = 1, \ldots, n.$$

Then

$$\mathbb{E}(X^2) = \sum_{x=1}^{n} x^2 \cdot P(X = x)$$

$$= \frac{1}{n} \sum_{x=1}^{n} x^2$$

$$= \frac{1}{n} \left\{ \frac{1}{6} n(n+1)(2n+1) \right\}$$

$$= \frac{1}{6}(n+1)(2n+1).$$

We previously found that $\mathbb{E}(X) = \frac{1}{2}(n+1)$. Using Proposition 2.4, this means that:

$$\text{var}(X) = \frac{1}{6}(n+1)(2n+1) - \frac{1}{4}(n+1)^2$$

$$= \frac{1}{12}(n+1)\{4n+2-3n-3\}$$

$$= \frac{1}{12}(n+1)\{n-1\}$$

$$= \frac{1}{12}(n^2-1).$$

The random variable $X + k$ follows the general discrete uniform distribution, with range space $k+1, k+2, \ldots, k+n$. It follows from Proposition 2.3, therefore, that the variance of this general distribution is also $\frac{1}{12}(n^2-1)$. We previously found, in agreement with Proposition 2.2, that its expected value is $\frac{1}{2}(n+1)+k$.

Example 2.9 – continued

When the continuous random variable X follows a Be(α, β) distribution, with

$$f_X(x) = \frac{x^{\alpha-1}(1-x)^{\beta-1}}{\mathbb{B}(\alpha, \beta)}, \qquad 0 \leq x \leq 1,$$

then

$$\mathbb{E}(X^2) = \int_0^1 x^2 f(x)\, dx$$

$$= \frac{1}{\mathbb{B}(\alpha, \beta)} \int_0^1 x^{\alpha+1}(1-x)^{\beta-1} dx$$

$$= \frac{1}{\mathbb{B}(\alpha, \beta)} \mathbb{B}(\alpha+2, \beta)$$

$$= \frac{\Gamma(\alpha+\beta)}{\Gamma(\alpha)\Gamma(\beta)} \frac{\Gamma(\alpha+2)\Gamma(\beta)}{\Gamma(\alpha+\beta+2)}$$

$$= \frac{\Gamma(\alpha+\beta)}{(\alpha+\beta+1)(\alpha+\beta)\Gamma(\alpha+\beta)} \frac{(\alpha+1)\alpha\Gamma(\alpha)}{\Gamma(\alpha)}$$

$$= \frac{(\alpha+1)\alpha}{(\alpha+\beta+1)(\alpha+\beta)}$$

and

$$\mathrm{var}(X) = \frac{(\alpha+1)\alpha}{(\alpha+\beta+1)(\alpha+\beta)} - \frac{\alpha^2}{(\alpha+\beta)^2}$$

$$= \frac{\alpha\beta}{(\alpha+\beta+1)(\alpha+\beta)^2}$$

Example 2.11

In some sampling situations, a finite population consists of just two distinct types of object. For example, voters in a local constituency might have to choose between two candidates in an opinion poll or consumers might be asked whether they enjoy or do not enjoy a particular product. This general situation can be represented using a bag that contains N beads, which are all identical apart from their colour. M of the beads are red and the remaining $N - M$ are black. A small number of beads, n, are sampled at random without replacement. For simplicity, we shall assume that $n < M \ (<N)$ and $n < N - M$. Let the random variable X be the number of red beads in the sample.

Altogether, there are $\binom{N}{n}$ different ways of choosing a sample of size n from the population of beads in the bag. Random sampling suggests that all of these samples are equally likely to be obtained.

$R_X = \{0, 1, \ldots, n\}$. For $x \in R_X$, there are $\binom{M}{x}$ different ways of obtaining x red beads from the M red beads in the population. For each of these ways of obtaining x red beads, there are $\binom{N-M}{n-x}$ different ways of obtaining $n - x$ black beads (to make up the total sample of n beads). So there are $\binom{M}{x}\binom{N-M}{n-x}$ ways of obtaining a random sample of n beads, of which exactly x are red and the remaining $n - x$ are black. This means that

$$P(X = x) = \binom{M}{x}\binom{N-M}{n-x} \bigg/ \binom{N}{n}, \qquad x = 0, 1, \ldots, n.$$

This is called a *hypergeometric* distribution. The hypergeometric identity (Result A1.11) shows that it is well defined.

In order to find $\mathbb{E}(X)$, we shall find it convenient to use the following result (proved in Exercise 18 below):

$$k \cdot \binom{K}{k} = K \cdot \binom{K-1}{k-1}.$$

Hence,

$$\mathbb{E}(X) = \sum_{x=0}^{n} x \cdot \binom{M}{x}\binom{N-M}{n-x} \bigg/ \binom{N}{n}$$

$$= \sum_{x=1}^{n} M \cdot \binom{M-1}{x-1}\binom{N-M}{n-x} \bigg/ \binom{N}{n}$$

(no contribution at $x = 0$)

$$= \frac{M}{\binom{N}{n}} \sum_{x=1}^{n} \binom{M-1}{x-1}\binom{(N-1)-(M-1)}{(n-1)-(x-1)}$$

(for the hypergeometric identity)

$$= \frac{M}{\binom{N}{n}} \sum_{y=0}^{n-1} \binom{P}{y}\binom{(N-1)-P}{(n-1)-y}$$

(setting $y = x - 1$ and $P = M - 1$)

$$= \frac{M}{\binom{N}{n}} \binom{N-1}{n-1}$$

$$= n\frac{M}{N}.$$

This means that, on average over all possible random samples, the proportion of red beads in the sample (x/n) is the same as the proportion in the population.

In order to find $\text{var}(X)$, it is easier to obtain $\mathbb{E}(X(X-1))$ than $\mathbb{E}(X^2)$. It can easily be shown that $\mathbb{E}(X(X-1)) = \mathbb{E}(X^2 - X) = \mathbb{E}(X^2) - \mathbb{E}(X)$, i.e. $\mathbb{E}(X^2) = \mathbb{E}(X(X-1)) + \mathbb{E}(X)$. We have

$$\mathbb{E}(X(X-1)) = \sum_{x=0}^{n} x(x-1) \cdot \binom{M}{x}\binom{N-M}{n-x} \bigg/ \binom{N}{n}$$

$$= \sum_{x=2}^{n} M(M-1) \cdot \binom{M-2}{x-2}\binom{N-M}{n-x} \bigg/ \binom{N}{n}$$

$$= \frac{M(M-1)}{\binom{N}{n}} \sum_{x=2}^{n} \binom{M-2}{x-2}\binom{(N-2)-(M-2)}{(n-2)-(x-2)}$$

$$= \frac{M(M-1)}{\binom{N}{n}} \sum_{z=0}^{n-2} \binom{Q}{z}\binom{(N-2)-Q}{(n-2)-z}$$

setting $z = x - 2$ and $Q = M - 2$

$$= \frac{M(M-1)}{\binom{N}{n}}\binom{N-2}{n-2}$$

$$= \frac{M(M-1)}{N(N-1)} n(n-1).$$

So,

$$\text{var}(X) = \mathbb{E}(X(X-1)) + \mathbb{E}(X) - (\mathbb{E}(X))^2$$

$$= \frac{M(M-1)n(n-1)}{N(N-1)} + \frac{Mn}{N} - \frac{M^2 n^2}{N^2}$$

$$= \frac{Mn\{(M-1)(n-1)N + N(N-1) - Mn(N-1)\}}{N^2(N-1)}$$

$$= \frac{Mn(N-M)(N-n)}{N^2(N-1)}$$

$$= n\frac{M}{N}\left(1 - \frac{M}{N}\right)\frac{N-n}{N-1}.$$

In Exercise 3 below, these results are compared with the corresponding results when the sample is drawn with replacement.

Notice that $\text{var}(X)$ is measured in the square of the units in which X itself is measured. For this reason, an alternative measure of spread is sometimes preferred that is measured in the same units as X. This measure, known as the *standard deviation* of X, is defined by

$$\sigma(X) = \sqrt{\text{var}(X)}.$$

Exercises

1 Suppose that the discrete random variable X follows the Ge(θ) distribution for some $0 < \theta < 1$:

$$P(X = x) = (1 - \theta)^{x-1}\theta, \qquad x = 1, 2, \ldots.$$

Find $\mathbb{E}(X)$, $\mathbb{E}(X(X - 1))$ and hence var(X). [*Hint*: some useful results on geometric series may be found in Appendix A1.]

2 Suppose that the discrete random variable X follows the Bi(n, θ) distribution for some positive integer n and some $0 < \theta < 1$:

$$P(X = x) = \binom{n}{x}\theta^x(1 - \theta)^{n-x}, \qquad x = 0, 1, \ldots, n.$$

Find $\mathbb{E}(X)$, $\mathbb{E}(X(X - 1))$ and hence var(X). [*Hint*: see Appendix A1.]

3 Consider again the problem of sampling from a finite distribution, which was introduced in Example 2.11 above. Suppose now that the sample is obtained with replacement. Show that the random variable X now follows a binomial distribution, and write down $\mathbb{E}(X)$ and var(X). Under what conditions on n, M and N do sampling with and without replacement give similar expected values and variances?

4 Repeat Exercise 2 for the Poisson distribution. Verify that $\mathbb{E}(X)$ and var(X) are equal in this case.

5 The *negative binomial distribution* NeBi(k, θ) is a generalization of the geometric distribution in the following sense. If a sequence of independent trials is conducted, on each of which there is success probability θ, then the number of the trial on which the kth success is obtained is a discrete random variable, $X \sim$ NeBi(k, θ). Show that

$$P(X = x) = \binom{x-1}{k-1}(1 - \theta)^{x-k}\theta^k, \qquad x = k, k+1, \ldots.$$

Use the results listed in Appendix A1 to show that this probability mass function is well defined. Also obtain $\mathbb{E}(X)$, $\mathbb{E}(X(X + 1))$ and hence var(X). [*Hint*: note well which expected values you are being asked to find here.]

6 The discrete random variable X has the following probability mass function:

$$p_X(x) = \frac{1}{x(x + 1)}, \qquad x = 1, 2, \ldots.$$

Show that this is a valid probability mass function. Show also that the expected value of X does not exist.

7 Suppose that the continuous random variable X follows a uniform distribution on the interval $[a, b]$, i.e. X has probability density function

$$f(x) = \frac{1}{b-a}, \qquad a < x < b.$$

Show that $\mathbb{E}(X) = \frac{1}{2}(a+b)$ and $\text{var}(X) = (b-a)^2/12$.

8 Suppose that the continuous random variable X follows the *Laplace distribution*

$$f(x) = \frac{\theta}{2} \exp(-\theta|x|), \qquad x \in \mathbb{R},$$

for some $\theta > 0$. Show that $\mathbb{E}(X) = 0$ and $\text{var}(X) = 2/\theta^2$.

9 The continuous random variable U follows a beta distribution with probability density function

$$f(u) = \frac{(m+n-1)!}{(m-1)!(n-1)!} u^{m-1}(1-u)^{n-1}, \qquad 0 < u < 1,$$

where m and n are positive integers. Find the expected value and variance of U.

10 Suppose that the continuous random variable X has a distribution that is symmetric around $x = c$. In other words, $f(x) = f(2c - x)$ for all possible values of x. Assuming that the expected value of X is finite, show that the expected value is equal to c.

11 Suppose that the continuous random variable X follows the *Cauchy distribution*

$$f(x) = \frac{\theta}{\pi} \cdot \frac{1}{\theta^2 + x^2}, \qquad x \in \mathbb{R},$$

for some $\theta > 0$. Although this distribution is symmetric about 0, show that $\mathbb{E}(X)$ does not exist because the appropriate integral does not converge.

12 Suppose that the continuous random variable X has a distribution that is symmetric around 0. In other words, $f(x) = f(-x)$ for all possible x. Show that:

(a) $F(x) = 1 - F(-x)$ for all possible x;

(b) $\int_{-c}^{c} F(x)\,dx = c$ for all real values, c.

13 Let X be a binary random variable with probability mass function

x	a	b
$p(x)$	$\frac{1}{2}$	$\frac{1}{2}$

(where $b > a$). Find $\mu = \mathbb{E}(X)$ and $\sigma^2 = \text{var}(X)$. Express a and b in terms of μ and σ^2.

14 Suppose that the binary random variable X has the following probability mass function

x	a	b
$p(x)$	θ	$1-\theta$

(where $b > a$ and $1 > \theta > 0$). Find $\mu = \mathbb{E}(X)$ and $\sigma^2 = \text{var}(X)$, and show that

$$a = \mu - \sigma\sqrt{\frac{1-\theta}{\theta}}, \qquad b = \mu + \sigma\sqrt{\frac{\theta}{1-\theta}}.$$

15 Suppose that X is a discrete random variable. Prove that

$$\text{var}(X) = \mathbb{E}(X^2) - \{\mathbb{E}(X)\}^2.$$

16 Suppose that the continuous random variable X has the gamma distribution

$$f(x) = \frac{\theta^\alpha x^{\alpha-1} e^{-\theta x}}{\Gamma(\alpha)}, \qquad x > 0,$$

for some $a > 0$ and $\theta > 0$.

(a) Show that, for any non-negative integer, k,

$$\mathbb{E}(X^k) = \frac{\Gamma(\alpha+k)}{\theta^k \Gamma(\alpha)}.$$

(b) Hence show that $\mathbb{E}(X) = \dfrac{\alpha}{\theta}$ and $\text{var}(X) = \dfrac{\alpha}{\theta^2}$. [*Hint*: see Appendix A2.]

(c) Show that the exponential distribution, $\text{Ex}(\theta)$, is a special case of the gamma distribution, and thus deduce its expected value and variance.

17 The continuous random variable X follows a *Weibull distribution* if it has the probability density function

$$f(x) = \alpha\theta x^{\alpha-1} \exp(-\theta x^\alpha), \qquad x > 0,$$

for some $a > 0$ and $\theta > 0$.

(a) Show that

$$\mathbb{E}(X) = \frac{1}{\theta^{1/\alpha}} \Gamma\left(1 + \frac{1}{\alpha}\right), \qquad \mathbb{E}(X^2) = \frac{1}{\theta^{2/\alpha}} \Gamma\left(1 + \frac{2}{\alpha}\right).$$

Find $\text{var}(X)$.

(b) The exponential distribution is a special case of the Weibull distribution. Verify that the above expressions give the correct values in this special case.

18 Suppose that k and K are integers such that $1 \le k \le K$. Show that

$$k\binom{K}{k} = K\binom{K-1}{k-1}.$$

19 If the discrete random variable X has the range space $\{0, 1, 2, \dots\}$, show that

$$\mathbb{E}(X) = \sum_{x=0}^{\infty} \{1 - F(x)\}.$$

[*Hint*: write out the terms of the usual expression for the expected value, and rearrange them.]

20 If the continuous random variable X has the range space $[0, \infty)$, show that

$$\mathbb{E}(X) = \int_0^{\infty} \{1 - F(x)\} \, dx.$$

[*Hint*: replace $1 - F(x)$ by the integral of $f(t)$ for t from x to infinity and change the order of integration.]
    Confirm that this result holds in the case where X follows an exponential distribution.

21 Let X be a continuous random variable with range space $[0, \infty)$, and let k be any positive constant. Show that

$$\mathbb{E}(X) \ge k \cdot P(X > k),$$

i.e.

$$P(X > k) \le \frac{\mathbb{E}(X)}{k}.$$

2.3 The moment-generating function

The expected value and variance of a random variable are examples of its moments. The *r*th *moment* of X ($r = 1, 2, \dots$), when it exists, is defined to be $\mu_r' = \mathbb{E}(X^r)$.
    The *r*th *central moment* of X ($r = 1, 2, \dots$), when it exists, is $\mu_r = \mathbb{E}\{[X - \mathbb{E}(X)]^r\}$. This means that $\text{var}(X)$ is the second central moment of X.

The *moment-generating function* (m.g.f.) of X is defined as follows:

$$M_X(t) = \mathbb{E}(e^{Xt}) = \begin{cases} \sum_{x \in R_X} e^{xt} p_X(x), & \text{when } X \text{ is discrete,} \\ \int_{R_X} e^{xt} f_X(x) \, dx, & \text{when } X \text{ is continuous.} \end{cases}$$

$M_X(t)$ is only well defined when the relevant series or integral is absolutely convergent. There are distributions for which the moment-generating function does not exist at all. In other cases, $M_X(t)$ exists for some but not all values of t. (For this reason, many authors prefer to introduce and use the *characteristic function* rather than the moment-generating function. This always exists, but is obtained by complex (as opposed to real) integration.)

Using the Maclaurin expansion of e^{xt}, it is possible to write:

$$M_X(t) = \int_{R_X} e^{xt} f_X(x) \, dx$$

$$= \int_{R_X} \left\{ 1 + xt + \frac{(xt)^2}{2!} + \frac{(xt)^3}{3!} + \cdots \right\} f_X(x) \, dx$$

$$= \int_{R_X} f_X(x) \, dx + t \int_{R_X} x \, f_X(x) \, dx + \frac{t^2}{2!} \int_{R_X} x^2 \, f_X(x) \, dx + \cdots$$

$$= 1 + t \, \mathbb{E}(X) + \frac{t^2}{2!} \, \mathbb{E}(X^2) + \cdots .$$

So

$$M'_X(t) = \frac{dM_X(t)}{dt} = \mathbb{E}(X) + t \, \mathbb{E}(X^2) + \frac{t^2}{2!} \, \mathbb{E}(X^3) + \cdots$$

and

$$M'_X(0) = \mathbb{E}(X)$$

Also

$$M''_X(t) = \frac{d^2 M_X(t)}{dt^2} = \mathbb{E}(X^2) + t\mathbb{E}(X^3) + \cdots$$

and

$$M''_X(0) = \mathbb{E}(X^2)$$

In general, the rth derivative $(r = 1, 2, 3, \ldots)$ is

$$M_X^{(r)}(t) = \frac{d^r}{dt^r} M_X(t) = \mathbb{E}(X^r) + t\mathbb{E}(X^{r+1}) + \cdots$$

Therefore

$$M_X^{(r)}(0) = \mathbb{E}(X^r).$$

So a general method of calculating the rth moment of X is to find its moment-generating function, differentiate r times and set $t = 0$. This is true for either continuous or discrete random variables.

Example 2.12 – The moment-generating function of the binomial distribution.

Suppose that $X \sim \text{Bi}(n, \theta)$. Then the m.g.f. of X can be found as follows:

$$M_X(t) = \mathbb{E}(e^{Xt})$$

$$= \sum_{x=0}^{n} e^{xt} \binom{n}{x} \theta^x (1-\theta)^{n-x}$$

$$= \sum_{x=0}^{n} \binom{n}{x} (\theta e^t)^x (1-\theta)^{n-x}$$

$$= \{\theta e^t + 1 - \theta\}^n \qquad \text{(binomial theorem)}.$$

Therefore,

$$M_X'(t) = n\{\theta e^t + 1 - \theta\}^{n-1} \theta e^t$$
$$M_X''(t) = [n(n-1)\{\theta e^t + 1 - \theta\}^{n-2} \theta e^t]\theta e^t + [n\{\theta e^t + 1 - \theta\}^{n-1}]\theta e^t.$$

So,

$$\mathbb{E}(X) = M_X'(0) = n\{\theta e^0 + 1 - \theta\}^{n-1} \theta e^0 = n\{\theta.1 + 1 - \theta\}^{n-1}.\theta.1 = n\theta,$$
$$\mathbb{E}(X^2) = M_X''(0) = n(n-1)\{\theta e^0 + 1 - \theta\}^{n-2} \theta e^0 \theta e^0 + n\{\theta e^0 + 1 - \theta\}^{n-1} \theta e^0$$
$$= n(n-1)\{\theta.1 + 1 - \theta\}^{n-2}.\theta.1.\theta.1 + n\{\theta.1 + 1 - \theta\}^{n-1}.\theta.1$$
$$= n(n-1)\theta^2 + n\theta,$$
$$\text{var}(X) = \mathbb{E}(X^2) - \{\mathbb{E}(X)\}^2 = n^2\theta^2 - n\theta^2 + n\theta - (n\theta)^2$$
$$= -n\theta^2 + n\theta = n\theta(1-\theta).$$

Example 2.13 – The moment-generating function of the normal distribution.

Suppose that $X \sim \text{N}(\mu, \sigma^2)$, for $\sigma > 0$. Then

$$M_X(t) = \mathbb{E}(e^{Xt}) = \int_{-\infty}^{\infty} e^{xt} \frac{1}{\sqrt{2\pi\sigma^2}} \exp\left\{-\frac{1}{2}\left(\frac{x-\mu}{\sigma}\right)^2\right\} dx.$$

The exponent (power of e) in this integral is

$$xt - \frac{1}{2}\left(\frac{x-\mu}{\sigma}\right)^2 = -\{(x-\mu)^2 - 2xt\sigma^2\}/2\sigma^2$$

$$= -\{[x - (\mu + t\sigma^2)]^2 - 2\mu t\sigma^2 - t^2\sigma^4\}/2\sigma^2$$

$$\text{[after completing the square]}$$

$$= -\{[x - (\mu + t\sigma^2)]^2\}/2\sigma^2 + \mu t + \tfrac{1}{2}t^2\sigma^2$$

Therefore

$$M_X(t) = \exp\left(t\mu + \frac{1}{2}t^2\sigma^2\right) \int_{-\infty}^{\infty} \frac{1}{\sqrt{2\pi\sigma^2}} \exp\left\{-\frac{1}{2}\left(\frac{x - [\mu + t\sigma^2]}{\sigma}\right)^2\right\} dx$$

$$= \exp\left(t\mu + \frac{1}{2}t^2\sigma^2\right).$$

The last step follows because the integrand is the p.d.f. of an $N(\mu + t\sigma^2, \sigma^2)$ random variable.

We can now use the moment-generating function to show that $\mathbb{E}(X) = \mu$ and $\text{var}(X) = \sigma^2$.

$$M_X'(t) = \exp\left(t\mu + \frac{1}{2}t^2\sigma^2\right) \cdot (\mu + t\sigma^2)$$

$$\therefore \quad \mathbb{E}(X) = M_X'(0) = \mu,$$

$$M_X''(t) = \exp\left(t\mu + \frac{1}{2}t^2\sigma^2\right) \cdot (\mu + t\sigma^2)^2 + \exp\left(t\mu + \frac{1}{2}t^2\sigma^2\right) \cdot \sigma^2$$

$$\therefore \quad \mathbb{E}(X^2) = M_X''(0) = \mu^2 + \sigma^2$$

$$\therefore \quad \text{var}(X) = \mathbb{E}(X^2) - \{\mathbb{E}(X)\}^2 = \sigma^2.$$

It might already appear from these two examples that the moment-generating function is useful since, in many cases, it allows us to obtain expected values and variances more conveniently than through direct summation or integration. The moment-generating function, however, is most useful because it characterizes a probability distribution uniquely. We state without proof the following important property.

Proposition 2.5 – The uniqueness property of moment-generating functions

Let X and Y be random variables. If the moment-generating functions $M_X(t)$ and $M_Y(t)$ exist and are equal for all values of t in an interval $(-h, h)$, where h is any positive constant, then the distribution functions F_X and F_Y are the same.

This uniqueness property of moment-generating functions will be used extensively to derive theoretical results in later chapters of this book. We will come to one important example after proving the following result.

Proposition 2.6

Suppose that X is a random variable with moment-generating function $M_X(t)$. Let $Y = aX + b$, for some constants a and b. Then Y has moment-generating function

$$M_Y(t) = e^{bt} M_X(at).$$

Proof.

$$M_Y(t) = \mathbb{E}\left(e^{Yt}\right) = \mathbb{E}\left(e^{(aX+b)t}\right) = \mathbb{E}\left(e^{bt}e^{Xat}\right) = e^{bt}\mathbb{E}\left(e^{Xat}\right) = e^{bt}M_X(at). \quad \blacksquare$$

Example 2.13 – continued

Suppose that $X \sim N(\mu, \sigma^2)$, for $\sigma > 0$. Then

$$M_X(t) = \exp\left(t\mu + \frac{1}{2}t^2\sigma^2\right).$$

Define the random variable Z by

$$Z = \frac{X - \mu}{\sigma} = \frac{1}{\sigma}X + \left(-\frac{\mu}{\sigma}\right).$$

Then, using Proposition 2.6, Z has moment-generating function

$$M_Z(t) = \exp\left(-\frac{\mu}{\sigma}t\right)M_X\left(\frac{1}{\sigma}t\right) = \exp\left(-\frac{\mu}{\sigma}t\right)\exp\left(\frac{1}{\sigma}t\mu + \frac{1}{2}\left(\frac{1}{\sigma}t\right)^2\sigma^2\right)$$

$$= \exp\left(\frac{1}{2}t^2\right).$$

This is the moment-generating function of the N(0, 1) or standard normal distribution. The uniqueness property (Proposition 2.5) tells us that $Z \sim N(0,1)$.

So, any normally distributed random variable can be transformed to a standard normal by first subtracting its expected value and then dividing by its standard deviation. This important property allows probabilities associated with an arbitrary normal distribution to be calculated using the standard normal distribution function that is tabulated in Table B2. The following example illustrates this.

Example 2.14

A large-scale survey has suggested that the age (in years) of pregnant women in the UK at the start of pregnancy is a normal random variable with expected value 28.7 and standard deviation 4.7. The probability that a randomly selected mother is greater than 35 years old at the start of pregnancy can be found as follows.

Let the random variable X be the age at the start of pregnancy. Then, $X \sim N(28.7, 4.7^2)$. Define the random variable Z by

$$Z = \frac{X - 28.7}{4.7}.$$

Then $Z \sim N(0, 1)$. Also,

$$P(X > 35) = P\left(\frac{X - 28.7}{4.7} > \frac{35 - 28.7}{4.7} = 1.34\right) = P(Z > 1.34)$$

$$= 1 - \Phi(1.34) = 0.0901.$$

Exercises

1 Suppose that the discrete random variable X follows the Po(θ) distribution, for $\theta > 0$. Find the moment-generating function of X and hence find its expected value and variance.

2 Suppose that the continuous random variable X follows the Ex(θ) distribution, for $\theta > 0$. Show that X has moment-generating function

$$M_X(t) = \frac{\theta}{\theta - t}, \qquad t < \theta.$$

Now define the new random variable Y by $Y = kX$, for some real constant $k > 0$. Use moment-generating functions to prove that Y also follows an exponential distribution.

3 An inventory of psychiatric illness has been devised which includes a scale for measuring the level of an individual's anxiety. This scale has been tested on the general, non-psychiatric population and the values obtained in this population are known to follow a normal distribution with expected value 50 and standard deviation 10. When used with a new subject, a score of 63 or more is taken to indicate that the individual might be suffering from 'case' levels of anxiety. Show that about 10% of the general population would mistakenly be considered as possible cases using this scale.

4 A disorder of the pancreas, known as acute pancreatitis, occurs in two distinct forms, Type A and Type G. Sixty per cent of all cases of acute pancreatitis are of Type A, the remainder are of Type G. The causes of the two forms of the disorder, and hence the required treatment, are quite different, but it is not easy to distinguish between them by clinical examination alone. Consequently, a biochemical test has been developed that is based on the continuous random variable X, which is the logarithm of the level of alkaline phosphate in the blood.

In patients with the Type A form of this disorder, X follows a normal distribution with expected value 5.2 and standard deviation 0.25. In patients with the Type G form, X follows a normal distribution with expected value 5.7 and standard deviation 0.20.

It has been decided to diagnose a patient suffering from acute pancreatitis as Type A whenever X is less than 5.5 and as Type G otherwise. Given that a patient suffering from acute pancreatitis is diagnosed as Type A, find the conditional probability that the patient is really of Type G.

Suppose now that a patient suffering from acute pancreatitis will be diagnosed as Type A whenever X is less than some value c and as Type G otherwise. Write down an expression for the probability that a randomly selected sufferer from acute pancreatitis is correctly diagnosed. By treating this probability as a function of c, show that it is maximized when $c = 5.5$ (approximately).

5 Suppose that the random variable X follows a normal distribution. Use moment-generating functions to prove that the random variable $Y = cX + d$ also follows a normal distribution (for any real constants $c \neq 0$ and d).

6 (a) Suppose that X is a random variable with moment-generating function $M_X(t)$. Define the new random variable Y by $Y = e^X$. Show that, for $k = 1, 2, \ldots,$

$$\mathbb{E}(Y^k) = M_X(k).$$

(b) Now suppose that $X \sim N(\mu, \sigma^2)$, for $\sigma > 0$. Then, $Y = e^X$ is said to follow a *lognormal* distribution. Find the expected value and variance of Y.

Summary

This chapter has introduced the main results about random variables that will be used throughout the rest of the book. The distinction between a discrete and a continuous random variable is particularly important. Key concepts include the distribution function, probability mass function and probability density function. Discussion of the expected value and variance led on to a brief treatment of moments and central moments in general. The moment-generating function has been introduced as a convenient way of obtaining the moments of a distribution, though its main use in this book will be to prove theoretical results about the distribution of a function of a random variable.

3
Transforming and simulating random variables

It is often necessary or useful to transform a random variable, for example multiplying it by a constant or taking its square. Transformation of a random variable sometimes allows us to rewrite a problem involving a random variable of unknown type as a problem involving a standard distribution, which greatly reduces the amount of work required to solve it. Functions of a random variable are themselves random variables, and we have already seen in Chapter 2 how the probability distribution of some simple (linear) functions of a random variable may be determined using moment-generating functions. In this chapter, we show how to use other methods to deal with more general functions of random variables. This chapter also begins to introduce simulation, many procedures for which depend on the idea of transforming random variables. Simulation allows us to investigate problems in probability that cannot be solved using analytical methods. Even when an analytical solution is available, greater insight may often be gained through simulation.

3.1 Functions of a random variable

Example 3.1

Suppose that a particle is moving along a straight line. At discrete time points, $1, 2, \ldots, n$, the particle may move forward one unit (with probability θ) or back one unit (with probability $1 - \theta$). At any time point, then, the further displacement of the particle is either $+1$ units, with probability θ, or -1 units, with probability $1 - \theta$. We will assume that moves made at different time points are independent.

Let the random variable X be the number of times the particle moves in a forward direction. Then $X \sim \text{Bi}(n, \theta)$, so that

$$P(X = x) = \binom{n}{x} \theta^x (1 - \theta)^{n-x} \qquad (x = 0, 1, \ldots, n),$$

and

$$\mathbb{E}(X) = n\theta, \qquad \text{var}(X) = n\theta(1 - \theta).$$

Now let the random variable Y be the total displacement of the particle from its starting position after n moves. $X = x$ if and only if the particle makes exactly x forward moves, and therefore $n - x$ backward moves, so $X = x$ if and only if Y takes the value $x - (n - x) = 2x - n$. Therefore, $Y = 2X - n$. Y must have range space $R_X = \{-n, -n + 2, \ldots, n - 2, n\}$. For $y \in R_Y$,

$$P(Y = y) = P(2X - n = y) = P(X = \tfrac{1}{2}(n + y))$$

$$= \binom{n}{(n + y)/2} \theta^{(n+y)/2} (1 - \theta)^{(n-y)/2}.$$

In order to determine $\mathbb{E}(Y)$ or $\mathrm{var}(Y)$ directly, we would need to sum series related to this probability distribution function, which could be quite challenging. However, we can exploit the relationship between X and Y to write immediately (using Propositions 2.2 and 2.3):

$$\mathbb{E}(Y) = \mathbb{E}(2X - n) = 2\mathbb{E}(X) - n = 2n\theta - n,$$
$$\mathrm{var}(Y) = \mathrm{var}(2X - n) = 4\mathrm{var}(X) = 4n\theta(1 - \theta).$$

In general, when X is a discrete random variable and the new random variable Y is defined by $Y = h(X)$, then Y must also be a discrete random variable; it cannot be continuous. The probability mass function of Y is found using $P(Y = y) = P(\{x: h(x) = y\})$. $\mathbb{E}(Y)$ and $\mathrm{var}(Y)$ may usually be found directly from this probability distribution or indirectly as functions of $\mathbb{E}(X)$ and $\mathrm{var}(X)$.

When X is a continuous random variable, and $Y = h(X)$ is defined by a continuous function of X, then the probability density function of Y must in general be derived using the distribution function of X.

Example 3.2

Suppose that X follows the standard normal distribution, so that X has probability density function

$$f_X(x) = \frac{1}{\sqrt{2\pi}} \exp\left\{-\frac{x^2}{2}\right\}, \qquad -\infty < x < \infty.$$

We have seen in Chapter 2 that X has a distribution function, denoted $\Phi(x)$, that cannot be written in closed algebraic form.

We shall derive the distribution function of $Y = 1/X$. The range space of Y is $R_Y = \{y: y \neq 0\}$. It is necessary to consider the two cases $y < 0$ and $y > 0$ separately. For $y < 0$,

$$F_Y(y) = P(Y \leq y) = P\left\{\frac{1}{X} \leq y\right\} = P\left\{\frac{1}{y} \leq X \leq 0\right\}$$

$$= P\{X \leq 0\} - P\left\{X \leq \frac{1}{y}\right\} = 0.5 - \Phi\left\{\frac{1}{y}\right\}.$$

For $y > 0$,

$$F_Y(y) = P(Y < 0) + P(0 < Y \le y) = P(X < 0) + P\left\{\frac{1}{y} \le X\right\}$$

$$= 0.5 + 1 - \Phi\left\{\frac{1}{y}\right\} = 1.5 - \Phi\left\{\frac{1}{y}\right\}.$$

In order to derive the probability density function of Y, we must use the fact that the derivative of $\Phi(y)$ is the corresponding probability density function,

$$f(y) = \frac{1}{\sqrt{2\pi}} \exp\left\{-\frac{y^2}{2}\right\}.$$

So

$$f_Y(y) = -f\left\{\frac{1}{y}\right\} \cdot \left\{-\frac{1}{y^2}\right\} = \frac{1}{\sqrt{2\pi y^2}} \exp\left\{-\frac{1}{2y^2}\right\}, \qquad \text{for } y < 0$$

and

$$f_Y(y) = -f\left\{\frac{1}{y}\right\} \cdot \left\{-\frac{1}{y^2}\right\} = \frac{1}{\sqrt{2\pi y^2}} \exp\left\{-\frac{1}{2y^2}\right\}, \qquad \text{for } y > 0,$$

i.e.

$$f_Y(y) = \frac{1}{\sqrt{2\pi y^2}} \exp\left\{-\frac{1}{2y^2}\right\}, \qquad y \ne 0.$$

Figure 3.1 shows a diagram of this general procedure for finding the probability density function of the continuous random variable Y, when Y can be expressed as a function of another continuous random variable X. In general, it is not possible to short-circuit this procedure, but in certain circumstances it

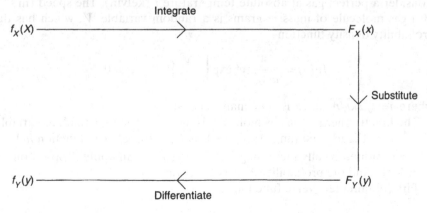

Figure 3.1 *Process for finding the p.d.f. of a function of a continuous random variable*

is possible to go from $f_X(x)$ to $f_Y(y)$ without having to work out the distribution functions explicitly. The two most important circumstances of this kind are described in Propositions 3.1 and 3.2.

Proposition 3.1

Suppose that X is a continuous random variable with range space $R_X = \{x : a < x < b\}$ and probability density function $f_X(x)$, $x \in R_X$. Define the random variable Y by $Y = h(X)$, where h is a strictly increasing or strictly decreasing, differentiable function on R_X. Then the range space of Y is $R_Y = \{y : \min(h(a), h(b)) < y < \max(h(a), h(b))\}$ and the p.d.f. of Y is:

$$f_Y(y) = f_X\{h^{-1}(y)\} \left| \frac{dx}{dy} \right|, \qquad y \in R_Y.$$

Proof. We prove the case where $h(x)$ is a strictly increasing function. It is clear that $R_Y = \{y : h(a) < y < h(b)\}$. For any y in this range,

$$F_Y(y) = P(Y \le y) = P\{h(X) \le y\} = P\{X \le h^{-1}(y)\} = F_X\{h^{-1}(y)\},$$

and so

$$f_Y(y) = \frac{d}{dy} F_X(h^{-1}(y)) = \frac{d}{dx} F_X(h^{-1}(y)) \frac{dx}{dy}$$

$$= f_X(h^{-1}(y)) \frac{dx}{dy}. \qquad \blacksquare$$

The proof for the case where $h(X)$ is a strictly decreasing function is left as an exercise.

Example 3.3

Consider a perfect gas at absolute temperature t (kelvin). The speed $(\mathrm{m\,s^{-1}})$ of a gas molecule of mass m grams is a random variable, V, which has the probability density function

$$f_V(v) = \frac{4}{a^3 \sqrt{\pi}} v^2 \exp\left\{ -\frac{v^2}{a^2} \right\}, \qquad v > 0,$$

where $a = \sqrt{2kt/m}$ and k is Boltzmann's constant.

The kinetic energy of a gas molecule is also a continuous random variable, $Y = \frac{1}{2}mV^2$. Clearly, the range space of Y is $R_Y = (0, \infty)$. The function $h(V) = \frac{1}{2}mV^2$ is monotonically increasing for $V > 0$, so we can apply Proposition 3.1 in order to find the probability density function of Y.

First, obtain the inverse function

$$Y = h(V) = \tfrac{1}{2}mV^2, \qquad \text{so } V = h^{-1}(Y) = \sqrt{\frac{2Y}{m}}.$$

Then

$$\frac{\mathrm{d}V}{\mathrm{d}Y} = \sqrt{\frac{1}{2mY}}$$

So

$$f_Y(y) = f_V\{h^{-1}(y)\} \left| \frac{\mathrm{d}v}{\mathrm{d}y} \right|$$

$$= \frac{4}{a^3\sqrt{\pi}} \frac{2y}{m} \exp\left\{-\frac{2y}{a^2m}\right\} \frac{1}{\sqrt{2my}}$$

$$= \frac{4\sqrt{2}}{a^3m\sqrt{m}\sqrt{\pi}} \sqrt{y} \exp\left\{-\frac{2y}{a^2m}\right\}.$$

Proposition 3.2

Suppose that the continuous random variable X has the range space $R_X = (-\infty, \infty)$, and probability density function $f_X(x)$ on R_X. Define the random variable Y by $Y = X^2$. Then Y has probability density function:

$$f_Y(y) = \frac{1}{2\sqrt{y}} \left[f_X\left(\sqrt{y}\right) + f_X\left(-\sqrt{y}\right) \right], \qquad y > 0.$$

Proof. Since Y is not a strictly increasing function of X on the whole of R_X, Proposition 3.1 does not apply. We proceed by the general method. $R_Y = \{y : 0 \le y\}$. For any $y \in R_Y$,

$$F_Y(y) = P(Y \le y) = P(X^2 \le y) = P(-\sqrt{y} \le X \le \sqrt{y}) = F_X(\sqrt{y}) - F_X(-\sqrt{y}).$$

Differentiating with respect to y gives

$$f_Y(y) = \frac{\mathrm{d}}{\mathrm{d}y} F_Y(y) = \frac{1}{2\sqrt{y}} f_X\left(\sqrt{y}\right) - \frac{-1}{2\sqrt{y}} f_X\left(-\sqrt{y}\right)$$

$$= \frac{1}{2\sqrt{y}} \left[f_X(\sqrt{y}) + f_X\left(-\sqrt{y}\right) \right]. \qquad \blacksquare$$

Example 3.4

Suppose that X has the standard normal distribution and let $Y = X^2$. The probability density function of X is

$$f_X(x) = \frac{1}{\sqrt{2\pi}} \exp\left(-\frac{x^2}{2}\right).$$

This is symmetric about $x = 0$. So, applying Proposition 3.2,

$$f_Y(y) = \frac{1}{2\sqrt{y}}\left[f_X(\sqrt{y}) + f_X(-\sqrt{y})\right] = \frac{1}{2\sqrt{y}}\left[2f_X(\sqrt{y})\right]$$

$$= \frac{1}{\sqrt{2\pi}\sqrt{y}}\exp\left(-\frac{y}{2}\right), \qquad y > 0.$$

This is the *chi-squared distribution* with one degree of freedom, written $Y \sim \chi_1^2$.

To end this section, we will discuss the relationship between one particular discrete distribution, the Poisson, and one continuous distribution, the exponential.

Example 3.5

Suppose that customers arrive at a bank according to a *Poisson process*. This means that the number of customers who arrive at the bank (during opening hours) in any time period of length t minutes, X_t say, is a Poisson random variable with expected value $t\theta$, for some $\theta > 0$. The numbers of customers arriving in non-overlapping time intervals are independent random variables.

The inter-arrival time is the length of time that elapses between consecutive arrivals at the bank. Suppose that a customer arrives and let the continuous random variable Y be the length of time (minutes) that elapses until the next customer arrives. Clearly, Y has range space $R_Y = (0, \infty)$. The distribution function of Y is

$$F_Y(y) = P(Y \leq y) = 1 - P(Y > y)$$

$$= 1 - P(\text{no customer arrives in the bank in the next } y \text{ minutes})$$

$$= 1 - P(X_y = 0)$$

$$= 1 - \frac{e^{-(\theta y)}(\theta y)^0}{0!}$$

$$= 1 - e^{-\theta y},$$

So that

$$f_Y(y) = \theta e^{-\theta y}, \qquad y > 0.$$

This is the probability density function of the Ex(θ) distribution. So when events occur at an average rate of θ events per unit time, according to a Poisson process, then the number of events in a time period of t minutes is a Po$(t\theta)$ random variable while the time that elapses between consecutive arrivals is an Ex(θ) random variable.

Exercises

1 In the context of Example 3.1, assume that, at each step, the particle moves forward two units with probability θ or back one unit with probability $1 - \theta$. Let the random variable Y be the position of the particle (units forward of its starting position) after n moves. Find the range space and probability distribution function of Y. Hence, or otherwise, find $\mathbb{E}(Y)$ and $\text{var}(Y)$.

2 The standard surgical procedure for removal of a certain type of skin lesion is successful in every case. The cost of one surgical treatment is £c_1. A new, non-surgical procedure is successful in a large proportion, θ, of cases $(0 < \theta < 1)$. The cost of this treatment is £c_2, where $c_2 < c_1$. A particular hospital decides that, in the future, every patient who requires treatment for this type of skin lesion will be treated by the non-surgical method first. Any patient for whom this treatment is not successful will later be treated surgically as well. Let the random variable C be the cost of treating a randomly selected patient. Show that the hospital's policy has a lower expected cost than treating everyone surgically, as long as $\theta > c_2/c_1$.

3 Each of a group of $10k$ people (for some $k \in \mathbb{Z}^+$) has to be tested for the presence of a rare disease. Blood samples are taken from all members of the group. Rather than test each sample separately, sub-samples of blood from 10 people are pooled and tested. This means that k pooled tests are conducted. If the test on a pooled sample is negative, then it may be assumed that none of these people has the disease. If the pooled test gives a positive result, then all the individual samples from that group of 10 people must be tested individually in order to determine who has the disease. (You may assume that the probability that an individual being tested has the disease is θ, independently for all the people being tested.)

   Let Y be the total number of tests that are carried out. Show how Y is related to a binomial random variable. Hence find the expected value and variance of Y.

4 If z is a real number, let $[z]$ represent the *integer part* of z, which is defined as follows:

$$\text{for } z \geq 0, [z] = m \text{ if and only if } m \leq z < m + 1,$$

$$\text{for } z < 0, [z] = m \text{ if and only if } m - 1 < z \leq m.$$

   Suppose that the continuous random variable X has a $\text{Un}(0,1)$ distribution and define the discrete random variable Y by $Y = [kX + 1]$, where k is a positive integer. Write down R_Y and find $P(Y = y)$ for all $y \in R_Y$. Deduce that Y has a discrete uniform distribution.

5 Suppose that the continuous random variable X follows a $\text{Un}(a, b)$ distribution, $b > a$. Find a function of X that follows a $\text{Un}(0, 1)$ distribution. Hence find $\mathbb{E}(X)$ and $\text{var}(X)$.

6 If a projectile is fired at an angle of X radians from the earth, at speed $v(\mathrm{m\,s^{-1}})$, then it travels a horizontal distance (metres) of $D = (v^2/g)\sin 2X$. (Here, g is the gravitational constant $9.81\ \mathrm{m\,s^{-2}}$.) Suppose that a hung-over gunner arrives on an army test range one morning. He is certain to fire a shell from his gun at speed v, but the angle at which he fires the gun is uniformly distributed on the range 0 to $\frac{1}{2}\pi$, i.e. $X \sim \mathrm{Un}(0, \frac{1}{2}\pi)$. Find the distribution function and hence the probability density function of D. Explain why Proposition 3.1 cannot be applied in this example.

7 Suppose that the continuous random variable X has an $\mathrm{N}(\mu, \sigma^2)$ distribution, and define the random variable Y by $Y = e^X$. Then Y is said to follow a *lognormal distribution*. Find the probability density function of Y and show that:

$$\mathbb{E}(Y) = \exp\{\mu + \tfrac{1}{2}\sigma^2\},$$
$$\mathrm{var}(Y) = \{\exp(\sigma^2) - 1\}\exp\{2\mu + \sigma^2\}.$$

(The expected value and variance of Y were obtained by a different method in Exercise 6 on Section 2.3.)

8 Suppose that the random variable X follows a standard normal distribution. Then Example 3.4 shows that X^2 follows a χ_1^2 distribution. Use the moment-generating function of X to obtain $\mathbb{E}(X^r)$ for $r = 1, 2, 3, 4$. Hence find the expected value and variance of the χ_1^2 distribution.

9 Suppose that $X \sim \mathrm{Ga}(\alpha, \theta)$. If k is a positive constant, show that $Y = kX \sim \mathrm{Ga}(\alpha, \theta/k)$.

10 Suppose that $X \sim \mathrm{We}(\alpha, \theta)$. Find the distribution of the following functions of X:

(a) $U = kX$ (where $k > 0$);

(b) $V = X^k$ (where $k > 0$).

Use (b) to find a function of X that follows an exponential distribution.

11 Prove Proposition 3.1 when $h(X)$ is a strictly decreasing function.

12 Suppose that X follows a $\mathrm{Be}(\alpha, \beta)$ distribution. Use the result of Exercise 11 to show that $Y = 1 - X$ follows a $\mathrm{Be}(\beta, \alpha)$ distribution. [*Hint*: the properties of the beta integral are listed in Appendix A2.]

13 Suppose that $X \sim F_{v_1, v_2}$. Show that $Y = 1/X \sim F_{v_2, v_1}$.

14 Suppose that the continuous random variable X follows a uniform distribution on the range space $(-\theta, \theta)$, where $\theta > 0$. Write down the probability density function of X. Find the probability density function of $Y = X^2$, and hence find $\mathbb{E}(Y)$ and $\mathrm{var}(Y)$.

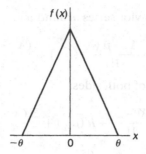

15 Suppose that the continuous random variable X has a probability density function of the form shown above on the range space $(-\theta, \theta)$, where $\theta > 0$.

Explain why $f(0) = 1/\theta$.

Write down the form of the probability density function explicitly. Find the probability density function of $Y = X^2$.

16 Suppose that X has a t_ν distribution. Show that $Y = X^2$ follows the $F_{1,\nu}$ distribution.

17 The continuous random variable X has the symmetric probability density function

$$f(x) = e^{-|x|}, \qquad x \in \mathbb{R}.$$

Find the probability density function of X^2.

3.2 Expected value and variance of a function of a random variable

When X is a continuous random variable, it can often be difficult or even impossible to determine the probability density function of $Y = h(X)$. The following results are often used, even in such cases, to find approximations to the expected value and variance of Y. Unfortunately, these results can give very misleading answers in particular cases.

Proposition 3.3

Let X be a continuous random variable with finite mean μ_X and finite standard deviation σ_X. Let $Y = h(X)$, where $h(\cdot)$ is a continuous function whose first and second derivatives exist at all points $x \in R_X$. Then

$$\mathbb{E}(Y) \approx h(\mu_X) + \frac{\sigma_X^2}{2} h''(\mu_X).$$

Proof. Expand $h(\cdot)$ as a Taylor series around μ_X:

$$Y = h(X) = h(\mu_X) + \frac{X - \mu_X}{1!} h'(\mu_X) + \frac{(X - \mu_X)^2}{2!} h''(\mu_X) + \cdots.$$

Now take expected values of both sides:

$$\mathbb{E}(Y) = h(\mu_X) + \frac{\mathbb{E}(X - \mu_X)}{1!} h'(\mu_X) + \frac{\mathbb{E}(X - \mu_X)^2}{2!} h''(\mu_X) + \cdots$$

$$= h(\mu_X) + \frac{0}{1!} h'(\mu_X) + \frac{\sigma_X^2}{2!} h''(\mu_X) + \cdots.$$

So

$$\mathbb{E}(Y) \approx h(\mu_X) + \frac{\sigma_X^2}{2} h''(\mu_X).$$

∎

Example 3.6

For an arbitrary continuous random variable X, let $Y = aX + b$ $(a \neq 0)$. Then

$$h(X) = aX + b, \qquad h'(X) = a, \qquad h''(X) = 0.$$

So

$$\mathbb{E}(Y) \approx h(\mu_X) + \frac{\sigma_X^2}{2} h''(\mu_X) = a\mathbb{E}(X) + b.$$

From Proposition 2.2, we already know that this result is exact, rather than approximate.

Example 3.7

Suppose that $X \sim N(\mu, \sigma^2)$ and let $Y = \exp(X)$. Then Y has the lognormal distribution discussed in Exercise 7 on Section 3.1. Here

$$h(X) = \exp(X), \qquad h'(X) = \exp(X), \qquad h''(X) = \exp(X)$$

So

$$\mathbb{E}(Y) \approx h(\mu) + \frac{\sigma^2}{2} h''(\mu) = e^\mu + \frac{\sigma^2}{2} e^\mu = \left(1 + \frac{\sigma^2}{2}\right) e^\mu.$$

In this case, the exact result is known to be

$$\mathbb{E}(Y) = \exp\left(\mu + \tfrac{1}{2}\sigma^2\right).$$

The diagram at the top of the next page shows the exact and approximate results for the case $\mu = 0$. The two are generally in good agreement for small values of σ, but diverge as σ increases.

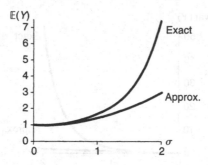

The following proposition gives the corresponding approximation for the variance of $Y = h(X)$. This approximate value is generally less accurate than that for the expected value of Y, since it is based on a first-order approximation.

Proposition 3.4

Let X be a continuous random variable with finite mean μ_X and finite standard deviation σ_X. Let $Y = h(X)$, where $h(\cdot)$ is a continuous function whose first and second derivatives exist at all points $x \in R_X$. Then

$$\text{var}(Y) \approx [h'(\mu_X)]^2\, \sigma_X^2.$$

Proof. Again expanding $h(\cdot)$ as a Taylor series around μ_X, but this time taking just the first two terms:

$$Y = h(X) \approx h(\mu_X) + \frac{(X - \mu_X)}{1!}\, h'(\mu_X).$$

Since $\mu_X, h(\mu_X)$ and $h'(\mu_X)$ are all constants, taking the variance of both sides gives

$$\text{var}(Y) \approx [h'(\mu_X)]^2\, \text{var}(X) = [h'(\mu_X)]^2\, \sigma_X^2. \qquad\blacksquare$$

Example 3.6 – continued

If $Y = h(X) = aX + b\ (a \neq 0)$, then

$$h'(X) = a$$

and

$$\text{var}(Y) \approx [h'(\mu_X)]^2 \sigma_X^2 = a^2\, \text{var}(X),$$

which we already know (Proposition 2.3) is an exact expression.

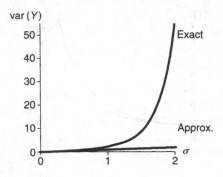

Example 3.7 – continued

$X \sim N(\mu, \sigma^2)$ and $Y = h(X) = \exp(X)$. So

$$h'(X) = \exp(X)$$

and

$$\text{var}(Y) \approx [h'(\mu_X)]^2 \sigma_X^2 = e^{2\mu}\sigma^2.$$

The exact result is

$$\text{var}(Y) = \{\exp(\sigma^2) - 1\} \exp\{2\mu + \sigma^2\}.$$

The diagram above shows a comparison of the exact and approximate standard deviations, in the case $\mu = 0$, for various values of σ. The two values diverge quickly.

Exercises

1 Let X be an arbitrary continuous random variable. Let $Y = h(X) = aX^2 + bX + c$, for some real constants a, b and c.

   (a) Use Proposition 3.3 to show that $\mathbb{E}(Y) \approx a(\mu_X^2 + \sigma_X^2) + b\mu_X + c$.

   (b) By direct integration of $(ax^2 + bx + c)f(x)$, show that (a) is an exact result.

   (c) Use Proposition 3.4 to find an approximation to $\text{var}(Y)$.

2 Suppose that X follows an $\text{Ex}(\theta)$ distribution, where $\theta \geq 2$. Find the probability density function of $U = e^X$. Hence find the expected value and variance of U. Compare these exact results with approximations obtained using Propositions 3.3 and 3.4.

3 Suppose that X follows a $\mathrm{Un}(0,1)$ distribution. Find the probability density function of $Y = \exp(X)$, and hence determine $\mathbb{E}(Y)$ and $\mathrm{var}(Y)$ exactly. Compare these results with approximations obtained using Propositions 3.3 and 3.4.

4 Suppose that X follows a $\mathrm{Ga}(\alpha, \theta)$ distribution, where $\alpha > 2$. By direct integration, find the expected value and variance of $Y = 1/X$. Compare these results with approximations obtained using Propositions 3.3 and 3.4.

3.3 Simulation using the inverse distribution function

So far in this book, we have restricted ourselves to investigating questions that have exact analytical solutions; in other words, we have been able to write down the solution as either a number or a formula. Many of the problems of interest in probability cannot be dealt with so simply, and such problems are solved by simulating the situation. This means that we try to mimic real life by an artificial process (often a computer program). We build into the mimicking process whatever probability distributions we think govern real life. We then run the program many times (typically 10,000 or more), on each occasion sampling from the basic probability distributions we have specified. Each simulation generates one value of the outcome we are interested in. We use the sample distribution of simulated results in order to estimate probabilities or parameters of interest.

We typically use simulation:

- when analytical answers are not available;

- when the analytical answers are very expensive or take a lot of time to collect;

- when we want to check an analytical result or gain some insight into its practical implication.

We begin by considering how to generate a sequence of random digits, i.e. a sequence $X_1, X_2, \ldots$ such that the X_i are independent and each is equally likely to take any of the values $0, 1, \ldots, 9$. These random digits are the basic building blocks for much work in simulation.

Perhaps the simplest methods are mechanical, based on tossing coins or dice or drawing cards from a pack or numbered cards or balls from a bag. These are very tedious, though, for generating the large number of digits required for most simulation studies.

Tables of random digits are widely available; there is one in Appendix B of this book. Although these are useful for small-scale studies, there is really no alternative to generating your own series of random digits when you are conducting a large study.

Computers are ideal for simulating random digits. Many standard software packages (such as spreadsheets) have simple built-in functions that give random digits. Alternatively, you could write your own computer program for this. It is important to realize, though, that the digits produced by most computer software are not truly 'random'; far from being independent, the digits follow each other in a predetermined fashion, and eventually a whole series of digits would appear in exactly the same order again. These are therefore called *pseudo-random* digits.

Most algorithms for generating pseudo-random digits are congruential generators. The digits are obtained using a recursion formula of the form

$$x_{n+1} = ax_n + b \pmod m,$$

where a, b and m are suitably chosen integer constants and the seed x_0 is an integer in the range 0 to $m - 1$ (and can be varied by the user). This formula means that x_{n+1} is the remainder when $ax_n + b$ is divided by m, and so each x_{n+1} is an integer in the range 0 to $m - 1$. We may then run all the x_i together in order to obtain our sequence of pseudo-random digits.

Example 3.8

Generate three pseudo-random integers in the range 0 to 9,999 using a congruential generator with $m = 10,000$, $a = 2,001$ and $b = 11$. Hence generate 12 pseudo-random digits.

Choose the seed $x_0 = 5$. Then

$$x_1 = ax_0 + b \pmod m = (2001 \times 5) + 11 \pmod{10,000}$$

$$= 10,016 \pmod{10,000} = 0016$$

$$x_2 = ax_1 + b \pmod m = (2001 \times 16) + 11 \pmod{10,000}$$

$$= 32,027 \pmod{10,000} = 2027$$

$$x_3 = ax_2 + b \pmod m = (2001 \times 2027) + 11 \pmod{10,000}$$

$$= 4,056,038 \pmod{10,000} = 6038$$

So the required pseudo-random integers are 0016, 2027, 6038 and the required pseudo-random digits are

$$0\ 0\ 1\ 6\ 2\ 0\ 2\ 7\ 6\ 0\ 3\ 8.$$

For efficiency of calculation in Example 3.8, m was chosen to be a power of 10. On a computer, it would be more efficient to use a power of 2 (since computers work in binary arithmetic).

Since m is the maximum possible number of different integers that can be generated, known as the *maximum cycle length*, m should be chosen to be large. For some choices of a and b, the number of distinct integers generated before the sequence begins to repeat will be much smaller than m. For $b > 0$, the maximum cycle length is obtained if and only if

- b and m have no common factors greater than 1;
- $a - 1$ is a multiple of every prime that divides m;
- $a - 1$ is a multiple of 4 when m is a multiple of 4.

So, when m is a power of 10, b should be set to be a prime number and a should be equal to $20c + 1$ for some c. When m is a power of 2, b should be set to an odd value and a should be equal to $4c + 1$ for some c.

Having obtained pseudo-random digits (or integers), it is very easy to obtain pseudo-random variates from the $Un(0, 1)$ distribution, as in the following example.

Example 3.8 – continued

Obtain three pseudo-random variates from the $Un(0, 1)$ distribution, working to four decimal places.

We previously obtained the pseudo-random integers 0016, 2027, 6038. From these we obtain the pseudo-random $Un(0, 1)$ variates

$$0.0016, \quad 0.2027, \quad 0.6038.$$

We can use a sequence of pseudo-random $Un(0, 1)$ variates to generate a sequence of pseudo-random variates from other distributions whose distribution function can be expressed in closed (algebraic) form. This is based on a particular result about functions of a random variable, as we shall soon see.

Suppose first that X is a discrete random variable with probability density function $p_X(x_i), i = 1, 2, \ldots$. We can generate a pseudo-random variate, x, from this distribution as follows:

- Generate a pseudo-random variate, u, from the $Un(0, 1)$ distribution.
- If $0 \le u < p_X(x_1)$, then set $x = x_1$;
- If $p_X(x_1) \le u < p_X(x_1) + p_X(x_2)$, then set $x = x_2$;
- etc.

In general, then, x is set to x_j if and only if

$$\sum_{i=1}^{j-1} p_X(x_i) \le u < \sum_{i=1}^{j} p_X(x_i)$$

i.e.

$$P(X \le x_{j-1}) \le u < P(X \le x_j)$$

or

$$F_X(x_{j-1}) \le u < F_X(x_j).$$

Since u is from the $\text{Un}(0, 1)$ distribution, it follows that

$$P(X = x_j) = P\{F_X(x_{j-1}) \leq u < F_X(x_j)\}$$
$$= P\{u < F_X(x_j)\} - P\{u < F_X(x_{j-1})\}$$
$$= F_X(x_j) - F_X(x_{j-1})$$
$$= p_X(x_j),$$

as required.

Example 3.9

The following numbers are five pseudo-random real numbers in the range from 0 to 1:

$$0.0421 \qquad 0.2333 \qquad 0.4253 \qquad 0.7225 \qquad 0.9830$$

We can use these values to generate five random variates from the $\text{Po}(2)$ distribution. We first require the terms of the probability mass function of the $\text{Po}(2)$ distribution. These may be found most easily using the recursive formula introduced in Chapter 2 (page 43):

$$p(0) = \exp(-\theta) = \exp(-2) = 0.1353,$$

$$p(1) = \frac{\theta}{1} p(0) = \frac{2}{1}(0.1353) = 0.2707,$$

$$p(2) = \frac{\theta}{2} p(1) = \frac{2}{2}(0.2707) = 0.2707,$$

etc.

The first few terms of the probability mass function and distribution function of the $\text{Po}(2)$ distribution are shown in Table 3.1.

Applying the above method gives

$$
\begin{aligned}
u_1 &= 0.0421 \quad \text{so} \qquad\quad u_1 < F(0) \quad \text{and } x_1 = 0, \\
u_2 &= 0.2333 \quad \text{so } F(0) \leq u_2 < F(1) \quad \text{and } x_2 = 1, \\
u_3 &= 0.4253 \quad \text{so } F(1) \leq u_3 < F(2) \quad \text{and } x_3 = 2, \\
u_4 &= 0.7225 \quad \text{so } F(2) \leq u_4 < F(3) \quad \text{and } x_4 = 3, \\
u_5 &= 0.9830 \quad \text{so } F(4) \leq u_5 < F(5) \quad \text{and } x_5 = 5.
\end{aligned}
$$

So we obtain the simulated Poisson variates: $0, 1, 2, 3, 5$.

An analogous procedure for simulating from a continuous distribution is justified by the following proposition.

Proposition 3.5

Suppose that F is a continuous distribution function. Let $U \sim \text{Un}(0, 1)$. Then the random variable $X = F^{-1}(U)$ has distribution function F.

Table 3.1 *Values of the Po(2) probability mass function and distribution function*

x	p(x)	F(x)
0	0.1353	0.1353
1	0.2707	0.4060
2	0.2707	0.6767
3	0.1804	0.8571
4	0.0902	0.9473
5	0.0361	0.9834
6	0.0120	0.9955
7	0.0034	0.9989
8	0.0009	0.9998
9	0.0002	1.0000
10	0.0000	1.0000

Proof.

$$F_X(x) = P(X \leq x) = P\{F^{-1}(U) \leq x\} = P\{U \leq F(x)\} = F(x). \qquad \blacksquare$$

This means we can generate a pseudo-random variate from a continuous distribution with distribution function F as follows:

- Generate a pseudo-random variate, u, from the $Un(0, 1)$ distribution.
- Set $x = F^{-1}(u)$.

Example 3.10

The following numbers are pseudo-random real numbers in the range from 0 to 1:

$$0.0421 \quad 0.2333 \quad 0.4253 \quad 0.7225 \quad 0.9830$$

We will use these values to generate four random variates from each of the following continuous distributions:

(a) the Ex(1) distribution;

(b) the standard normal distribution;

(c) the normal distribution with expected value -5 and variance 0.36;

(d) the chi-squared distribution with one degree of freedom.

(a) The Ex(1) distribution has probability density function $f(x) = \exp(-x)$, $x > 0$. So,

$$F(x) = \int_0^x \exp(-u)\, du = 1 - \exp(-x).$$

In order to use the inverse distribution function method, we set

$$u = F(x) = 1 - \exp(-x),$$

i.e.

$$x = -\log_e(1 - u).$$

So

$$x_1 = -\log_e(1 - u_1) = -\log_e(1 - 0.0421) = 0.043,$$
$$x_2 = -\log_e(1 - u_2) = -\log_e(1 - 0.2333) = 0.266,$$
$$x_3 = -\log_e(1 - u_3) = -\log_e(1 - 0.4253) = 0.554,$$
$$x_4 = -\log_e(1 - u_4) = -\log_e(1 - 0.7225) = 1.282,$$
$$x_5 = -\log_e(1 - u_5) = -\log_e(1 - 0.9830) = 4.075.$$

So five pseudo-random variates from the Ex(1) distribution are:

$$0.043, \quad 0.266, \quad 0.554, \quad 1.282, \quad 4.075.$$

(b) Using the inverse distribution function method, we obtain the pseudo-random variate z from the equation $z = \Phi^{-1}(u)$. From Table B2,

$$z_1 = \Phi^{-1}(u_1) = \Phi^{-1}(0.0421) = -\Phi^{-1}(0.9579) = -1.73,$$
$$z_2 = \Phi^{-1}(u_2) = \Phi^{-1}(0.2333) = -\Phi^{-1}(0.7667) = -0.73,$$
$$z_3 = \Phi^{-1}(u_3) = \Phi^{-1}(0.4253) = -\Phi^{-1}(0.5747) = -0.19,$$
$$z_4 = \Phi^{-1}(u_4) = \Phi^{-1}(0.7225) = 0.59,$$
$$z_5 = \Phi^{-1}(u_5) = \Phi^{-1}(0.9830) = 2.12.$$

So five pseudo-random variates from the standard normal distribution are:

$$-1.73, \quad -0.73, \quad -0.19, \quad 0.59, \quad 2.12.$$

(c) The $N(\mu, \sigma^2)$ distribution is related to the standard normal by the equation $X = \mu + \sigma Z$. So, random variates from the $N(-5, 0.36)$ distribution may be obtained from the standard normal variates obtained in (b) using the relationship $x_i = \mu + \sigma z_i = 0.6z_i - 5$. This yields the simulated values

$$-6.04, \quad -5.44, \quad -5.11, \quad -4.65, \quad -3.73.$$

(d) This distribution is related to the standard normal by the equation $Y = Z^2$. Squaring the variates obtained from the standard normal in part (b) yields the following simulated values from the χ_1^2 distribution:

$$2.98, \quad 0.53, \quad 0.04, \quad 0.35, \quad 4.49.$$

Example 3.11

The continuous random variable X has probability density function $f_X(x) = e^{-2|x|}$, for all real x. This is a symmetric distribution. With a little care, it is possible to invert its distribution function and simulate from it in the way described above. However, there is another possibility.

Let $Y = |X|$. Then the distribution function of Y is

$$F_Y(y) = P(Y \le y) = P(-y \le X \le y) = \int_{-y}^{0} e^{2x} dx + \int_{0}^{y} e^{-2x} dx$$

$$= 1 - e^{-2y}, \quad y > 0.$$

So $Y \sim \text{Ex}(2)$.

We can simulate from the distribution of Y using its inverse distribution function. A simulated value, y, for Y must correspond to either $X = y$ or $X = -y$. The symmetry of the distribution of X means that these two possible values should be equally likely given $Y = y$. We could generate an (independent) $\text{Un}(0, 1)$ random variate, u_2, and decide to take $X = -y$, if $U < \frac{1}{2}$ and $X = y$, if $U > \frac{1}{2}$. This allows us to simulate from the original distribution.

Exercises

1 (a) Use the congruential generator with $m = 10{,}000$, $a = 2{,}001$ and $b = 11$ to generate 20 pseudo-random digits. [You may choose an appropriate seed, x_0.]

(b) From these pseudo-random digits, obtain four pseudo-random real numbers in the range 0 to 1. Work to five decimal places.

2 (a) Use the random digits table provided (Table B1) to generate ten pseudo-random integers in the range 0 to 999.

(b) Hence obtain ten pseudo-random real numbers in the range 0 to 1, each to three decimal places.

3 Suppose that we wish to simulate from the $\text{Bi}(n, \theta)$ distribution, where the parameters n and θ are known. Show how the recursive formula for the probabilities of this distribution can be utilized to set up a method of simulation. Use this formula to generate ten random variates from the $\text{Bi}(5, 0.3)$ distribution.

4 The following numbers are a random sample of real numbers in the range from 0 to 1.

$$0.1281 \quad 0.2735 \quad 0.3163 \quad 0.5840 \quad 0.8640$$

Use these values to generate five random variates from each of the following discrete distributions:

(a) hypergeometric, Hyp(3, 10, 6);

(b) binomial, Bi(10, 0.25);

(c) geometric, Ge$\left(\frac{5}{6}\right)$;

(d) Poisson, Po(3.5).

5 The following numbers are a random sample of real numbers in the range from 0 to 1:

$$0.0816 \quad 0.3724 \quad 0.4646 \quad 0.7924 \quad 0.9581$$

Use these values to generate five random variates from each of the following continuous distributions.

(a) exponential, Ex(1);

(b) exponential, Ex(3);

(c) Weibull, We$\left(\frac{1}{2}, 2\right)$.

[*Hint*: for (b) and (c), consider transformations to the Ex(1) distribution.]

6 Describe how to simulate from the We(α, θ) distribution, for known values of α and θ, using a method based on the inverse distribution function.

7 A motorist is trying to collect a set of six different tokens from her petrol station. On each visit to the petrol station, the motorist receives one token, which is equally likely to be any of the six available types. The types of tokens received on different visits are independent.

(a) Using the table of random digits provided (Table B1), simulate the number of times the motorist will require to visit the petrol station in order to collect a full set of tokens. Repeat the simulation at least 20 times.

(b) The prize for collecting a full set of six different tokens is a child's toy. The motorist requires two of these toys. Simulate the number of times she must visit the petrol station in order to collect two full sets of tokens. Repeat the simulation at least 20 times.

8 A rural bus route connects six towns, A, B, C, D, E and F (in that order). The time (in minutes) taken to travel between any pair of consecutive towns is believed to be an N(10, 1) random variable. The times taken on different stages of the same journey can be assumed to be independent.

The timetable allows 12 minutes to cover the journey between each pair of consecutive towns. If the bus is late in arriving at B, C, D or E, then it departs on the next stage of the journey immediately. Otherwise, it waits until its scheduled departure time. The bus always leaves A on time.

Starting with random digits from the table provided (Table B1), simulate a bus journey from A to F on this route. Repeat the simulation at least 20 times. Record whether or not the bus is late each time.

9 Suppose that X follows a χ^2_n distribution and define the random variable Y by $Y = \sqrt{2X}$. Find the probability density function of Y, and hence show that:

$$\mathbb{E}(Y) = \frac{\Gamma\left(\dfrac{n+1}{2}\right)}{\Gamma\left(\dfrac{n}{2}\right)} \qquad \mathbb{E}\left(Y^2\right) = 2n.$$

It follows that, as $n \to \infty$, $\mathbb{E}(Y) \to 1$ and $\operatorname{var}(Y) \to 2n - 1$. Use a random number generator on your computer to simulate 1,000 times from each of a range of χ^2 distributions, with different values of n. In each case, form the values $\sqrt{2x}$ and plot them on a histogram. You should be able to convince yourself that, for large values of n, the distribution of Y is very similar to the $N(1, 2n - 1)$ distribution.

10 Suppose that X follows the standard normal distribution and let $Y = |X|$. Show that Y has probability density function

$$f_X(x) = \sqrt{\frac{2}{\pi}}\, e^{-x^2/2}, \qquad x > 0.$$

Explain how to simulate values of Y.

Summary

In this chapter, we have looked at some general methods for dealing with functions of a random variable. Sometimes it is possible to obtain the distribution function, and hence the probability density function, of a continuous function of a continuous random variable. When the function in question is strictly increasing or decreasing on the range space of X, and when the function is the square, it is possible to circumvent the full process using Proposition 3.1 or 3.2. Even when the density function of the transformed variable may not be obtained, it is still often possible to find the expected value or variance, though the results usually quoted for this can be misleading. Simulation has also been introduced, using methods based on the inverse distribution function.

4
Bivariate distributions

Up until now, we have implicitly assumed that each outcome of a stochastic experiment may be adequately represented by just one numerical value. This has allowed us to consider just one random variable at a time. In general, though, we must allow for the possibility that each replicate of an experiment generates more than one piece of numerical information, so that more than one random variable is required to represent each outcome. In this chapter, we take the simplest case of this and begin to consider how to deal with two random variables at a time. A key issue that arises in this context is how to describe the relationship between two random variables; in order to address this, we introduce various measures of association between random variables and also the idea of regression.

4.1 Joint and marginal distributions

Example 4.1

Two identical machines, A and B, are due for a refit. Two identical electronic components will have to be fitted to each of them. The maintenance department has a stock of ten spare components of which, unknown to them, two are defective. Let the discrete random variables X and Y, respectively, be the numbers of defective components that are fitted to machines A and B.

Consider X first of all, ignoring Y. X is a discrete random variable with range space, $R_X = \{0, 1, 2\}$. Indeed, X is a hypergeometric random variable, $X \sim \text{Hyp}(2, 10, 2)$, with probability mass function

$$p_X(x) = \frac{\binom{2}{x}\binom{8}{2-x}}{\binom{10}{2}}, \qquad x = 0, 1, 2.$$

Now consider Y, ignoring X. The symmetry of the situation requires Y to follow the same probability distribution as X, i.e. Y has range space $R_Y = \{0, 1, 2\}$ and $Y \sim \text{Hyp}(2, 10, 2)$.

However, it is really not very informative to consider X and Y separately. For one thing, X and Y may not simultaneously take every possible value in their range spaces. Since there are only two defective components in total, it is not possible that both $X = 2$ and $Y = 2$, for example. The set of all possible pairs of values for (X, Y) is called the *joint range space* of (X, Y); in this example, the joint range space is

$$R_{XY} = \{(0,0), (0,1), (0,2), (1,0), (1,1), (2,0)\}.$$

The *joint probability mass function* of (X, Y) is defined to be

$$p_{XY}(x, y) = P(X = x, Y = y), \qquad (x, y) \in \mathbb{R}^2.$$

Clearly, $p_{XY}(x,y) = 0$ except when $(x, y) \in R_{XY}$. In order to obtain the joint probability mass function on R_{XY}, note first that there are $10!/2!\,2!\,6!$ different ways to choose two of the spare components for machine A and two different components for machine B (leaving six components in the store). Assuming that the choice is made at random, then each of these possible outcomes is equally likely.

Take any point $(x, y) \in R_{XY}$. There are $2!/x!\,y!\,(2 - x - y)!$ different ways to choose x defective components for machine A and y (different) defective components for machine B (leaving $2 - x - y$ defectives in store). Similarly, there are $8!/(2 - x)!\,(2 - y)!\,(4 + x + y)!$ different ways to choose $2 - x$ perfect components for machine A and $2 - y$ (different) perfect components for machine B (leaving $4 + x + y$ perfect components in store).

This means that, for $(x, y) \in R_{XY}$,

$$p_{XY}(x, y) = P(X = x, Y = y)$$

$$= \frac{2!}{x!\,y!\,(2 - x - y)!} \frac{8!}{(2 - x)!\,(2 - y)!\,(4 + x + y)!} \bigg/ \frac{10!}{2!\,2!\,6!}$$

This joint probability mass function is tabulated below:

		x			
$p_{XY}(x, y)$		0	1	2	$p_Y(y)$
y	0	$\frac{15}{45}$	$\frac{12}{45}$	$\frac{1}{45}$	$\frac{28}{45}$
	1	$\frac{12}{45}$	$\frac{4}{45}$	—	$\frac{16}{45}$
	2	$\frac{1}{45}$	—	—	$\frac{1}{45}$
$p_X(x)$		$\frac{28}{45}$	$\frac{16}{45}$	$\frac{1}{45}$	1

The probability mass function of X alone, known in this context as the *marginal probability mass function* of X, can be recovered from the joint probability mass function by summing down the columns of the above table. For example,

$$p_X(0) = P(X = 0) = p(0, 0) + p(0, 1) + p(0, 2) = 15/45 + 12/45 + 1/45 = 28/45.$$

In general, for $x \in R_X$,

$$p_X(x) = P(X = x) = \sum_{y:(x,y) \in R_{XY}} p_{XY}(x, y).$$

These values are shown on the bottom margin of the table, and can be shown to agree with the values of the hypergeometric distribution for X previously discussed. Moments of X can be determined as usual from the marginal distribution of X, for example,

$$\mathbb{E}(X) = \left(0 \times \tfrac{28}{45}\right) + \left(1 \times \tfrac{16}{45}\right) + \left(2 \times \tfrac{1}{45}\right) = \tfrac{18}{45} = 0.4,$$

$$\mathbb{E}(X^2) = \left(0^2 \times \tfrac{28}{45}\right) + \left(1^2 \times \tfrac{16}{45}\right) + \left(2^2 \times \tfrac{1}{45}\right) = \tfrac{20}{45},$$

$$\text{var}(X) = \tfrac{20}{45} - \left(\tfrac{18}{45}\right)^2 = \tfrac{576}{2025} = \tfrac{64}{225}.$$

In a similar way, the marginal probability mass function of Y can be found by summing along the rows of the table of the joint probability mass function. For $y \in R_Y$,

$$p_Y(y) = \sum_{x:(x,y)\in R_{XY}} p_{XY}(x,y).$$

This marginal distribution is shown on the right-hand margin of the table of the joint probability mass function. In this example, it is the same as the marginal distribution of X.

In general, when X and Y are discrete random variables, their joint probability mass function must satisfy the following conditions:

(a) $0 \le p_{XY}(x,y) \le 1$, since $p_{XY}(x,y)$ is a probability.

(b) $\sum\sum_{(x,y)\in R_{XY}} p_{XY}(x,y) = 1$, since all these terms are probabilities of mutually exclusive outcomes (x,y) that partition the sample space.

If $g(X,Y)$ is any real-valued function of the discrete random variables X and Y, then its expected value is defined as follows:

$$\mathbb{E}\{g(X,Y)\} = \sum_{(x,y)\in R_{XY}}\sum g(x,y)p_{XY}(x,y).$$

This value is well defined (or *finite*) if and only if $\sum\sum_{(x,y)\in R_{XY}}|g(x,y)|p_{XY}(x,y)$ converges to a finite limit. Several functions $g(X,Y)$ are of particular importance.

First of all, g might be a function of X only (e.g. X or X^2). In this case, the definition of the expected value given above agrees with the usual definition based on the marginal distribution of X, for

$$\mathbb{E}\{g(X)\} = \sum_{(x,y)\in R_{XY}}\sum g(x)p_{XY}(x,y) = \sum_{x\in R_X} g(x) \sum_{y:(x,y)\in R_{XY}} p_{XY}(x,y)$$

$$= \sum_{x\in R_X} g(x)p_X(x).$$

A similar argument shows that expected values of functions of Y alone may continue to be obtained from the marginal probability mass function of Y.

Another important case arises when $g(X, Y) = X + Y$. Using the general definition of expected value,

$$\mathbb{E}\{X + Y\} = \sum_{(x,y) \in R_{XY}} \sum (x + y) p_{XY}(x, y)$$

$$= \sum_{(x,y) \in R_{XY}} \sum x p_{XY}(x, y) + \sum_{(x,y) \in R_{XY}} \sum y p_{XY}(x, y)$$

$$= \mathbb{E}(X) + \mathbb{E}(Y).$$

This result can be generalized by a process of induction to the sum of any finite sequence of random variables.

The variance of $X + Y$ is

$$\text{var}(X + Y) = \mathbb{E}\left\{[X + Y - \mathbb{E}(X + Y)]^2\right\}$$

$$= \mathbb{E}\left\{[X + Y - \mathbb{E}(X) - \mathbb{E}(Y)]^2\right\}$$

$$= \mathbb{E}\left\{[X - \mathbb{E}(X)]^2 + [Y - \mathbb{E}(Y)]^2 + 2[X - \mathbb{E}(X)][Y - \mathbb{E}(Y)]\right\}$$

$$= \mathbb{E}\left\{[X - \mathbb{E}(X)]^2\right\} + \mathbb{E}\left\{[Y - \mathbb{E}(Y)]^2\right\}$$

$$+ 2\mathbb{E}\left\{[X - \mathbb{E}(X)][Y - \mathbb{E}(Y)]\right\}$$

$$= \text{var}(X) + \text{var}(Y) + 2\text{cov}(X, Y).$$

In the last step, we have written $\text{cov}(X, Y)$ for $\mathbb{E}\left\{[X - \mathbb{E}(X)][Y - \mathbb{E}(Y)]\right\}$. This is called the *covariance* between the random variables X and Y. The covariance may take any real value and, as we shall see later, gives some indication of how X and Y are related. Notice that $\text{cov}(Y, X) = \text{cov}(X, Y)$.

Notice the similarity with the definition of the variance, $\text{var}(X) = \mathbb{E}\{[X - \mathbb{E}(X)]^2\}$. As with the variance, there is another expression for the covariance that is often easier to evaluate.

$$\mathbb{E}\left\{[X - \mathbb{E}(X)][Y - \mathbb{E}(Y)]\right\}$$

$$= \sum_{(x,y) \in R_{XY}} \sum [x - \mathbb{E}(X)][y - \mathbb{E}(Y)] p_{XY}(x, y)$$

$$= \sum_{(x,y) \in R_{XY}} \sum xy\, p_{XY}(x, y) - \mathbb{E}(X) \sum_{(x,y) \in R_{XY}} \sum y p_{XY}(x, y)$$

$$- \mathbb{E}(Y) \sum_{(x,y) \in R_{XY}} \sum x p_{XY}(x, y) + \mathbb{E}(X)\mathbb{E}(Y) \sum_{(x,y) \in R_{XY}} \sum p_{XY}(x, y)$$

$$= \mathbb{E}(XY) - \mathbb{E}(X)\mathbb{E}(Y) - \mathbb{E}(Y)\mathbb{E}(X) + \mathbb{E}(X)\mathbb{E}(Y).1$$

$$= \mathbb{E}(XY) - \mathbb{E}(X)\mathbb{E}(Y).$$

Example 4.1 – continued

We begin by finding $\mathbb{E}(XY)$. In R_{XY}, the function XY takes a non-zero value only at $(1,1)$, so

$$\mathbb{E}(XY) = 1 \times 1 \times \tfrac{4}{45} = \tfrac{4}{45}.$$

Therefore,

$$\text{cov}(X,Y) = \mathbb{E}(XY) - \mathbb{E}(X)\mathbb{E}(Y) = \tfrac{4}{45} - \tfrac{18}{45} \cdot \tfrac{18}{45} = -\tfrac{16}{225}$$

The negative value of the covariance here indicates that larger values of X tend to occur along with smaller values of Y (and vice versa). This is intuitively reasonable in the present case; since there are only a fixed number of defective components in total, the larger the number of defective items fitted into machine A the smaller the number that will be available to be fitted into machine B (and vice versa).

$X + Y$ represents the total number of defective components fitted to machines A and B at their refit. Using the results obtained up till now,

$$\mathbb{E}\{X + Y\} = \mathbb{E}(X) + \mathbb{E}(Y) = \frac{4}{5}$$

$$\text{var}(X + Y) = \text{var}(X) + \text{var}(Y) + 2\text{cov}(X,Y) = \frac{64 + 64 - 32}{225} = \frac{96}{225}.$$

The *joint distribution function* of two random variables X and Y is defined to be

$$F_{XY}(x,y) = P(X \leq x, Y \leq y), \qquad (x,y) \in \mathbb{R}^2.$$

Notice that the marginal d.f. of X or Y can be recovered from the joint d.f. For example:

$$F_X(x) = P(X \leq x) = P(X \leq x, Y \leq \infty) = F_{XY}(x,\infty), \quad x \in \mathbb{R}.$$

Similarly,

$$F_Y(y) = F_{XY}(\infty, y), \qquad y \in \mathbb{R}.$$

Example 4.1 – continued

The table below shows the values of the joint distribution function of X and Y at points $(x, y) \in R_{XY}$. $F_{XY}(x, y)$ only changes on this lattice of points; it is constant at intermediate values.

		x		
$F_{XY}(x, y)$		**0**	**1**	**2**
y	**0**	$\frac{15}{45}$	$\frac{27}{45}$	$\frac{28}{45}$
	1	$\frac{27}{45}$	$\frac{43}{45}$	$\frac{44}{45}$
	2	$\frac{28}{45}$	$\frac{44}{45}$	1

Suppose that X and Y are (absolutely) continuous random variables, and that their joint distribution function can be written in the form

$$F_{XY}(x,y) = \int_{-\infty}^{x} \int_{-\infty}^{y} f_{XY}(u,v)\, dv\, du.$$

Then X and Y are said to be jointly continuous and f_{XY} is their *joint probability density function*. This means that f_{XY} is a second-order partial derivative of $F_{XY}(x,y)$:

$$f_{XY}(x,y) = \frac{\partial^2}{\partial x \partial y} F_{XY}(x,y), \qquad (x,y) \in \mathbb{R}^2.$$

Since $F_{XY}(x,y)$ must be an increasing function of both x and y, this definition means that

$$f_{XY}(x,y) \geq 0, \quad (x,y) \in \mathbb{R}^2.$$

Clearly, $f_{XY}(x,y) > 0$ only for $(x,y) \in R_{XY}$. Also

$$\int_{-\infty}^{\infty} \int_{-\infty}^{\infty} f_{XY}(x,y)\, dy\, dx = F_{XY}(\infty,\infty) = 1.$$

It has already been pointed out that $F_X(x) = F_{XY}(x,\infty)$, so

$$F_X(x) = F_{XY}(x,\infty) = \int_{-\infty}^{x} \int_{-\infty}^{\infty} f_{XY}(u,y)\, dy\, du.$$

Given certain regularity conditions that almost always apply in practical cases, this means that

$$f_X(x) = \frac{d}{dx} F_X(x) = \frac{d}{dx} \int_{-\infty}^{x} \left\{ \int_{-\infty}^{\infty} f_{XY}(u,y)\, dy \right\} du = \int_{-\infty}^{\infty} f_{XY}(x,y)\, dy.$$

In other words, we can recover the marginal p.d.f. of X from the joint p.d.f. by integrating out y. In a similar way,

$$f_Y(y) = \int_{-\infty}^{\infty} f_{XY}(x,y)\, dx.$$

Example 4.2

Suppose that the continuous random variables X and Y have the joint range space

$$R_{XY} = \{(x,y) : 0 < x < b;\ 0 < y < c\}, \qquad b > 0 \text{ and } c > 0.$$

We will suppose that X and Y have a joint uniform distribution over R_{XY}. This means that there is a positive real constant k such that the joint probability density function of X and Y is of the form $f_{XY}(x, y) = k$, $(x, y) \in R_{XY}$. In order to determine the constant k we must integrate the joint density function over the joint range space and set the result to 1:

$$1 = \int_{-\infty}^{\infty} \int_{-\infty}^{\infty} f_{XY}(x, y) \, dy \, dx = \int_{0}^{b} \int_{0}^{c} k \, dy \, dx = kbc,$$

i.e.

$$k = \frac{1}{bc} \quad \left(= \frac{1}{\text{area of } R_{XY}} \right)$$

and

$$f_{XY}(x, y) = \frac{1}{bc}, \quad 0 < x < b, \quad 0 < y < c.$$

Therefore

$$f_X(x) = \int_{0}^{c} f_{XY}(x, y) \, dy = \int_{0}^{c} \frac{1}{bc} \, dy = \left[\frac{y}{bc} \right]_{0}^{c} = \frac{c}{bc} = \frac{1}{b}, \quad 0 < x < b.$$

Marginally, then, X follows a Un(0, b) distribution, with $\mathbb{E}(X) = b/2$ and $\text{var}(X) = b^2/12$.

The marginal distribution of Y is Un(0, c), with $\mathbb{E}(Y) = c/2$ and $\text{var}(X) = c^2/12$.

When X and Y are jointly continuous random variables, and $g(X, Y)$ is a real-valued function of X and Y, then the expected value of $g(X, Y)$ (if it exists) is defined to be

$$\mathbb{E}\{g(X, Y)\} = \int_{-\infty}^{\infty} \int_{-\infty}^{\infty} g(x, y) f_{XY}(x, y) \, dy \, dx.$$

As in the discrete case, when $g(\cdot)$ is just a function of X, then this general definition is in agreement with the marginal definition of $\mathbb{E}\{g(X)\}$ that was introduced previously:

$$\mathbb{E}\{g(X)\} = \int_{-\infty}^{\infty} \int_{-\infty}^{\infty} g(x) f_{XY}(x, y) \, dy \, dx = \int_{-\infty}^{\infty} g(x) \left\{ \int_{-\infty}^{\infty} f_{XY}(x, y) \, dy \right\} dx$$

$$= \int_{-\infty}^{\infty} g(x) f(x) \, dx.$$

Example 4.2 – continued

We have

$$\mathbb{E}(XY) = \int_{0}^{b} \int_{0}^{c} xy \frac{1}{bc} \, dy \, dx = \frac{1}{bc} \int_{0}^{b} x \, dx \int_{0}^{c} y \, dy = \frac{bc}{4}.$$

Therefore

$$\text{cov}(X, Y) = \mathbb{E}(XY) - \mathbb{E}(X)\mathbb{E}(Y) = \frac{bc}{4} - \frac{b}{2} \cdot \frac{c}{2} = 0.$$

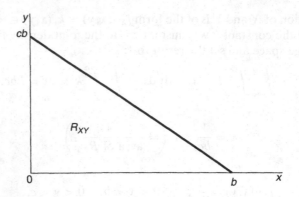

Example 4.3

Consider another uniform bivariate distribution, this time on a triangular range space (see diagram above):

$$f(x, y) = k, \qquad 0 < x < b; \quad 0 < y < c(b - x).$$

The area of R_{XY} is $\frac{1}{2}cb^2$, so it is easily shown that

$$f(x, y) = \frac{2}{cb^2}, \qquad (x, y) \in R_{XY}.$$

Hence,

$$f_X(x) = \int_0^{c(b-x)} f(x, y)\, dy = \left[\frac{2y}{cb^2} \right]_0^{c(b-x)} = \frac{2c(b-x)}{cb^2} = \frac{2(b-x)}{b^2}, \qquad 0 < x < b,$$

and

$$f_Y(y) = \int_0^{b-y/c} f(x, y)\, dx = \left[\frac{2x}{cb^2} \right]_0^{b-y/c} = \frac{2(cb - y)}{c^2 b^2}, \qquad 0 < y < bc.$$

In this case, although the joint distribution of X and Y is uniform, neither of the marginal distributions is uniform. This is a result of the non-rectangular form of the joint range space R_{XY}. More will be said about this later.

We can find $\mathbb{E}(X)$ and $\mathrm{var}(X)$ in the usual way. For $r = 1, 2, \ldots$,

$$\mathbb{E}(X^r) = \frac{2}{b^2} \int_0^b x^r (b - x)\, dx$$

$$= \frac{2}{b^2} \int_0^1 b^r u^r b(1 - u) b\, du \qquad \text{putting } u = x/b$$

$$= 2b^r \mathbb{B}(r + 1, 2)$$

$$= 2b^r \frac{r!1!}{(r+2)!} \qquad \text{(Result A2.6)}$$

$$= \frac{2b^r}{(r+2)(r+1)}$$

$$\therefore \qquad \mathbb{E}(X) = \frac{b}{3}, \qquad \mathbb{E}(X^2) = \frac{b^2}{6}, \qquad \text{var}(X) = \frac{b^2}{18}.$$

In a similar way, we can show that $\mathbb{E}(Y) = cb/3$ and $\text{var}(Y) = c^2b^2/18$.
Futhermore,

$$\mathbb{E}(XY) = \frac{2}{cb^2} \int_0^b \int_0^{c(b-x)} xy \, dy \, dx$$

$$= \frac{1}{cb^2} \int_0^b x \left[y^2\right]_{y=0}^{y=c(b-x)} dx$$

$$= \frac{c}{b^2} \int_0^b x(b-x)^2 \, dx$$

$$= cb^2 \int_0^1 u(1-u)^2 \, du \qquad \text{putting } u = x/b.$$

$$= cb^2 \, \mathbb{B}(2,3)$$

$$= \frac{cb^2}{12}.$$

So

$$\text{cov}(X,Y) = \mathbb{E}(XY) - \mathbb{E}(X)\mathbb{E}(Y) = -\frac{cb^2}{36}.$$

Exercises

1 X and Y are discrete random variables with the joint probability mass
function shown below.

$p_{XY}(x, y)$		x	
	−1	**0**	**1**
y **−1**	$\frac{1}{4}(1-\theta)$	0	$\frac{1}{4}(1-\theta)$
0	0	θ	0
1	$\frac{1}{4}(1-\theta)$	0	$\frac{1}{4}(1-\theta)$

(a) Derive the marginal probability mass functions of X and Y.

(b) Find $\mathbb{E}(X), \mathbb{E}(Y), \text{var}(X)$ and $\text{var}(Y)$.

(a) Find $\mathbb{E}(XY)$ and, hence, $\text{cov}(X, Y)$.

2 Repeat Exercise 1 when X and Y have the following joint probability mass function (for some real constant $k > 0$):

$p_{XY}(x, y)$	x 0	k
y 0	θ	$\frac{1}{2}(1 - \theta)$
k	$\frac{1}{2}(1 - \theta)$	0

3 The discrete random variables X and Y have the joint probability mass function shown below.

$p_{XY}(x, y)$	x 0	1	2
y 0	0.2	0.1	0.1
2	0.1	0.1	0.1
4	0.1	0.1	0.1

(a) Find $\mathbb{E}(X)$ and $\mathbb{E}(Y)$ without first calculating the marginal probability mass functions. Check that the same answer is obtained from the marginal distributions.

(b) Find $\mathbb{E}(X + Y)$ and check that $\mathbb{E}(X + Y) = \mathbb{E}(X) + \mathbb{E}(Y)$.

(c) Find $\mathbb{E}(X - Y)$ and check that $\mathbb{E}(X - Y) = \mathbb{E}(X) - \mathbb{E}(Y)$.

4 The continuous random variables X and Y have the following joint probability density function, for some real constant k:

$$f_{XY}(x, y) = k(x + y), \qquad 0 < x < 1, 0 < y < 1,$$

(a) Find the value of k.

(b) Find (i) $P(X \leq \frac{1}{2})$; (ii) $P(Y \leq \frac{1}{2})$; (iii) $P(X \leq \frac{1}{2}, Y \leq \frac{1}{2})$.

(c) Derive the marginal probability density functions of X and Y.

(d) Find $\mathbb{E}(X), \mathbb{E}(Y), \text{var}(X)$ and $\text{var}(Y)$.

(e) Find $\mathbb{E}(XY)$ and, hence, $\text{cov}(X, Y)$.

5 The continuous random variables X and Y have joint probability density function

$$f_{XY}(x, y) = \frac{1}{4}(x^2 + y^2) \exp\{-(x + y)\}, \qquad x > 0, y > 0.$$

(a) Check that this is a valid (joint) probability density function.

(b) Find (i) $P(X > 1)$; (ii) $P(Y > 1)$, (iii) $P(X > 1, Y > 1)$.

(c) Derive the marginal probability density functions of X and Y.

(d) Find $\mathbb{E}(X), \mathbb{E}(Y)$ and $\mathbb{E}(XY)$. Hence find $\text{cov}(X, Y)$.

6 Let X and Y be discrete random variables, and let a and b be real numbers.

(a) Show that $\mathbb{E}\{aX + bY\} = a\,\mathbb{E}(X) + b\,\mathbb{E}(Y)$.

(b) By induction, result (a) can be extended to a sum of any finite number of random variables. Use this to show that

$$\text{var}\{aX + bY\} = a^2\,\mathbb{E}(X) + b^2\,\mathbb{E}(Y) + 2ab\,\text{cov}(X, Y).$$

7 Repeat Exercise 6 for the case where X and Y are continuous random variables.

8 Let X and Y be arbitrary random variables, and let a, b, c and d be real constants. Use the definition of covariance, and the results of Exercises 6 and 7, to prove that

$$\text{cov}(a + bX, c + dY) = bd\,\text{cov}(X, Y).$$

4.2 Conditional distributions and independence

Example 4.4

Deadman and MacDonald (2004) describe the results of a UK survey into criminal activity among young people (aged 12 to 30 years) in inner city areas in England and Wales. Two discrete random variables, X and Y, might be defined as follows:

$$X = \begin{cases} 0, & \text{if an individual has never committed a criminal offence,} \\ 1, & \text{if an individual has ever committed a criminal offence,} \end{cases}$$

$$Y = \begin{cases} 0, & \text{if an individual has never been a victim of crime,} \\ 1, & \text{if an individual has ever been a victim of crime.} \end{cases}$$

The joint probability mass function of these two random variables for the population investigated by Deadman and MacDonald might be as shown below.

		x		
$p_{XY}(x, y)$		0	1	$p_Y(y)$
y	0	0.44	0.30	0.74
	1	0.11	0.15	0.26
$p_X(x)$		0.55	0.45	1

Suppose we were interested in *conditional* probabilities associated with Y *given* that $X = 1$. Using the definition of conditional probability,

$$P(Y = 0 \mid X = 1) = \frac{P(X = 1, Y = 0)}{P(X = 1)} = \frac{p_{XY}(1,0)}{p_X(1)} = \frac{0.30}{0.45} = \frac{2}{3},$$

$$P(Y = 1 \mid X = 1) = \frac{P(X = 1, Y = 1)}{P(X = 1)} = \frac{p_{XY}(1,1)}{p_X(1)} = \frac{0.15}{0.45} = \frac{1}{3}.$$

These probabilities define the *conditional* probability mass function of Y *given* that $X = 1$. In general, whenever $p_X(x) > 0$, the *conditional probability mass function* of Y given $X = x$ is defined by:

$$p_{Y \mid X}(y \mid x) = \frac{p_{XY}(x,y)}{p_X(x)}, \qquad y: (x, y) \in R_{XY}.$$

The conditional probability mass functions of Y given $X = 0$ and $X = 1$ are:

y	0	1
$p_{Y \mid X}(y \mid X = 0)$	$\frac{4}{5}$	$\frac{1}{5}$
$p_{Y \mid X}(y \mid X = 1)$	$\frac{2}{3}$	$\frac{1}{3}$

Notice that these conditional probability mass functions are quite different from one another and both are different from the marginal probability mass function of Y. Overall, 26% of the population has been the victim of a criminal act ($Y = 1$), but this includes 33% of those who have themselves committed criminal acts and only 20% of those who have not. Thus, information about X modifies the probabilities associated with different values of Y. Similarly, it can be shown that information about Y modifies the probabilities associated with different values of X.

The idea of a *conditional* distribution of Y *given* $X = x$, or of X *given* $Y = y$, leads to the idea of the independence of X and Y. Intuitively, we would think of X and Y as independent if the *conditional* distribution of Y *given* $X = x$ was equal to the marginal distribution of Y for all possible values of x. This would mean that information about X did not change the probabilities associated with Y. The following definition, which mirrors the definition of independent events, leads to that intuitively appealing result.

Two discrete random variables, X and Y, are said to be *independent* if

$$p_{XY}(x, y) = p_X(x)p_Y(y), \qquad \text{for all } (x, y) \in \mathbb{R}^2.$$

Example 4.4 – continued

It is easy to show that X and Y are not independent. We need only find one pair (x, y) such that $p_{XY}(x, y) \neq p_X(x)p_Y(y)$. Now,

$$p_{XY}(0,0) = 0.44$$

while

$$p_X(0)p_Y(0) = 0.55 \times 0.74 = 0.407.$$

For jointly continuous random variables, we approach the problem of identifying the conditional distribution of Y given X by first defining an appropriate distribution function. At first sight, it might appear that a conditional distribution function of Y given $X = x$ could be defined by $P(Y \leq y | X = x)$. However, since the event $X = x$ has probability 0, this would not be in agreement with our usual restrictions on a conditional probability. Assuming that $x \in R_X$, and letting dx be a small positive value,

$$
\begin{aligned}
P(Y \leq y | x \leq X \leq x + dx) &= \frac{P(Y \leq y, x \leq X \leq x + dx)}{P(x \leq X \leq x + dx)} \\
&= \frac{\int_x^{x+dx} \int_{-\infty}^y f_{XY}(u, v) \, dv \, du}{\int_x^{x+dx} du} \\
&\approx \frac{dx \int_{-\infty}^y f_{XY}(x, v) \, dv}{dx \, f_X(x)} \\
&= \int_{-\infty}^y \frac{f_{XY}(x, v)}{f_X(x)} \, dv.
\end{aligned}
$$

Letting $dx \to 0^+$, the left-hand side of this expression tends to the conditional probability that we are trying to obtain. So, for $x \in R_X$, the *conditional distribution function* of Y given $X = x$ is

$$F_{Y|X}(y|x) = \int_{-\infty}^y \frac{f_{XY}(x, v)}{f_X(x)} \, dv.$$

Assuming certain regularity conditions that almost always hold in practice, then the *conditional probability density function* of Y given X can be defined by

$$f_{Y|X}(y|x) = \frac{d}{dy} F_{Y|X}(y|x) = \frac{d}{dy} \int_{-\infty}^y \frac{f_{XY}(x, v)}{f_X(x)} \, dv = \frac{f_{XY}(x, y)}{f_X(x)}$$

$F_{X|Y}(x|y)$ is defined in an analogous manner.

Example 4.2 – continued

Here, X and Y are continuous random variables with

$$f(x, y) = \frac{1}{cb}, \qquad 0 < x < b, \quad 0 < y < c.$$

Previously, we showed that

$$f_X(x) = \frac{1}{b}, \qquad 0 < x < b,$$

$$f_Y(y) = \frac{1}{c}, \qquad 0 < y < c.$$

For $0 < y < c$, X may take any value between 0 and b. The conditional probability density function of X given $Y = y$ is

$$f(x|y) = \frac{f(x,y)}{f(y)} = \frac{1/cb}{1/c} = \frac{1}{b}, \qquad 0 < x < b.$$

This shows that the conditional distribution of X given $Y = y$, for any $y \in R_Y$, is Un$(0, b)$. This is the same as the marginal distribution of X. Similarly, for any $x \in R_X$, the conditional distribution of Y given $X = x$ is Un$(0, c)$, which is just the marginal distribution of Y.

In Example 4.2, the conditional distribution of X given $Y = y$ is equal to the marginal distribution of X for all possible values of y, and the conditional distribution of Y given $X = x$ is equal to the marginal distribution of Y for all possible values of x. This is in accordance with an intuitive concept of X and Y being independent.

Formally, the jointly continuous random variables, X and Y, are said to be *independent* if and only if

$$F_{XY}(x, y) = F_X(x)F_Y(y), \qquad \text{for all } (x, y) \in \mathbb{R}^2.$$

This is, in fact, a general definition of the independence of two random variables; it can be shown that the definition of the independence of discrete random variables in terms of probability mass functions is equivalent to this definition in terms of distribution functions. However, since we usually prefer to work with probability density functions rather than distribution functions when dealing with continuous random variables, this is not the most helpful definition. It is left as an exercise to prove that it is equivalent to say that jointly continuous random variables X and Y are independent if and only if

$$f_{XY}(x, y) = f_X(x)f_Y(y), \qquad \text{for all } (x, y) \in \mathbb{R}^2.$$

Example 4.2 – continued

We have already shown that

$$f_{XY}(x, y) = f_X(x) f_Y(y), \qquad \text{for all } x \text{ and } y.$$

So X and Y are independent.

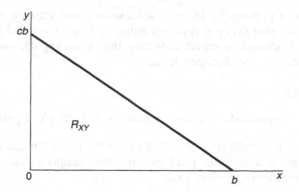

Example 4.3 – continued

As shown in the diagram above,

$$f(x,y) = \frac{2}{cb^2}, \qquad 0 < x < b, \quad 0 < y < c(b-x),$$

$$f_X(x) = \frac{2(b-x)}{b^2} \qquad 0 < x < b,$$

$$f_Y(y) = \frac{2(cb-y)}{c^2b^2} \qquad 0 < y < bc.$$

This means that X and Y are not independent, since generally $f_{XY}(x,y) \neq f_X(x)f_Y(y)$.

The conditional distribution of X given $Y = y$ must depend on the value of y in this case, since the non-rectangular nature of the joint range space means that the conditional range space of X given $Y = y$ changes with the value of y. Formally,

$$f(x|y) = \frac{f(x,y)}{f(y)} = \frac{2/cb^2}{2(bc-y)/c^2b^2} = \frac{c}{bc-y}, \qquad 0 < x < \frac{bc-y}{c}.$$

This means that, conditional on $Y = y$, X has a uniform distribution on the range 0 to $(bc-y)/c$. This is not the same as the marginal distribution of X, another indication that X and Y are not independent.

Also, the conditional expected value of X given $Y = y$ is

$$E(X|Y = y) = \frac{bc-y}{2c} = \frac{b}{2} - \frac{y}{2c},$$

which depends on the value of y.

Contrasting Examples 4.2 and 4.3, it is clear that the shape of the joint range space must have an influence on whether or not the random variables are independent. In order for X and Y to be independent, their joint range space must be rectangular. A non-rectangular range space means that the random variables cannot possibly be independent, since there must be values $x \in R_X$ and $y \in R_Y$ such that $(x, y) \notin R_{XY}$, meaning that $f_X(x) > 0$ and $f_Y(y) > 0$ but $f_{XY}(x, y) = 0$. It should be noted, however, that a rectangular range space by itself does not guarantee independence.

Proposition 4.1

If X and Y are independent random variables, then $\text{cov}(X, Y) = 0$.

Proof. This is generally true. We shall prove the result for the case when X and Y are jointly continuous random variables. Suppose that X and Y are independent random variables. Then

$$
\begin{aligned}
\mathbb{E}(XY) &= \iint_{(x,y) \in R_{XY}} xy \cdot f_{XY}(x, y) \, dy \, dx \\
&= \iint_{(x,y) \in R_{XY}} xy \cdot f_X(x) f_Y(y) \, dy \, dx \\
&= \int_{x \in R_X} x \cdot f_X(x) \, dx \int_{y \in R_Y} y \cdot f_Y(y) \, dy \\
&= \mathbb{E}(X)\mathbb{E}(Y).
\end{aligned}
$$
∎

The converse of this proposition does not hold. It is possible to have $\text{cov}(X, Y) = 0$ when X and Y are not independent. We shall see examples of this later.

Example 4.2 – continued

We have already seen that X and Y are independent, and we previously (page 91) showed that $\text{cov}(X, Y) = 0$.

Example 4.3 – continued

In this case, X and Y are not independent, and we previously (page 93) showed that $\text{cov}(X, Y) = -cb^2/36 \neq 0$.

When X and Y are not independent, we would like some way of indicating the strength of the relationship between them. A numerical value that does this is called a *measure of association*. It has proved very difficult to propose general measures of association. The covariance itself is one possibility but it has various undesirable features, one of which is indicated in the following proposition.

Proposition 4.2

Suppose that X and Y are random variables with finite covariance and let a, b, c and d be real constants such that b and d are non-zero. Then

$$
\text{cov}(a + bX, c + dY) = bd \, \text{cov}(X, Y).
$$

The proof of this is left as an exercise. It means, however, that covariance is scale-dependent. For example, if we wish to know how height and weight are associated in a certain population, then the covariance between them would be different depending on whether height was measured in inches or centimetres and on whether weight was measured in pounds or kilograms. It is for this reason that a covariance by itself is hard to interpret as a measure of association. Other measures are required.

In the discrete case, our search for a measure of association begins with the *odds ratio*. Suppose that X and Y are binary variables, so each takes exactly two different values which, rescaling if required, we may take to be 0 and 1. In general, we have the sort of joint probability mass function shown below:

$$\begin{array}{c|cc} & \multicolumn{2}{c}{x} \\ p_{XY}(x, y) & 0 & 1 \\ \hline y \quad 0 & p_{00} & p_{10} \\ 1 & p_{01} & p_{11} \end{array}$$

Given $X = 0$, the *odds* on $Y = 1$ are

$$\frac{P(Y = 1 | X = 0)}{P(Y \neq 1 | X = 0)} = \frac{P(Y = 1 | X = 0)}{P(Y = 0 | X = 0)} = \frac{p_{01}/(p_{00} + p_{01})}{p_{00}/(p_{00} + p_{01})} = \frac{p_{01}}{p_{00}}.$$

Given $X = 1$, the odds on $Y = 1$ are

$$\frac{P(Y = 1 | X = 1)}{P(Y \neq 1 | X = 1)} = \frac{P(Y = 1 | X = 1)}{P(Y = 0 | X = 1)} = \frac{p_{11}/(p_{10} + p_{11})}{p_{10}/(p_{10} + p_{11})} = \frac{p_{11}}{p_{10}}.$$

The *odds ratio* is the ratio of these values:

$$OR = \frac{p_{11}/p_{10}}{p_{01}/p_{00}} = \frac{p_{00}p_{11}}{p_{01}p_{10}}.$$

The odds ratio can take any positive value. $OR = 1$ if the conditional odds on $Y = 1$ given X is the same for both $X = 0$ and $X = 1$. But, this only happens when the conditional probability of $Y = 1$ given X is the same for both $X = 0$ and $X = 1$, i.e. when X and Y are independent.

Notice that the odds ratio is symmetric in X and Y, in the following sense. Given $Y = 0$, the odds on $X = 1$ is

$$\frac{p_{10}/(p_{00} + p_{10})}{p_{00}/(p_{00} + p_{10})} = \frac{p_{10}}{p_{00}}.$$

Given $Y = 1$, the odds on $X = 1$ is

$$\frac{p_{11}/(p_{01} + p_{11})}{p_{01}/(p_{01} + p_{11})} = \frac{p_{11}}{p_{01}}.$$

So, the odds ratio is the ratio of these two values as well.

Example 4.4 – continued

The joint probability mass function is shown below.

	$p_{XY}(x, y)$	x 0	1	$p_Y(y)$
y	0	0.44	0.30	0.74
	1	0.11	0.15	0.26
	$p_X(x)$	0.55	0.45	1

Given $X = 0$, the odds on $Y = 1$ is

$$\frac{p_{01}}{p_{00}} = \frac{0.11}{0.44} = \frac{1}{4}.$$

Given $X = 1$, the odds on $Y = 1$ is

$$\frac{p_{11}}{p_{10}} = \frac{0.15}{0.30} = \frac{1}{2}.$$

The odds ratio is

$$OR = \frac{1/2}{1/4} = 2.$$

Given $Y = 0$, the odds on $X = 1$ is

$$\frac{p_{10}}{p_{00}} = \frac{0.30}{0.44}.$$

Given $Y = 1$, the odds on $X = 1$ is

$$\frac{p_{11}}{p_{01}} = \frac{0.15}{0.11}.$$

Again the odds ratio is

$$OR = \frac{15/11}{30/44} = 2.$$

We can interpret this result by saying that, in this population, the odds on being a victim of crime are twice as great for those who have themselves committed a criminal act as they are for those who have never committed a criminal act. This is a fairly clear statement of the strength of the association that exists between these two random variables.

Example 4.1 – continued

This joint probability mass function of X and Y in this example is tabulated below. Neither X nor Y is a binary random variable, so the odds ratio cannot be calculated. A more general measure of association is required.

$p_{XY}(x, y)$		x 0	1	2	$p_Y(y)$
y	0	$\frac{15}{45}$	$\frac{12}{45}$	$\frac{1}{45}$	$\frac{28}{45}$
	1	$\frac{12}{45}$	$\frac{4}{45}$	–	$\frac{16}{45}$
	2	$\frac{1}{45}$	–	–	$\frac{1}{45}$
$p_X(x)$		$\frac{28}{45}$	$\frac{16}{45}$	$\frac{1}{45}$	1

Goodman and Kruskal (1954) proposed a measure of association that is usually denoted by the Greek letter *gamma* (γ). In order to see how this is calculated, we need to consider making two observations, (X_1, Y_1) and (X_2, Y_2), from this distribution. These observations are called *concordant* if either $(X_2 > X_1$ and $Y_2 > Y_1)$ or $(X_2 < X_1$ and $Y_2 < Y_1)$. The observations are called *discordant* if either $(X_2 > X_1$ and $Y_2 < Y_1)$ or $(X_2 < X_1$ and $Y_2 > Y_1)$. Note that it is possible for the two observations to be neither concordant nor discordant, but this possibility is ignored in the calculation of γ.

Concordance is indicative of positive association between X and Y, in other words that larger values of X tend to be observed along with larger values of Y while smaller values of X tend to be observed along with smaller values of Y. Discordance is indicative of negative association between X and Y, in other words that larger values of X tend to be observed along with smaller values of Y while smaller values of X tend to be observed along with larger values of Y. Overall, γ indicates the extent to which concordance is more likely to occur than discordance. It is defined to be

$$\gamma = \frac{P-Q}{P+Q} = \frac{P}{P+Q} - \frac{Q}{P+Q},$$

where P is the probability of concordance and Q is the probability of discordance. This means that γ is the difference between the conditional probability of concordance, given that there is either concordance or discordance, and the conditional probability of discordance, given that there is either concordance or discordance. This means that γ must lie in the range -1 to 1. $\gamma = 0$ when the random variables are independent.

P is found by multiplying every probability in the joint probability mass function by the probability of every cell that is both to the right and below it. In this example, there is only one cell that has a cell both to the right and below it, that is p_{00}, which has p_{11} to its lower right. So

$$P = p_{00}p_{11} = \frac{15}{45}\frac{4}{45} = \frac{60}{2025}.$$

Q is found by multiplying every probability in the joint probability mass function by the probability of every cell that is to its left and below it.

In this example,

$$Q = p_{10}(p_{01} + p_{02}) + p_{20}(p_{01} + p_{02} + p_{11}) + p_{11}p_{02}$$

$$= \frac{12}{45} \left(\frac{12}{45} + \frac{1}{45} \right) + \frac{1}{45} \left(\frac{12}{45} + \frac{1}{45} + \frac{4}{45} \right) + \frac{4}{45} \frac{1}{45}$$

$$= \frac{177}{2025}$$

Therefore,

$$\gamma = \frac{P - Q}{P + Q} = \frac{60 - 177}{60 + 177} = -\frac{39}{79} = -0.494.$$

A negative value of γ indicates that X and Y are negatively associated. A value of about -0.5 suggests that there is moderately strong association in this example.

In the case of two binary random variables, considered previously, $P = p_{00}p_{11}$ and $Q = p_{01}p_{10}$. This means that

$$\gamma = \frac{P - Q}{P + Q} = \frac{p_{00}p_{11} - p_{01}p_{10}}{p_{00}p_{11} + p_{01}p_{10}} = \frac{p_{00}p_{11}/p_{01}p_{10} - p_{01}p_{10}/p_{01}p_{10}}{p_{00}p_{11}/p_{01}p_{10} + p_{01}p_{10}/p_{01}p_{10}} = \frac{OR - 1}{OR + 1}.$$

This is a monotonically increasing function of the odds ratio. As OR increases from 0 towards ∞, γ increases from -1 towards 1. It is an often-stated principle of dealing with binary random variables that all sensible measures of association are monotonic functions of OR.

Example 4.4 – continued

Here

$$P = 0.44 \times 0.15,$$

$$Q = 0.30 \times 0.11,$$

so

$$\gamma = \frac{P - Q}{P + Q} = \frac{0.44 \times 0.15 - 0.30 \times 0.11}{0.44 \times 0.15 + 0.30 \times 0.11} = \frac{0.033}{0.099} = \frac{1}{3}.$$

This is indicative of positive association, i.e. X and Y tend to take the same values (either 0 and 0 or 1 and 1).

We will now consider the most popular measure of association for two continuous random variables, namely the correlation. Some readers might be surprised that this has not already been proposed as a measure of association in the discrete case. In fact, it really is not a good option at all for dealing with discrete random variables and is only really helpful in particular cases of continuous random variables, two truths generally ignored in statistics textbooks.

If X and Y are random variables with finite expected values and finite variances, then the *correlation* between X and Y is defined to be:

$$\rho_{XY} = \frac{\text{cov}(X, Y)}{\sqrt{\text{var}(X)\text{var}(Y)}}.$$

The properties of the covariance mean that $\rho_{XY} = 0$ when X and Y are independent.

Like the covariance, the correlation is a measure of association between two random variables. Its main advantage over the covariance is that the absolute value of the correlation is invariant to changes in scale. This is expressed in the following proposition.

Proposition 4.3

Let $U = a + bX$ and $V = c + dY$, where a, b, c and d are real constants such that b and d are non-zero. Then

$$\rho_{UV} = \begin{cases} \rho_{XY}, & \text{if } bd \geq 0, \\ -\rho_{XY}, & \text{if } bd < 0. \end{cases}$$

Proof. Using Propositions 2.3 and 4.2,

$$\text{var}(U) = b^2 \, \text{var}(X),$$
$$\text{var}(V) = d^2 \, \text{var}(Y),$$
$$\text{cov}(U, V) = bd \, \text{cov}(X, Y).$$

So

$$\rho_{UV} = \frac{bd \, \text{cov}(X, Y)}{\sqrt{b^2 \, \text{var}(X) d^2 \, \text{var}(Y)}} = \frac{bd}{|b||d|} \frac{\text{cov}(X, Y)}{\sqrt{\text{var}(X).\text{var}(Y)}}$$

$$= \begin{cases} \rho_{XY}, & \text{if } bd \geq 0, \\ -\rho_{XY}, & \text{if } bd < 0. \end{cases} \qquad \blacksquare$$

This means that, for example, if we were interested in measuring the association between height and weight, the correlation would be the same whether we measured height in inches or centimetres and the weight in pounds or kilograms; as we have already established, the covariance would be different for different units of measurement.

Another advantage of the correlation as a measure of association is that, like γ, it can only take values in a limited range, as we shall now show.

Proposition 4.4

Let X and Y be random variables with finite expected values and finite, non-zero variances. Then:

(a) $-1 \leq \rho_{XY} \leq 1$;

(b) $\rho_{XY} = 1 \Leftrightarrow Y = a + bX$ for some $b > 0$;

(c) $\rho_{XY} = -1 \Leftrightarrow Y = a + bX$ for some $b < 0$.

Proof. Define the standardized random variables U and V by

$$U = \frac{X - \mathbb{E}(X)}{\sqrt{\operatorname{var}(X)}} = -\frac{\mathbb{E}(X)}{\sqrt{\operatorname{var}(X)}} + \frac{1}{\sqrt{\operatorname{var}(X)}}X,$$

$$V = \frac{Y - \mathbb{E}(Y)}{\sqrt{\operatorname{var}(Y)}} = -\frac{\mathbb{E}(Y)}{\sqrt{\operatorname{var}(Y)}} + \frac{1}{\sqrt{\operatorname{var}(Y)}}Y.$$

Then $\mathbb{E}(U) = \mathbb{E}(V) = 0$ and $\operatorname{var}(U) = \operatorname{var}(V) = 1$. So

$$\mathbb{E}(U^2) = \operatorname{var}(U) + \{\mathbb{E}(U)\}^2 = 1,$$
$$\mathbb{E}(V^2) = \operatorname{var}(V) + \{\mathbb{E}(V)\}^2 = 1.$$

Also

$$\operatorname{cov}(U, V) = \mathbb{E}(UV) - \mathbb{E}(U)\mathbb{E}(V) = \mathbb{E}(UV)$$

and

$$\rho_{UV} = \mathbb{E}(UV).$$

By Proposition 4.3, $\rho_{XY} = \rho_{UV}$, since

$$\frac{1}{\sqrt{\operatorname{var}(X)}} \frac{1}{\sqrt{\operatorname{var}(Y)}} \geq 0.$$

Now consider the random variable $(U - V)^2$. Since this takes only non-negative values, it must follow that $\mathbb{E}\left\{(U - V)^2\right\} \geq 0$. But

$$\mathbb{E}\left\{(U - V)^2\right\} = \mathbb{E}\left\{U^2 - 2UV + V^2\right\}$$
$$= \mathbb{E}\left\{U^2\right\} + \mathbb{E}\left\{V^2\right\} - 2\mathbb{E}\{UV\}$$
$$= 2(1 - \rho_{XY}),$$

so

$$1 - \rho_{XY} \geq 0,$$

i.e.

$$\rho_{XY} \leq 1.$$

Similarly, by considering $\mathbb{E}\left\{(U+V)^2\right\}$, it can be shown that $\rho_{XY} \geq -1$. This establishes property (a).

Notice that

$$\rho_{XY} = 1 \Leftrightarrow 1 - \rho_{XY} = 0$$
$$\Leftrightarrow \mathbb{E}\left\{(U-V)^2\right\} = 0$$
$$\Leftrightarrow U - V = 0 \text{ (with probability 1)}$$
$$\Leftrightarrow X = \left[\mathbb{E}(X) - \frac{\sqrt{\mathrm{var}(X)}}{\sqrt{\mathrm{var}(Y)}}\mathbb{E}(Y)\right] + \frac{\sqrt{\mathrm{var}(X)}}{\sqrt{\mathrm{var}(Y)}}Y \text{ (with probability 1)}$$

This establishes property (b).

A similar argument using $\mathbb{E}\left\{(U+V)^2\right\}$ establishes property (c). ∎

Values of ρ_{XY} that are close to $+1$ indicate strong positive correlation (high values of X tend to occur along with high values of Y), while values of ρ_{XY} that are close to -1 indicate strong negative correlation (high values of X tend to occur along with low values of Y).

Example 4.5

The jointly continuous random variables X and Y have joint probability density function

$$f_{XY}(x,y) = \begin{cases} \theta^2 e^{-\theta y}, & y > x > 0, \\ 0, & \text{otherwise.} \end{cases}$$

We will first of all derive the marginal probability density functions of X and Y, and their expected values and variances. We will then calculate the correlation between X and Y.

We have

$$f_X(x) = \int_x^\infty \theta^2 e^{-\theta y}\,\mathrm{d}y \qquad\qquad f_Y(y) = \int_0^y \theta^2 e^{-\theta y}\,\mathrm{d}x$$
$$= \theta e^{-\theta x}, \qquad x > 0, \qquad\qquad = \theta^2 y e^{-\theta y}, \qquad y > 0.$$

This means that $X \sim \mathrm{Ex}(\theta)$ and $Y \sim \mathrm{Ga}(2, \theta)$. So, using results obtained in Chapter 2,

$$\mathbb{E}(X) = \frac{1}{\theta}, \qquad \mathrm{var}(X) = \frac{1}{\theta^2},$$
$$\mathbb{E}(Y) = \frac{2}{\theta}, \qquad \mathrm{var}(Y) = \frac{2}{\theta^2}$$

Also,

$$\mathbb{E}(XY) = \int_0^\infty \int_0^y xy\theta^2 e^{-\theta y} \, dx \, dy$$

$$= \int_0^\infty y\theta^2 e^{-\theta y} \left\{ \int_0^y x \, dx \right\} dy$$

$$= \frac{\theta^2}{2} \int_0^\infty y^3 e^{-\theta y} \, dy$$

$$= \frac{\theta^2}{2} \frac{1}{\theta^4} \Gamma(4)$$

$$= \frac{3}{\theta^2}.$$

Therefore,

$$\mathrm{cov}(X, Y) = \mathbb{E}(XY) - \mathbb{E}(X)\mathbb{E}(Y) = \frac{1}{\theta^2}$$

and

$$\rho_{XY} = \frac{1}{\sqrt{2}} = 0.707$$

This positive value of ρ indicates positive association, i.e. large values of X tend to go with large values of Y. The value of 0.707 indicates that this positive association is moderately strong.

It is very important to emphasize that the correlation is not a general measure of association. It is a measure of the *linear* association between two random variables. When the relationship between X and Y is not linear, ρ_{XY} is a very poor guide to the strength of association between the variables, as the following example shows.

Example 4.6

Suppose that X is an N(0, 1) random variable, and let $Y = X^2$. Then, X and Y are perfectly associated; if we know the value of one of them, then we can immediately find the value of the other. Since $X \sim \mathrm{N}(0, 1)$ and $Y \sim \chi^2(1)$, then

$$\mathbb{E}(X) = 0 \qquad \mathrm{var}(X) = 1,$$

$$\mathbb{E}(Y) = 1 \qquad \mathrm{var}(Y) = 2.$$

So

$$\mathrm{cov}(X, Y) = \mathbb{E}(XY) - \mathbb{E}(X)\mathbb{E}(Y) = \mathbb{E}(X^3) - \mathbb{E}(X)\mathbb{E}(Y) = \mathbb{E}(X^3).$$

Now, the probability density function of X is symmetric around the value $x = 0$, so $x^3 f_X(x)$ is an odd function and $\mathbb{E}(X^3)$ must equal 0. Hence, $\mathrm{cov}(X, Y) = 0$. So $\rho_{XY} = 0$ even though X and Y are perfectly related by the parabola $y = x^2$.

We have already shown that, if X and Y are independent, then $\operatorname{cov}(X,Y)=0$ and so $\rho_{XY}=0$. The above example shows that ρ_{XY} can be equal to zero even when X and Y are not independent. So, a correlation of 0 does not prove, by itself, that two random variables are independent. On the other hand, a correlation that is not equal to zero does always imply that the random variables are not independent.

Example 4.7

A scientist was investigating the damage that can be caused to the stomach by long-term (possibly excessive) use of aspirin and related drugs. In the course of a study of rats, the scientist found two measures of degradation that could be recorded from microscopic examination of a specimen of stomach tissue. In control (i.e. untreated) rats, these random variables, X and Y, both marginally followed exponential distributions. X and Y were not, however, independent. How might we find a plausible bivariate model for X and Y?

Write $f(x)$ and $g(y)$ for the marginal p.d.f.s, and $F(x)$ and $G(y)$ for the marginal distribution functions, and consider the class of bivariate models with joint distribution function

$$F_{XY}(x,y) = F(x)G(y)\{1 - k(1 - F(x))(1 - G(y))\},$$

where k is some value between -1 and 1. This class of model has the appropriate marginal distribution functions, since

$$F_{XY}(x,\infty) = F(x)G(\infty)\{1 - k(1 - F(x))(1 - G(\infty))\} = F(x)$$

and

$$F_{XY}(\infty,y) = F(\infty)G(y)\{1 - k(1 - F(\infty))(1 - G(y))\} = G(y).$$

Differentiating F_{XY} with respect to x and then with respect to y gives

$$f_{XY}(x,y) = f(x)g(y)\{1 - k(1 - 2F(x))(1 - 2G(y))\}.$$

In this particular example,

$$f(x) = \theta \exp(-\theta x), \quad F(x) = 1 - \exp(-\theta x), \quad x > 0,$$

and

$$g(y) = \phi \exp(-\phi y), \quad G(y) = 1 - \exp(-\phi y), \quad y > 0$$

so

$$f_{XY}(x,y) = \theta \exp(-\theta x)\phi \exp(-\phi y)$$
$$\times \{1 - k[2\exp(-\theta x) - 1][2\exp(-\phi y) - 1]\}, \quad x > 0, \ y > 0.$$

From this it follows that

$$E(XY) = \int_0^\infty \theta x e^{-\theta x} \int_0^\infty \phi y e^{-\phi y}(1 - k + 2ke^{-\theta x} + 2ke^{-\phi y} - 4ke^{-\theta x}e^{-\phi y})\, dy\, dx.$$

But the inner integral is

$$\int_0^\infty \phi y e^{-\phi y}(1 - k + 2ke^{-\theta x})\, dy + \int_0^\infty \phi y e^{-2\phi y} 2k(1 - 2e^{-\theta x})\, dy$$

$$= (1 - k + 2ke^{-\theta x})\frac{1}{\phi} + k(1 - 2e^{-\theta x})\frac{1}{2\phi}$$

$$= \frac{1}{2\phi}(2 - k) + \frac{k}{\phi}e^{-\theta x}.$$

Hence,

$$E(XY) = \frac{2 - k}{2\phi\theta} + \frac{k}{4\phi\theta}$$

$$\text{cov}(X, Y) = \frac{2 - k}{2\phi\theta} + \frac{k}{4\phi\theta} - \frac{1}{\theta}\cdot\frac{1}{\phi} = -\frac{k}{4\theta\phi}$$

$$\rho_{XY} = \frac{-k/4\theta\phi}{\sqrt{\theta^{-2}\phi^{-2}}} = -\frac{k}{4}.$$

Since, as originally specified, k must lie between -1 and 1, it follows that this bivariate model for X and Y only allows a correlation between $-\frac{1}{4}$ and $\frac{1}{4}$. This seems an unreasonable restriction, so this might not be a good class of bivariate model to consider. Some other extension to two dimensions might be required.

Exercises

1 Consider again the discrete random variables X and Y described in Exercise 1 on Section 4.1.

(a) Find the conditional distribution $p_{X|Y}(x|y)$ for all possible values of y. Compare these with one another and with the marginal distribution of X.

(b) Find the covariance between X and Y.

(c) Calculate Goodman and Kruskal's γ.

(d) How do you know that X and Y are not independent?

2 Consider again the discrete random variables X and Y described in Exercise 2 on Section 4.1.

(a) Find the conditional distribution $p_{Y|X}(y|x)$ for all possible values of x. Compare these with one another and with the marginal distribution of Y.

(b) Find the covariance between X and Y.

(c) Calculate Goodman and Kruskal's γ.

(d) How do you know that X and Y are not independent?

3 The discrete random variables X and Y have the following (marginal) probability mass functions:

x	1	2	3	4
$p_X(x)$	0.1	0.2	0.3	0.4

y	−1	0	1
$p_Y(y)$	0.25	0.5	0.25

(a) Assuming that X and Y are independent, find their joint probability mass function.

(b) Confirm that the conditional probability mass function of X given $Y = y$ is the same as the marginal probability mass function of X for all possible values of y.

(c) Show that $\gamma = 0$.

4 Sweet peas have flowers that are either purple or red. The colour of a plant's flowers is determined by one gene which can be denoted (F, f). Suppose that two hybrid plants are crossed to produce offspring. Each parent plant has one F gene and one f gene. Each parent independently is equally likely to donate either of its genes to an offspring.

(a) Find the probability that an offspring has (i) two F genes, (ii) one F and one f gene, (iii) two f genes.

(b) Offspring with at least one F gene have purple flowers; other offspring have red flowers. Find the probability that an offspring has (i) purple, (ii) red flowers.

5 The sweet peas described in Exercise 4 also have either long or round pollen. The shape of a plant's pollen is determined by a gene which can be denoted (G, g), which is inherited according to the same rules that govern inheritance of the (F, f) gene discussed in the previous exercise. Any plant with at least one G gene has long pollen. Suppose that two parent plants, known as dihybrids, have one G and one g gene as well as one F and one f gene.

(a) If the (F, f) and (G, g) genes are inherited independently of one another, what is the probability that an offspring has purple flowers and long pollen?

(b) In fact, these genes are not inherited independently. Suppose there is probability θ ($0.5 < \theta < 0.75$) that an offspring has purple flowers and long pollen. Complete the following table of joint probabilities.

joint probability	long pollen	round pollen	
purple flowers	θ		0.75
red flowers			0.25
	0.75	0.25	1

6 Let A and B be two events in the sample space S. Define the binary random variables I_A and I_B as follows:

$$I_A = \begin{cases} 1, & \text{if } A \text{ occurs,} \\ 0, & \text{otherwise,} \end{cases} \qquad I_B = \begin{cases} 1, & \text{if } B \text{ occurs,} \\ 0, & \text{otherwise.} \end{cases}$$

I_A and I_B are known as *indicator variables*. Show that I_A and I_B are independent random variables if and only if A and B are independent events. Show also that A and B are independent events if and only if the odds ratio is equal to one.

7 Strachan *et al.* (1996) investigated the population of 11-year-old children in the UK. They were interested in two binary random variables:

$$X = \begin{cases} 0, & \text{if a child does not suffer from hay fever,} \\ 1, & \text{if a child does suffer from hay fever,} \end{cases}$$

$$Y = \begin{cases} 0, & \text{if a child does not suffer from eczema,} \\ 1, & \text{if a child does suffer from eczema.} \end{cases}$$

The joint probability mass function indicated by their survey is shown below.

		x	
$p_{XY}(x, y)$		0	1
y	0	0.90	0.06
	1	0.03	0.01

Calculate and interpret the odds ratio and Goodman and Kruskal's γ.

8 Consider again the random variables X and Y defined in Exercise 4 on Section 4.1. Find and interpret the correlation between X and Y.

9 Consider again the random variables X and Y defined in Exercise 5 on Section 4.1. Find and interpret the correlation between X and Y.

10 The continuous random variables X and Y have joint probability density function

$$f(x,y) = 6x, \qquad 0 < x < y < 1.$$

Derive the marginal probability density functions of X and Y, and their expected values and variances. Find and interpret the correlation between X and Y.

11 The continuous random variables X and Y have the triangular joint range space bounded by the x-axis and the lines $y = c(b - x)$ and $y = c(b + x)$ (for some positive constant c). For each of the following joint probability density functions, find the constant k, the marginal probability density functions of X and Y, the expected value and variance of X and Y, the conditional probability density of Y given $X = x$, and the correlation between X and Y.

(a) $f(x,y) = k$;

(b) $f(x,y) = k|x|$;

(c) $f(x,y) = ky$.

12 The continuous random variables X and Y have the joint probability density function

$$f_{XY}(x,y) = \frac{2}{c^2}, \qquad 0 < x < c, \; 0 < y < c, \; 0 < x + y < c,$$

where c is a positive constant. Find and interpret the correlation between X and Y.

13 The continuous random variables X and Y have the joint probability density function

$$\frac{\Gamma(\alpha + \beta + \gamma)}{\Gamma(\alpha)\Gamma(\beta)\Gamma(\gamma)} x^{\alpha-1} y^{\beta-1} (1 - x - y)^{\gamma-1}, \quad 0 < x < 1, \; 0 < y < 1, \; 0 < x + y < 1,$$

where $\alpha > 0, \beta > 0, \gamma > 0$ and $\Gamma(\cdot)$ is the gamma function.

(a) Let r and s be non-negative integers. Show that the expected value of $X^r Y^s$ is

$$\mathbb{E}(X^r Y^s) = \frac{\Gamma(\alpha + r)}{\Gamma(\alpha)} \cdot \frac{\Gamma(\beta + s)}{\Gamma(\beta)} \cdot \frac{\Gamma(\alpha + \beta + \gamma)}{\Gamma(\alpha + \beta + \gamma + r + s)}$$

(b) Hence determine the expected value and variance of X.

(c) Find the correlation between X and Y.

14 Suppose that the continuous random variables X and Y are independent with the following (marginal) probability density functions:

$$f_X(x) = \exp(-x), \qquad x > 0,$$
$$f_Y(y) = \exp(-y), \qquad y > 0.$$

Write down the joint probability density function of X and Y. Show that the correlation between X and Y is 0.

15 Suppose that the continuous random variable X follows the $Ex(\theta)$ distribution. Then we may write $X = U + V$, where U is the integer part of X (i.e. the largest integer that is not greater than X) and $V = X - U$ is the remainder.

    (a) Show that the discrete random variable U follows a geometric distribution.

    (b) Find the conditional distribution of the continuous random variable V given that $U = u(\geq 0)$.

    (c) Hence show that U and V are independent random variables and identify the marginal distribution of V.

16 X and Y are continuous random variables.

    (a) Suppose that X and Y are independent, i.e. $F_{XY}(x, y) = F_X(x)F_Y(y)$ for all x and y. Show that $f_{XY}(x, y) = f_X(x)f_Y(y)$ for all x and y.

    (b) Now suppose instead that $f_{XY}(x, y) = f_X(x)f_Y(y)$ for all x and y. Show that X and Y are independent.

17 X and Y are continuous random variables. Show that, if X and Y are independent, then $f_{XY}(x, y) = f_{X|Y}(x|y)f_Y(y)$ for any $(x, y) \in R_{XY}$.

18 Prove Proposition 4.1 when X and Y are both discrete random variables.

19 X and Y are independent random variables. Let $g(X)$ be any function of X and $h(Y)$ be any function of Y. By considering the distribution functions of $g(X)$ and $h(Y)$, show that these are independent random variables. Deduce that

$$\mathbb{E}\{g(X)h(Y)\} = \mathbb{E}\{g(X)\}\mathbb{E}\{h(Y)\}.$$

20 Suppose that X is a constant, i.e. $P(X = c) = 1$ for some $c \in \mathbb{R}$. Prove that the covariance between X and any other random variable must be 0. What happens to the correlation between X and any other random variable?

4.3 Iterated expectation and variance

Example 4.8

A board game is played with an unbiased, six-sided die whose sides are marked $1, 2, \ldots, 6$. Whenever a player scores a 6, then he rolls the die again. The score for the player's turn is the total of the scores on the die on all the times it is rolled.

If the random variable Y is the total score on a turn, then Y has the range space

$$R_Y = \{1, 2, \ldots, 5, 7, 8, \ldots, 11, 13, 14, \ldots, 17, 19, \ldots\}$$

and associated probabilities

$$p(1) = p(2) = \ldots = p(5) = \tfrac{1}{6},$$
$$p(7) = p(8) = \ldots = p(11) = \tfrac{1}{36}, \ldots .$$

We could obtain $\mathbb{E}(Y)$ and $\mathrm{var}(Y)$ directly from this probability mass function, but that is extremely tedious. An easier way can be found, using conditional expected values and variances, if the random variable X is defined to be the number of times the die is rolled in the course of a turn. The following result is required.

Proposition 4.5 – Iterated expectation and variance

Suppose that X and Y are random variables. Then, assuming that the required expected values and variances exist,

(a) $\mathbb{E}(Y) = \mathbb{E}\{\mathbb{E}(Y|X)\}$;

(b) $\mathrm{var}(Y) = \mathbb{E}\{\mathrm{var}(Y|X)\} + \mathrm{var}\{\mathbb{E}(Y|X)\}$.

Proof. We prove this proposition for the case where X and Y are both discrete random variables.

(a) We have

$$\mathbb{E}(Y|X = x) = \sum_{y:(x,y)\in R_{XY}} y\, p(y|x) = \sum_{y:(x,y)\in R_{XY}} y \frac{p_{XY}(x,y)}{p_X(x)}.$$

Now, this is a function of x, but not of y, and so it makes sense to take its expected value with respect to X.

$$\mathbb{E}\{\mathbb{E}(Y|X)\} = \sum_{x\in R_X}\left\{ \sum_{y:(x,y)\in R_{XY}} y\cdot\frac{p_{XY}(x,y)}{p_X(x)} \right\} p_X(x) = \sum\sum_{(x,y)\in R_{XY}} y\, p(x,y) = \mathbb{E}(Y).$$

(b) We have

$$\mathrm{var}\{\mathbb{E}(Y|X)\} = \mathbb{E}\left\{[\mathbb{E}(Y|X)]^2\right\} - \left[\mathbb{E}\left\{\mathbb{E}(Y|X)\right\}\right]^2 = \mathbb{E}\left\{[\mathbb{E}(Y|X)]^2\right\} - [\mathbb{E}(Y)]^2,$$

$$\mathbb{E}\{\mathrm{var}(Y|X)\} = \mathbb{E}\{\mathbb{E}(Y^2|X)\} - \mathbb{E}\left\{[\mathbb{E}(Y|X)]^2\right\} = \mathbb{E}(Y^2) - \mathbb{E}\left\{[\mathbb{E}(Y|X)]^2\right\},$$

and we sum these results to obtain

$$\mathbb{E}\{\mathrm{var}(Y|X)\} + \mathrm{var}\{\mathbb{E}(Y|X)\} = \mathbb{E}(Y^2) - [\mathbb{E}(Y)]^2 = \mathrm{var}(Y). \quad\blacksquare$$

Example 4.8 – continued

If $X = x$, then the die has been rolled exactly x times, a 6 being obtained on the first $x - 1$ rolls but not on the xth roll. This means that X is a $Ge(\theta)$ random variable, with $\theta = 5/6$. So, $\mathbb{E}(X) = 6/5$ and $\mathrm{var}(X) = 6/25$.

Given that $X = x$, the conditional distribution of Y is the discrete uniform distribution on the five values

$$6(x - 1) + 1, \ldots, 6(x - 1) + 5.$$

Using results for the uniform discrete distribution that were derived in Chapter 2,

$$\mathbb{E}(Y|X = x) = 6(x - 1) + 3 = 6x - 3,$$
$$\mathrm{var}(Y|X = x) = (5^2 - 1)/12 = 2.$$

Using Proposition 4.5(a),

$$\mathbb{E}(Y) = \mathbb{E}\{\mathbb{E}(Y|X)\} = \mathbb{E}(6X - 3) = 6\,\mathbb{E}(X) - 3 = 6 \times 6/5 - 3 = 4.2.$$

Using Proposition 4.5(b),

$$\mathbb{E}\{\mathrm{var}(Y|X)\} = \mathbb{E}(2) = 2,$$
$$\mathrm{var}\{\mathbb{E}(Y|X)\} = \mathrm{var}\{6X - 3\} = 36\,\mathrm{var}(X) = 36 \times 6/25 = 8.64$$

and

$$\mathrm{var}(Y) = \mathbb{E}\{\mathrm{var}(Y|X)\} + \mathrm{var}\{\mathbb{E}(Y|X)\} = 2 + 8.64 = 10.64.$$

Proposition 4.6

Suppose that X and Y are random variables, and let $g(X)$ be any real function of X. Then, assuming that the required expected values exist,

$$\mathbb{E}\{g(X)Y\} = \mathbb{E}\{g(X)\,\mathbb{E}(Y|X)\}.$$

This result is an extension of Proposition 4.5, and its proof is left as an exercise.

Example 4.9 – The trinomial distribution

When many people are injured in an emergency, for example an explosion or a train crash, it is common for medical personnel to categorize casualties according to the triage system – as either minor, moderate or critical. Suppose that, in incidents of this kind, the proportions of minor and moderate casualties are θ_1 and θ_2. In a particular incident, there are a total of n casualties of which X are minor and Y are moderate (so that $n - X - Y$ are critical). Then the random variables X and Y jointly follow a *trinomial* distribution with joint probability mass function

$$p_{XY}(x,y) = \frac{n!}{x!y!(n-x-y)!}\,\theta_1^x\theta_2^y(1-\theta_1-\theta_2)^{n-x-y}$$

on the range space $R_{XY} = \{x=0,1,\ldots,n; y=0,1,\ldots,n-x\}$. In general, we write $(X,\ Y) \sim \mathrm{Tri}(n,\theta_1,\theta_2)$, where $0 \le \theta_1 \le 1$, $0 \le \theta_2 \le 1$, $0 \le \theta_1+\theta_2 \le 1$. This distribution is an extension of the binomial to a situation where there are three (rather than two) categories. In Chapter 5, we shall extend this further to the *multinomial* distribution, where there is no restriction on m, the number of mutually exclusive categories.

Marginally, it is easy to show that X and Y both follow binomial distributions. For $x=0,1,\ldots,n$,

$$p_X(x) = \sum_{y=0}^{n-x} p_{XY}(x,y)$$

$$= \sum_{y=0}^{n-x} \frac{n!}{x!y!(n-x-y)!}\theta_1^x\theta_2^y(1-\theta_1-\theta_2)^{n-x-y}$$

$$= \frac{n!}{x!(n-x)!}\theta_1^x \sum_{y=0}^{n-x} \binom{n-x}{y}\theta_2^y(1-\theta_1-\theta_2)^{(n-x)-y}$$

$$= \binom{n}{x}\theta_1^x\{\theta_2+(1-\theta_1-\theta_2)\}^{n-x} \qquad \text{(Result A1.8)}$$

$$= \binom{n}{x}\theta_1^x\{1-\theta_1\}^{n-x}.$$

So, $X \sim \mathrm{Bi}(n,\theta_1)$. This is a reasonable result, since any casualty is either minor (with probability θ_1) or not minor (with probability $1-\theta_1$). Similarly, $Y \sim \mathrm{Bi}(n,\theta_2)$.

Since R_{XY} is not a 'rectangular' region, it is immediately clear that X and Y are not independent. In order to prove this formally, consider the point (n,n). Since this point does not lie in R_{XY}, then $p_{XY}(n,n)=0$. But $p_X(n)$ and $p_Y(n)$ are both non-zero.

The conditional distribution of Y given $X == x$ (for $x \in R_X$) is

$$p(y|x) = \frac{p_{XY}(x,y)}{p_X(x)}$$

$$= \frac{n!}{x!y!(n-x-y)!}\theta_1^x\theta_2^y(1-\theta_1-\theta_2)^{n-x-y}\frac{x!(n-x)!}{n!\theta_1^x\{1-\theta_1\}^{n-x}}$$

$$= \binom{n-x}{y}\left(\frac{\theta_2}{1-\theta_1}\right)^y\left(1-\frac{\theta_2}{1-\theta_1}\right)^{(n-x)-y}$$

for $y = 0, 1, \ldots, n - x$. So, conditional on $X = x, Y$ follows the $\mathrm{Bi}(n - x, \theta_2 / (1 - \theta_1))$ distribution. This is a reasonable result: given that x of the casualties are minor, then each of the remaining $n - x$ casualties is either moderate (with conditional probability $\theta_2/(1 - \theta_1)$) or critical. Similarly, conditional on $Y = y, X$ follows the $\mathrm{Bi}(n - y, \theta_1/(1 - \theta_2))$ distribution.

We will now use Proposition 4.6 in order to find $\mathbb{E}(XY)$ and hence $\mathrm{cov}(X, Y)$ and ρ_{XY}. We have

$$\mathbb{E}(XY) = \mathbb{E}\{X\,\mathbb{E}(Y|X)\}$$

$$= \mathbb{E}\left\{X(n - X)\frac{\theta_2}{1 - \theta_1}\right\}$$

$$= \frac{\theta_2}{1 - \theta_1}\left\{\mathbb{E}(nX) - \mathbb{E}(X^2)\right\}$$

$$= \frac{\theta_2}{1 - \theta_1}\left\{n\,\mathbb{E}(X) - \mathrm{var}(X) - [\mathbb{E}(X)]^2\right\}$$

$$= \frac{\theta_2}{1 - \theta_1}\left\{nn\theta_1 - n\theta_1(1 - \theta_1) - n^2\theta_1^2\right\}$$

$$= \frac{\theta_2}{1 - \theta_1}\left\{n^2\theta_1(1 - \theta_1) - n\theta_1(1 - \theta_1)\right\}$$

$$= n(n - 1)\theta_1\theta_2.$$

Hence

$$\mathrm{cov}(X, Y) = \mathbb{E}(XY) - \mathbb{E}(X)\,\mathbb{E}(Y) = -n\theta_1\theta_2.$$

Notice the negative sign here, which indicates that there is a negative relationship between X and Y. This makes sense intuitively, since the more minor casualties there are, the fewer moderate casualties there are likely to be (and vice versa). Now,

$$\rho_{XY} = \frac{\mathrm{cov}(X, Y)}{\sqrt{\mathrm{var}(X)\,\mathrm{var}(Y)}} = \frac{-n\theta_1\theta_2}{\sqrt{n\theta_1(1 - \theta_1)\,n\theta_2(1 - \theta_2)}} = -\sqrt{\frac{\theta_1\theta_2}{(1 - \theta_1)(1 - \theta_2)}}.$$

If $\theta_1 + \theta_2 = 1$, i.e. $\theta_2 = 1 - \theta_1$, then this correlation is -1. In this case, $X + Y$ is certain to take the value n, i.e. there are no critical casualties.

Exercises

1 Let Y be the number of vehicles that travel along a certain stretch of motorway in a period of 5 minutes. At peak times, $Y \sim \mathrm{Po}(\lambda)$. A proportion θ of all the vehicles that travel along this road at peak times then leave the motorway at the next interchange. Given $Y = y$, then X, the number of vehicles that go off at the interchange in a period of 5 minutes, follows a $\mathrm{Bi}(y, \theta)$ distribution.

(a) Use the relationship $p_{XY}(x,y) = p_{X|Y}(x|y)p_Y(y)$ to find the joint probability mass function of X and Y.

(b) Show that, marginally, X follows a Poisson distribution.

(c) Verify that, in this example,

$$\mathbb{E}(Y) = \mathbb{E}\{\mathbb{E}(Y|X)\},$$

$$\text{var}(Y) = \mathbb{E}\{\text{var}(Y|X)\} + \text{var}\{\mathbb{E}(Y|X)\}.$$

2 The discrete random variable Y follows a Bi(n, θ) distribution. Given that $Y = y$, the discrete random variable X follows a Bi$(n-y, \phi)$ distribution, where $0 < \phi < 1 - \theta$. Show that (X, Y) jointly follow a trinomial distribution.

3 On a certain factory production line, k electronic components are manufactured each working day (where $k \geq 1$). The number of defective items from a randomly selected day's production, X, is a Bi(k, y) random variable, where y varies from day to day in accordance with a beta distribution,

$$f_Y(y) = \frac{(m+n-1)!}{(m-1)!\,(n-1)!} y^{m-1}(1-y)^{n-1}, \qquad 0 < y < 1,$$

where m and n are positive integers.

(a) Sketch the joint range space for (X, Y).

(b) Find the marginal probability mass function for X using the relationship

$$p_X(x) = \int_0^1 p_{X|Y}(x|y)f_Y(y)\,\mathrm{d}y.$$

Recognize this distribution and hence write down $\mathbb{E}(X)$ and $\text{var}(X)$.

(c) Verify that, in this example,

$$\mathbb{E}(X) = \mathbb{E}\{\mathbb{E}(X|Y)\},$$

$$\text{var}(X) = \mathbb{E}\{\text{var}(X|Y)\} + \text{var}\{\mathbb{E}(X|Y)\}.$$

4 (a) A sequence of independent trials is carried out. Each trial may result in a success, with probability ϕ (where $0 < \phi < 1$), or a failure, with probability $1 - \phi$. The discrete random variable Y is the number of consecutive failures recorded before the first success. Write down the probability mass function of Y, and hence find its expected value and variance.

(b) The discrete random variable Y and the continuous random variable X are jointly distributed. Marginally, X has the probability density function

$$f_X(x) = \theta e^{-\theta x}, \qquad x > 0$$

where $\theta > 0$. Conditional on $X = x$, Y follows a Poisson distribution with expected value x. Show that the marginal distribution of Y is the distribution described in part (a), for a particular value of ϕ.

(c) Confirm that, for X and Y defined as in part (b),

$$\mathbb{E}(Y) = \mathbb{E}\{\mathbb{E}(Y|X)\},$$

$$\mathrm{var}(Y) = \mathbb{E}\{\mathrm{var}(Y|X)\} + \mathrm{var}\{\mathbb{E}(Y|X)\}.$$

5 The continuous random variable Y follows an $\mathrm{Ex}(\theta)$ distribution. Conditional on $Y = y$, X follows an $\mathrm{Ex}(1/y)$ distribution. Use Proposition 4.5 to show that $\mathbb{E}(X) = 1/\theta$ and $\mathrm{var}(X) = 3/\theta^2$.

6 A certain train service is at least 5 minutes late on 3% of occasions, and late by less than 5 minutes on a further 12% of occasions. On the remaining occasions, the service runs on time. In the course of one week, you use this service ten times. Find:

(a) the probability that you are at least 5 minutes late at least once;

(b) the probability that you are on time on at least eight occasions;

(c) the probability that you are on time on exactly eight occasions and at least 5 minutes late on exactly one occasion;

(d) the expected number of times you will be at least 5 minutes late.

7 Of the fine china ornaments manufactured in a particular factory, 75% are perfect, 20% are seconds and the other 5% are rejects. In a batch of 20 ornaments, find

(a) the probability that at least 17 are perfect;

(b) the probability that none is a reject;

(c) the probability that exactly 17 are perfect and exactly three are seconds;

(d) the expected number of ornaments that are rejects.

8 It is believed that 6% of all children in the UK suffer moderately from dyslexia and a further 4% suffer severely from the same disorder. Of 30 children in a primary school class, find the probability that:

(a) at least one child is severely dyslexic;

(b) at least two children are moderately dyslexic;

(c) at least one child is severely dyslexic and at least another two are moderately dyslexic;

(d) at most one child has severe dyslexia, given that exactly three children have moderate dyslexia.

9 The discrete random variables X and Y jointly follow the $\text{Tri}(3, \frac{1}{4}, \frac{1}{2})$ distribution. Tabulate the joint probability mass function of X and Y on their joint range space. Evaluate Goodman and Kruskal's γ for this distribution.

10 The discrete random variables X and Y have the joint range space

$$R_{XY} = \{(x, y) : x = 0, 1, \ldots, n; y = 0, 1, \ldots, n; x + y \le n\}.$$

The joint probability mass function of X and Y on R_{XY} is

$$p_{XY}(x, y) = \frac{\binom{M_1}{x}\binom{M_2}{y}\binom{N - M_1 - M_2}{n - x - y}}{\binom{N}{n}},$$

where N, M_1 and M_2 are integers such that $n \le M_1$, $n \le M_2$, $n \le N - M_1 - M_2$.

(a) Obtain the marginal probability mass functions of X and Y. Recognize these as hypergeometric distributions and hence write down the marginal expected values and variances of X and Y.

(b) For $y \in R_Y$, find the conditional probability mass function of X given $Y = y$. By recognizing this too as a hypergeometric distribution, write down the conditional expected value and variance of X given $Y = y$.

11 $N (\ge 1)$ small beads are mixed together in a bag. The beads are all identical apart from their colour; M_1 of the beads are red, M_2 are yellow and the remaining $N - M_1 - M_2$ are blue. A sample of n beads is to be drawn, where $n \le M_1, n \le M_2, n \le N - M_1 - M_2$. Let the random variable X be the number of red beads in the sample and the random variable Y be the number of yellow beads in the sample.

(a) Suppose that the sample is obtained 'with replacement'. This means that a bead is drawn from the bag at random, its colour is recorded and the bead is returned to the bag and thoroughly mixed in with the other beads again. This process is repeated n times. Show that X and Y jointly follow a trinomial distribution.

(b) Suppose now that the sample is obtained 'without replacement'. This means that sampled beads are drawn consecutively from the bag, but a sampled bead is not returned to the bag. Write down the joint probability mass function of X and Y in this case. [*Hint*: it should be of the form given in Exercise 10 above.]

(c) Explain, without doing any more algebra, why the probability distribution obtained in (b) must converge to a trinomial distribution as N is increased in such a way that the proportions of red, yellow and blue beads in the bag remain the same.

12 Suppose that a sequence of independent trials is to be conducted, where each trial has three possible outcomes (E_1, E_2 and E_3). The outcomes of the trials are all independent and, on each trial, $P(E_i) = \theta_i$ (where $0 < \theta_i < 1$ and $\theta_1 + \theta_2 + \theta_3 = 1$). The trials are numbered sequentially 1, 2, ... Trials will be conducted until at least one E_1 outcome and at least one E_2 outcome are observed. Let the random variables X and Y, respectively, be the serial number of the trial on which the first E_1 and the first E_2 are observed.

(a) Explain why X and Y cannot be equal. Hence write down the joint range space of X and Y. Find $p_{XY}(x, y)$. [*Hint*: you might wish to consider the cases $X > Y$ and $X < Y$ separately.]

(b) Show that, marginally, X and Y follow geometric distributions. Explain this result.

4.4 The bivariate normal distribution and regression

Example 4.10 – The bivariate normal distribution

The bivariate normal distribution is a generalization to two dimensions of the normal distribution. Two continuous random variables, X and Y, are said to follow a (non-singular) bivariate normal distribution if their joint probability density function is of the form

$$\frac{1}{2\pi\sqrt{\sigma_1^2\sigma_2^2(1 - \rho^2)}} \exp\left\{ -\frac{\begin{bmatrix} \sigma_2^2(x - \mu_1)^2 - 2\rho\sigma_1\sigma_2(x - \mu_1) \\ \times (y - \mu_2) + \sigma_1^2(y - \mu_2)^2 \end{bmatrix}}{2\sigma_1^2\sigma_2^2(1 - \rho^2)} \right\}, \quad (x, y) \in \mathbb{R}^2.$$

It will be shown later that X has expected value μ_1 and variance σ_1^2, that Y has expected value μ_2 and variance σ_2^2, and that the covariance between X and Y is $\rho\sigma_1\sigma_2$ (so that ρ is the correlation between X and Y).

The joint p.d.f. can be written much more neatly if we define the following matrix:

$$\Sigma = \begin{bmatrix} \sigma_1^2 & \rho\sigma_1\sigma_2 \\ \rho\sigma_1\sigma_2 & \sigma_2^2 \end{bmatrix}.$$

Later (Chapter 5), we shall call this the *covariance matrix* of X and Y. Then

$$f_{XY}(x,y) = \frac{1}{2\pi\sqrt{|\Sigma|}} \exp\left\{-\tfrac{1}{2}[x-\mu_1 \quad y-\mu_2]\, \Sigma^{-1}\, [x-\mu_1 \quad y-\mu_2]^{\mathrm{T}}\right\},$$

where the determinant of the matrix Σ is

$$|\Sigma| = \sigma_1^2\sigma_2^2 - (\rho\sigma_1\sigma_2)^2 = \sigma_1^2\sigma_2^2(1-\rho^2).$$

$f_{XY}(x,y)$ is well defined unless Σ is a singular matrix. This occurs when

$$|\Sigma| = 0$$
$$\Leftrightarrow \sigma_1^2\sigma_2^2 - (\rho\sigma_1\sigma_2)^2 = 0$$
$$\Leftrightarrow \rho = \pm 1$$
$$\Leftrightarrow X \text{ and } Y \text{ are just linear transformations of one another.}$$

This issue will be discussed in general later (Chapter 8).

The following result shows one way in which the bivariate normal is an extension of the univariate normal distribution.

Proposition 4.7

Suppose that X and Y follow a (non-singular) bivariate normal distribution. Then, marginally, X and Y both follow normal distributions.

Proof. The marginal distribution of X is found from the usual formula:

$$f_X(x) = \int_{-\infty}^{\infty} f_{XY}(x,y)\,\mathrm{d}y.$$

By completing the square in the exponent in the joint probability density function, we obtain:

$$-\frac{1}{2\sigma_2^2(1-\rho^2)}\left\{\left[(y-\mu_2) - \frac{\rho\sigma_2}{\sigma_1}(x-\mu_1)\right]^2 + \frac{\sigma_2^2(1-\rho^2)}{\sigma_1^2}(x-\mu_1)^2\right\}$$

$$= -\frac{1}{2\sigma_2^2(1-\rho^2)}\left[(y-\mu_2) - \frac{\rho\sigma_2}{\sigma_1}(x-\mu_1)\right]^2 - \frac{1}{2\sigma_1^2}(x-\mu_1)^2.$$

This means that $f_X(x)$ can be written as:

$$\frac{1}{2\pi\sqrt{\sigma_1^2\sigma_2^2(1-\rho^2)}} \exp\left\{-\frac{1}{2\sigma_1^2}(x-\mu_1)^2\right\}$$

$$\times \int_{-\infty}^{\infty} \exp\left\{-\frac{\left[(y-\mu_2)-\frac{\rho\sigma_2}{\sigma_1}(x-\mu_1)\right]^2}{2\sigma_2^2(1-\rho^2)}\right\} dy$$

$$= \frac{1}{\sqrt{2\pi\sigma_1^2}} \exp\left\{-\frac{1}{2\sigma_1^2}(x-\mu_1)^2\right\}$$

$$\times \int_{-\infty}^{\infty} \frac{1}{\sqrt{2\pi\sigma_2^2(1-\rho^2)}} \exp\left\{-\frac{\left[y-\left[\mu_2+\frac{\rho\sigma_2}{\sigma_1}(x-\mu_1)\right]\right]^2}{2\sigma_2^2(1-\rho^2)}\right\} dy.$$

The integrand in this expression is the probability density function of a normal distribution with expected value

$$\mu_2 + \frac{\rho\sigma_2}{\sigma_1}(x-\mu_1)$$

and variance

$$\sigma_2^2(1-\rho^2).$$

Hence, the integral equals 1 and

$$f_X(x) = \frac{1}{\sqrt{2\pi\sigma_1^2}} \exp\left\{-\frac{1}{2\sigma_1^2}(x-\mu_1)^2\right\}.$$

In other words, $X \sim N(\mu_1, \sigma_1^2)$.
   Similarly, it can be shown that $Y \sim N(\mu_2, \sigma_2^2)$. ∎

   It is left as an exercise to show that $\rho_{XY} = \rho$.
   The following proposition states an important property of the bivariate normal distribution, which (as we have seen) does not hold in general for bivariate distributions.

Proposition 4.8

Suppose that X and Y follow a (non-singular) bivariate normal distribution. Then X and Y are independent if and only if X and Y are uncorrelated.

Proof. If X and Y are independent, then it necessarily follows that they are also uncorrelated (see Proposition 4.1).

Now suppose that X and Y are uncorrelated, i.e. $\rho = 0$. Then

$$f_{XY}(x,y) = \frac{1}{2\pi\sqrt{\sigma_1^2\sigma_2^2}} \exp\left\{-\frac{\sigma_2^2(x-\mu_1)^2 + \sigma_1^2(y-\mu_2)^2}{2\sigma_1^2\sigma_2^2}\right\}$$

$$= \frac{1}{2\pi\sqrt{\sigma_1^2}} \exp\left\{-\frac{(x-\mu_1)^2}{2\sigma_1^2}\right\} \frac{1}{2\pi\sqrt{\sigma_2^2}} \exp\left\{-\frac{(y-\mu_2)^2}{2\sigma_2^2}\right\}$$

$$= f_X(x)f_Y(y).$$

Since the joint p.d.f. factorizes, X and Y are independent. ■

Proposition 4.9

Suppose that X and Y follow a (non-singular) bivariate normal distribution. Then the conditional distribution of Y, *given* $X = x$, is a normal distribution (for all possible values of x).

Proof. In the proof of Proposition 4.7, we showed that the joint probability density function of X and Y can be written in the form

$$\frac{1}{\sqrt{2\pi\sigma_1^2}} \exp\left\{-\frac{1}{2\sigma_1^2}(x-\mu_1)^2\right\}$$

$$\times \frac{1}{\sqrt{2\pi\sigma_2^2(1-\rho^2)}} \exp\left\{-\frac{\left[(y-\mu_2) - \frac{\rho\sigma_2}{\sigma_1}(x-\mu_1)\right]^2}{2\sigma_2^2(1-\rho^2)}\right\}$$

Now,

$$f_{Y|X}(y|x) = \frac{f_{XY}(x,y)}{f_X(x)} = \frac{1}{\sqrt{2\pi\sigma_2^2(1-\rho^2)}} \exp\left\{-\frac{\left[(y-\mu_2) - \frac{\rho\sigma_2}{\sigma_1}(x-\mu_1)\right]^2}{2\sigma_2^2(1-\rho^2)}\right\}.$$

As remarked previously, this is the p.d.f. of a normal distribution with expected value

$$\mathbb{E}(Y|x) = \mu_2 + \frac{\rho\sigma_2}{\sigma_1}(x-\mu_1)$$

and variance

$$\mathrm{var}(Y|x) = \sigma_2^2(1-\rho^2). \qquad\blacksquare$$

In proving the above proposition, we have shown that the conditional expected value of Y, given $X = x$, depends on x (except in the very special case when $\rho = 0$ and X and Y are independent). An equation that expresses the conditional expected value of Y, given $X = x$, as a function of x is called the *regression* of Y on X. Similarly, there is a regression of X on Y. For the bivariate normal distribution, both regressions are linear functions, which is not generally the case for arbitrary random variables. We can exploit this result in order to find ρ_{XY} in a simpler way, using the following general result.

Proposition 4.10

(a) Suppose that X and Y are random variables such that the regression of Y on X is linear for all possible x, i.e.

$$\mathbb{E}(Y|X = x) = \alpha + \beta x.$$

Then

$$\alpha = \mathbb{E}(Y) - \frac{\rho_{XY}\sqrt{\mathrm{var}(Y)}}{\sqrt{\mathrm{var}(X)}}\mathbb{E}(X) \quad \text{and} \quad \beta = \frac{\rho_{XY}\sqrt{\mathrm{var}(Y)}}{\sqrt{\mathrm{var}(X)}}$$

(b) If, in addition, the conditional variance of Y given X is constant for all possible x, i.e.

$$\mathrm{var}(Y|X = x) = \sigma^2,$$

then

$$\sigma^2 = \left(1 - \rho_{XY}^2\right)\mathrm{var}(Y).$$

Proof.

(a) $$\mathbb{E}(Y) = \mathbb{E}\{\mathbb{E}(Y|X)\} = \mathbb{E}(\alpha + \beta X) = \alpha + \beta\mathbb{E}(X),$$

$$\mathbb{E}(XY) = \mathbb{E}\{X\mathbb{E}(Y|X)\} = \mathbb{E}(\alpha X + \beta X^2) = \alpha\mathbb{E}(X) + \beta\mathbb{E}(X^2).$$

It follows from the usual formula for the correlation that

$$\rho_{XY} = \frac{\{\alpha\mathbb{E}(X) + \beta\mathbb{E}(X^2)\} - \mathbb{E}(X)\{\alpha + \beta\mathbb{E}(X)\}}{\sqrt{\mathrm{var}(X)\mathrm{var}(Y)}}$$

$$= \frac{\beta\sqrt{\mathrm{var}(X)}}{\sqrt{\mathrm{var}(Y)}}$$

So

$$\beta = \frac{\rho_{XY}\sqrt{\mathrm{var}(Y)}}{\sqrt{\mathrm{var}(X)}}$$

and

$$\alpha = \mathbb{E}(Y) - \beta\mathbb{E}(X) = \mathbb{E}(Y) - \frac{\rho_{XY}\sqrt{\mathrm{var}(Y)}}{\sqrt{\mathrm{var}(X)}}\mathbb{E}(X).$$

(b) Assuming in addition that $\mathrm{var}(Y|X) = \sigma^2$, then

$$\mathrm{var}(Y) = \mathbb{E}\{\mathrm{var}(Y|X)\} + \mathrm{var}\{\mathbb{E}(Y|X)\} = \sigma^2 + \beta^2\mathrm{var}(X)$$

so that

$$\sigma^2 = \mathrm{var}(Y) - \beta^2\mathrm{var}(X)$$

$$= \mathrm{var}(Y) - \rho_{XY}^2\frac{\mathrm{var}(Y)}{\mathrm{var}(X)}\mathrm{var}(X)$$

$$= \left(1 - \rho_{XY}^2\right)\mathrm{var}(Y). \qquad\blacksquare$$

What has been called the conditional variance, σ^2, in this example is some-times called the *residual variance* in statistical inference. We have just shown that, as long as X and Y are not uncorrelated ($\rho_{XY} \neq 0$), then the residual vari-ance of Y given x must be smaller than the unconditional variance of Y. This means that information about X reduces our uncertainty about Y.

In fact, Proposition 4.10(b) has shown rather more about the extent to which the variance of Y is reduced by knowledge of X. Rearranging this result, we obtain

$$\rho_{XY}^2 = 1 - \frac{\sigma^2}{\mathrm{var}(Y)} = \frac{\mathrm{var}(Y) - \mathrm{var}(Y|X)}{\mathrm{var}(Y)}$$

So, in the special circumstances described in this proposition, the square of the correlation can be interpreted as the proportional reduction in $\mathrm{var}(Y)$ due to the regression on X. In Chapter 8, we will see how this can be generalized to the multiple correlation.

We will now use Proposition 4.10 to find the correlation between X and Y, when they jointly follow a bivariate normal distribution.

Example 4.10 – continued

In the case of bivariate normal random variables, X and Y, we have already seen that

$$\mathbb{E}(Y|X = x) = \alpha + \beta x,$$

where

$$\alpha = \mu_2 - \frac{\rho\sigma_2}{\sigma_1}\mu_1 \quad \text{and} \quad \beta = \frac{\rho\sigma_2}{\sigma_1}.$$

Therefore, using Proposition 4.10(a),

$$\frac{\rho_{XY}\sqrt{\mathrm{var}(Y)}}{\sqrt{\mathrm{var}(X)}} = \beta = \frac{\rho\sigma_2}{\sigma_1},$$

i.e.

$$\rho_{XY} = \rho.$$

Example 4.8 – continued

In this example, we have already seen (page 116) that the regression of Y on X is linear in x,

$$\mathbb{E}(Y|X = x) = -3 + 6x,$$

and that the conditional variance of Y given $X = x$ is a constant,

$$\text{var}(Y|X = x) = 2.$$

We also found that

$$\mathbb{E}(X) = \tfrac{6}{5}, \qquad \text{var}(X) = \tfrac{6}{25},$$
$$\mathbb{E}(Y) = \tfrac{21}{5}, \qquad \text{var}(Y) = \tfrac{266}{25}$$

Using Proposition 4.10,

$$\beta = \frac{\rho_{XY}\sqrt{\text{var}(Y)}}{\sqrt{\text{var}(X)}},$$

so

$$\rho_{XY} = \frac{\beta\sqrt{\text{var}(X)}}{\sqrt{\text{var}(Y)}} = \frac{6\sqrt{6}}{\sqrt{266}} = 0.901.$$

Exercises

1 Suppose that X and Y are continuous random variables with joint probability density function $f(x,y)$. Let $g(X)$ be any real-valued function of X. Show that:

$$\mathbb{E}\{g(X)Y\} = \mathbb{E}\{g(X)\mathbb{E}(Y|X)\}.$$

2 Suppose that X and Y follow a bivariate normal distribution. Find the covariance between X and Y and hence show that their correlation is ρ. [*Hint*: use the definition of $\text{cov}(X, Y)$.]

3 When designing a plant to convert wind energy into usable electricity, it is necessary to try to model the annual energy obtainable using data about wind speeds. However, it is often the case that the only available data for wind speeds refer to an altitude lower than the altitude of the proposed plant. Kaminsky and Kirchhoff (1988) propose the following joint model for X, the wind speed at the lower height, and Y, the wind speed at the operational height.

Marginally, X has a Rayleigh distribution with parameter α,

$$f_X(x) = \frac{\pi}{2} \cdot \frac{x}{\alpha^2} \cdot \exp\left\{-\frac{\pi}{4} \cdot \left(\frac{x}{\alpha}\right)^2\right\}, \qquad x > 0.$$

Conditional on $X = x$, Y follows a normal distribution with expected value $c + dx$ and variance σ^2.

(a) Show that $\mathbb{E}(X) = \alpha$ and $\text{var}(X) = (4/\pi - 1)\alpha^2$.

(b) Hence find $\mathbb{E}(Y)$ and $\text{var}(Y)$. Find also the correlation between X and Y. [*Hint*: use the fact that the regression of Y on X is linear.]

4 The continuous random variables X and Y have the following joint probability density function:

$$f(x, y) = e^{-y}, \qquad 0 < x < y.$$

(a) Find the marginal distributions of X and Y.

(b) Find the conditional distribution of Y given $X = x$. Hence find $\mathbb{E}(Y|X = x)$.

(c) Find the conditional distribution of X given $Y = y$. Hence find $\mathbb{E}(X|Y = y)$.

(d) Find the correlation between X and Y.

(e) Verify that the general relationship between the regression parameters, α and β, and the correlation, ρ, proved in Proposition 4.10, holds in this particular case.

5 Suppose that X and Y are jointly distributed such that the regressions of Y on x and X on y are both linear. Show that these linear regressions must intersect at the point $(\mathbb{E}(X), \mathbb{E}(Y))$. Verify that this is true for the particular example in Exercise 4.

6 The continuous random variables X and Y have the following joint probability density function:

$$f(x, y) = 2e^{-(x+y)}, \qquad 0 < y < x.$$

(a) Find the marginal distributions of X and Y.

(b) Obtain the correlation between X and Y.

(c) Find the conditional distribution of X given $Y = y$.

(d) Show that $\mathbb{E}(X|Y = y) = 1 + y$, and that $\text{var}(X|Y = y) = 1$ for all values of x.

(e) Find the correlation between X and Y, using Proposition 4.10.

7 The continuous random variables X and Y have the joint probability density function

$$f_{XY}(x, y) = 12y^2, \qquad 0 < y < x < 1.$$

Show that the regression of X on y and the regression of Y on x are both linear. Hence find the correlation between X and Y.

8 Repeat Exercise 7 when the random variables have joint probability density function

$$f_{XY}(x,y) = 2, \qquad 0 < x < 1, \quad 0 < y < 1 - x.$$

9 Example 4.9 and Exercise 2 on Section 4.3 show that a trinomial distribution for the random variables (X,Y) is fully characterized as follows:

(i) X follows a $\text{Bi}(n, \theta_1)$ distribution;

(ii) given that $X = x$, Y follows a $\text{Bi}(n - x, \theta_2/(1 - \theta_1))$ distribution.

Simulate (x,y) values from a trinomial distribution of your choice, using the following two-stage procedure:

- simulate a value x from the marginal distribution of X;

- simulate y from the conditional distribution of Y given the simulated value of x.

10 Repeat Exercise 9 for the bivariate normal distribution, making use of Propositions 4.7 and 4.9.

Summary

In this chapter, we have discussed joint distributions of two random variables where the variables are either both discrete or jointly continuous. We have introduced basic concepts, such as the joint range space, distribution function, probability mass function and probability density function. We have considered how to obtain marginal and conditional distributions from the joint distribution. We have also discussed what it means for random variables to be independent and how to measure the strength of the association between variables that are not independent. Two very important distributions have been introduced – the trinomial (a two-dimensional analogue of the binomial) and the bivariate normal (a two-dimensional analogue of the normal). The idea of regression has also been introduced. In the next chapter, we will extend these ideas to collections of more than two random variables.

5
Random vectors

In the previous chapter, we discussed stochastic experiments where the outcome consisted of a pair of random variables. We emphasized that the principal reason for studying the bivariate probability distribution, rather than the two marginal distributions, is that it tells us about the association between the two variables. In many situations, we might wish to analyse patterns of association among more than two random variables, in which case it is neater to group the (two or more) random variables together as a random vector. In this chapter, we shall discuss some basic results about random vectors, such as the definitions of the joint distribution function and probability density function and the joint, marginal and conditional distributions. The problem of visualizing a multivariate distribution is more acute. It also becomes more difficult to describe the relationships among individual components of the random vector, and new measures of association, such as the partial correlation coefficient, are required.

5.1 Joint and marginal distributions

Example 5.1

Children are to be recruited into a longitudinal study of nutritional status. Each child's height and weight will be measured at $1, 2, 3, \ldots, 15$ years of age and these results will allow the child to be classified as underweight, in the normal range of weights or overweight at each age, in relation to the general population of children of the same age. The discrete random variable X_i indicates a child's status at age i years, with $X_i = -1$ (when a child is underweight), $X_i = 0$ (in the normal range), $X_i = 1$ (overweight). There is likely to be correlation between pairs of these 15 random variables, with children who are overweight at age i more likely to remain overweight the next year.

Example 5.2

In a study of blood pressure, an individual's age (X_1), height (X_2), weight (X_3), systolic blood pressure (X_4) and diastolic blood pressure (X_5) are all to be measured. It is of interest to investigate relationships among these random variables in the population of people with high blood pressure, for example how strongly systolic blood pressure and diastolic blood pressure are correlated. Now that there are more than two random variables, it is possible to consider more complicated patterns of relationship, such as the possibility that the strength of relationship between systolic and diastolic blood pressure is different for people of different ages.

Example 5.3

As part of the quality control procedures adopted in a factory that makes batteries, $X_1, \ldots, X_{20}$, the lifetimes of 20 different batteries, are noted. The lifetimes of different batteries might be independent. Interest here is likely to focus on properties of the whole sample that might indicate important properties of the output of the factory, for example the average lifetime of the sample.

Let $X_1, \ldots, X_m$ $(m \geq 1)$ be random variables. Then the m-dimensional vector $\mathbf{X} = [X_1 \ldots X_m]^T$ is called a *random vector*.

Suppose that, in the blood pressure study discussed in Example 5.2, the same five measurements are to be made on each of 20 subjects. Then the study is concerned with 100 random variables in all, which can be conveniently arranged in the matrix

$$X = \begin{bmatrix} X_{11} & X_{12} & \ldots & X_{15} \\ X_{21} & X_{22} & \ldots & X_{25} \\ \vdots & \vdots & & \vdots \\ X_{20,1} & X_{20,2} & \ldots & X_{20,5} \end{bmatrix}$$

Here, X_{ij} is a random variable describing the jth measurement on the ith subject. In general, if $X_{11}, \ldots, X_{1m}, X_{21}, \ldots, X_{2m}, \ldots, X_{nm}$ are random variables, then

$$X = \begin{bmatrix} X_{11} & \ldots & X_{1m} \\ X_{21} & \ldots & X_{2m} \\ \vdots & & \vdots \\ X_{n1} & \ldots & X_{nm} \end{bmatrix}$$

is a *random matrix* (of order $n \times m$).

Multivariate probability typically deals with random vectors. It is always true that a random vector is just a particular case of a random matrix. In practice, though, it is often more convenient to think of the elements of a random matrix of order $n \times m$ forming an mn-dimensional random vector

$$[X_{11} \cdots X_{1m} X_{21} \cdots X_{2m} \cdots X_{nm}].$$

The *joint range space* of the m-dimensional random vector $\mathbf{X}$ is a region $R_\mathbf{X} \subseteq \mathbb{R}^m$ such that $\mathbf{X}$ takes values outside $R_\mathbf{X}$ with probability 0.

The *joint distribution function* of the random vector $\mathbf{X}$ is the real-valued function $F_\mathbf{X}: \mathbb{R}^m \to [0, 1]$ defined by:

$$F_\mathbf{X}(\mathbf{x}) = P(X_1 \leq x_1, X_2 \leq x_2, \ldots, X_m \leq x_m), \qquad \text{for all } \mathbf{x} = [x_1 \ldots x_m]^T \in \mathbb{R}^m.$$

It will sometimes be convenient to write this in the alternative form

$$F_{\mathbf{X}}(\mathbf{x}) = P(\mathbf{X} \le \mathbf{x}),$$

where the vector inequality is implicitly taken to apply to the vectors element by element.

When all the random variables $X_1, X_2, \dots, X_m$ are discrete, then we shall refer to $\mathbf{X}$ as a *discrete random vector*. In that case, the *joint probability mass function* of $\mathbf{X}$ is

$$p_{\mathbf{X}}(\mathbf{x}) = P(X_1 = x_1, X_2 = x_2, \dots, X_m = x_m), \qquad \mathbf{x} = [x_1 \, x_2 \, \dots \, x_m]^{\mathrm{T}} \in \mathbb{R}^m.$$

As usual, the values of $p_{\mathbf{X}}(\mathbf{x})$ must sum to 1 when added over the whole of the joint range space of $\mathbf{X}$.

Example 5.4

Suppose that the discrete random vector $\mathbf{X}$ consists of three discrete random variables, X_1, X_2 and X_3, so that the joint range space of $\mathbf{X} = [X_1 \, X_2 \, X_3]^{\mathrm{T}}$ consists of eight points,

$$R_{\mathbf{X}} = \{(0,0,0), (0,0,1), (0,1,0), (0,1,1), (1,0,0), (1,0,1), (1,1,0), (1,1,1)\}.$$

The joint probability mass function of $\mathbf{X}$ is tabulated below. The possible values of X_1 are used to define the columns, the values of X_2 the rows and the values of X_3 the layers of the table.

$x_3 = 0$				$x_3 = 1$		
		x_1				x_1
$p_{\mathbf{X}}(x_1, x_2, 0)$		0	1	$p_{\mathbf{X}}(x_1, x_2, 1)$		0	1
x_2	0	0.10	0.15	x_2	0	0.15	0.10
	1	0.15	0.10		1	0.10	0.15

The marginal range space of X_1 is $R_{X_1} = \{0, 1\}$. A marginal probability related to X_1 is obtained by summing over all the possible values of X_2 and X_3; for example,

$$P(X_1 = 0) = p(0,0,0) + p(0,1,0) + p(0,0,1) + p(0,1,1) = 0.50,$$
$$P(X_1 = 1) = p(1,0,0) + p(1,1,0) + p(1,0,1) + p(1,1,1) = 0.50.$$

These two marginal probabilities make up the marginal probability mass function of X_1, which is tabulated below.

x_1	0	1
$p_1(x_1)$	0.5	0.5

In general, when the random vector $\mathbf{X}$ consists of m discrete random variables, the marginal probability mass function of X_i is defined as follows.

$$p_i(x_i) = \sum_{x_1} \cdots \sum_{x_{i-1}} \sum_{x_{i+1}} \cdots \sum_{x_m} p_{\mathbf{X}}(\mathbf{x}).$$

This is just the same kind of marginal distribution discussed in Chapter 4 in the context of bivariate distributions. From it, we can obtain the moments and central moments of X_i. In this example, for instance,

$$\mu_1 = \mathbb{E}(X_1) = 0.5, \qquad \sigma_1^2 = \operatorname{var}(X_1) = 0.25.$$

It is easily shown that X_2 and X_3 each have the same marginal distribution as X_1 in this example. The expected values of the three random variables may be grouped together in a vector, known as the *vector of expected values* (or *mean vector*),

$$\mathbb{E}(\mathbf{X}) = \begin{bmatrix} \mathbb{E}(X_1) \\ \mathbb{E}(X_2) \\ \mathbb{E}(X_3) \end{bmatrix} = \begin{bmatrix} \mu_1 \\ \mu_2 \\ \mu_3 \end{bmatrix} = \begin{bmatrix} 0.5 \\ 0.5 \\ 0.5 \end{bmatrix}.$$

Higher-order marginal distributions are also of interest, for example the joint (marginal) probability mass function of X_1 and X_2, which is denoted $p_{12}(x_1, x_2)$. In Example 5.4, this is easily obtained by combining the two layers of the original table:

$p_{12}(x_1, x_2)$		x_1 0	1
x_2	0	0.25	0.25
	1	0.25	0.25

In general, when there are m discrete random variables, the (joint) marginal probability mass function of X_i and X_j ($i < j$) is defined as follows:

$$p_{ij}(x_i, x_j) = \sum_{x_1} \cdots \sum_{x_{i-1}} \sum_{x_{i+1}} \cdots \sum_{x_{j-1}} \sum_{x_{j+1}} \cdots \sum_{x_m} p_{\mathbf{X}}(\mathbf{x})$$

From the joint (marginal) probability mass function of X_i and X_j, we can in general obtain the covariance, $\operatorname{cov}(X_i, X_j)$, which is often denoted σ_{ij}. In Example 5.4,

$$\sigma_{12} = \operatorname{cov}(X_1, X_2) = 0.$$

The symmetry of the distribution allows us to infer that

$$\sigma_{13} = \text{cov}(X_1, X_3) = 0,$$
$$\sigma_{23} = \text{cov}(X_2, X_3) = 0.$$

All the marginal variances and the pairwise covariances can be put together in the *covariance matrix* (or *variance–covariance matrix*) of the random vector **X**,

$$\text{cov}(\mathbf{X}) = \begin{bmatrix} \text{var}(X_1) & \text{cov}(X_1, X_2) & \text{cov}(X_1, X_3) \\ \text{cov}(X_2, X_1) & \text{var}(X_2) & \text{cov}(X_2, X_3) \\ \text{cov}(X_3, X_1) & \text{cov}(X_3, X_2) & \text{var}(X_3) \end{bmatrix}$$

$$= \begin{bmatrix} \sigma_1^2 & \sigma_{12} & \sigma_{13} \\ \sigma_{12} & \sigma_2^2 & \sigma_{23} \\ \sigma_{13} & \sigma_{23} & \sigma_3^2 \end{bmatrix}$$

$$= \begin{bmatrix} 0.25 & 0 & 0 \\ 0 & 0.25 & 0 \\ 0 & 0 & 0.25 \end{bmatrix}.$$

Since $\text{cov}(X_i, X_j) = \text{cov}(X_j, X_i)$, it follows that $\text{cov}(\mathbf{X})$ must always be a symmetric matrix.

Example 5.5 – The multinomial distribution

The multinomial distribution is the generalization to an arbitrary number of dimensions of the binomial and trinomial distributions considered in earlier chapters. Suppose that n objects are each independently to be placed in one of $m+1$ different categories, each object having probability θ_i of being placed in the ith category $(i = 1, 2, \ldots, m+1)$. This means that $0 \leq \theta_i \leq 1$ and that $\theta_1 + \cdots + \theta_{m+1} = 1$.

Let the random variable X_i denote the total number of objects placed in the ith category $(i = 1, 2, \ldots, m+1)$. Notice that $X_1 + X_2 + \cdots + X_{m+1} = n$, since every object must be placed in one and only one of the available categories. This means that only m of the random variables need to be considered explicitly, say $X_1, X_2, \ldots, X_m$, since the value of X_{m+1} may be inferred from the values of the other random variables. The random vector $\mathbf{X} = [X_1 X_2 \ldots X_m]^T$ is said to follow a multinomial distribution, often written $\mathbf{X} \sim \text{Mu}(n, \theta_1, \theta_2, \ldots, \theta_m)$. Note the restriction $\theta_1 + \cdots + \theta_m \leq 1$.

X has joint range space

$$R_{\mathbf{X}} = \{(x_1, \ldots, x_m) : x_1, \ldots, x_m = 0, 1, \ldots, n; \, x_1 + \cdots + x_m \leq n\}.$$

X has joint probability mass function

$$p_{\mathbf{X}}(x_1,\ldots,x_m) = \frac{n!}{x_1! \cdots x_m!\,(n - x_1 - \cdots - x_m)!}\,\theta_1^{x_1}\ldots\theta_m^{x_m}$$

$$\times\,(1 - \theta_1 - \cdots - \theta_m)^{n - x_1 - \cdots - x_m},\qquad (x_1,\ldots,x_m) \in R_{\mathbf{X}}.$$

The binomial distribution is the special case of the multinomial when $m = 1$, so $\mathrm{Bi}(n, \theta)$ is the same as $\mathrm{Mu}(n, \theta)$. The trinomial distribution is the special case when $m = 2$, so $\mathrm{Tri}(n, \theta_1, \theta_2)$ is the same as $\mathrm{Mu}(n, \theta_1, \theta_2)$.

Marginal probability mass functions can be obtained recursively. We first find the marginal distribution of $X_1,\ldots,X_{m-1}$, by summing out X_m. Notice that $(X_1,\ldots,X_{m-1})$ has the joint range space

$$\{(x_1,\ldots,x_{m-1}): x_1,\ldots,x_{m-1} = 0, 1,\ldots,n;\; x_1 + \cdots + x_{m-1} \leq n\}.$$

On this range space, the marginal probability mass function is

$$p_{12\ldots(m-1)}(x_1,\ldots,x_{m-1})$$

$$= \sum_{x_m=0}^{n - x_1 - \cdots - x_{m-1}} p_{\mathbf{X}}(\mathbf{x})$$

$$= \frac{n!}{x_1!\ldots x_{m-1}!}\,\theta_1^{x_1}\ldots\theta_{m-1}^{x_{m-1}}$$

$$\times \sum_{x_m=0}^{n - x_1 - \cdots - x_{m-1}} \frac{1}{x_m!(n - x_1 - \cdots - x_m)!}\,\theta_m^{x_m}(1 - \theta_1 - \cdots - \theta_m)^{n - x_1 - \cdots - x_m}.$$

Aiming to use the binomial theorem (Appendix A1), we rewrite the constant inside the summation as a binomial coefficient, making the required changes outside the summation:

$$p_{12\ldots(m-1)}(x_1,\ldots,x_{m-1})$$

$$= \frac{n!}{x_1!\ldots x_{m-1}!(n - x_1 - \ldots - x_{m-1})!}\,\theta_1^{x_1}\ldots\theta_{m-1}^{x_{m-1}}$$

$$\times \sum_{x_m=0}^{n - x_1 - \cdots - x_{m-1}} \frac{(n - x_1 - \cdots - x_{m-1})!}{x_m!(n - x_1 - \cdots - x_m)!}\,\theta_m^{x_m}(1 - \theta_1 - \cdots - \theta_m)^{n - x_1 - \cdots - x_m}$$

$$= \frac{n!}{x_1! \ldots x_{m-1}!(n - x_1 - \cdots - x_{m-1})!} \theta_1^{x_1} \cdots \theta_{m-1}^{x_{m-1}}$$

$$\times \sum_{x_m=0}^{n-x_1-\cdots-x_{m-1}} \binom{n - x_1 - \cdots - x_{m-1}}{x_m} \theta_m^{x_m} (1 - \theta_1 - \cdots - \theta_m)^{n-x_1-\cdots-x_m}$$

$$= \frac{n!}{x_1! \ldots x_{m-1}!(n - x_1 - \cdots - x_{m-1})!} \theta_1^{x_1} \cdots \theta_{m-1}^{x_{m-1}}$$

$$\times \{\theta_m + (1 - \theta_1 - \cdots - \theta_m)\}^{n-x_1-\cdots-x_{m-1}}$$

$$= \frac{n!}{x_1! \ldots x_{m-1}!(n - x_1 - \cdots - x_{m-1})!} \theta_1^{x_1} \cdots \theta_{m-1}^{x_{m-1}}$$

$$\times (1 - \theta_1 - \cdots - \theta_{m-1})^{n-x_1-\cdots-x_{m-1}}.$$

In other words, the marginal distribution of $[X_1 \ldots X_{m-1}]$ is $\mathrm{Mu}(n, \theta_1, \theta_2, \ldots, \theta_{m-1})$. This can be explained intuitively as follows. We began with $m + 1$ categories of objects and explicitly modelled the counts in categories $1, 2, \ldots, m$; category $m + 1$ may be considered as a 'miscellaneous' category which is only considered implicitly. When we restrict our attention to the marginal distribution of $X_1, \ldots, X_{m-1}$, we are implicitly combining categories m and $m + 1$ into a new 'miscellaneous' category.

Since we have now established that $[X_1 \ldots X_{m-1}]$ follows a $\mathrm{Mu}(n, \theta_1, \theta_2, \ldots, \theta_{m-1})$ distribution, we may use the above result to infer that $[X_1 \ldots X_{m-2}]$ follows a $\mathrm{Mu}(n, \theta_1, \theta_2, \ldots, \theta_{m-2})$ distribution. This argument may be continued until we establish that (X_1, X_2) marginally follows a $\mathrm{Mu}(n, \theta_1, \theta_2)$ or $\mathrm{Tri}(n, \theta_1, \theta_2)$ distribution (Chapter 4) and X_1 marginally follows a $\mathrm{Mu}(n, \theta_1)$ or $\mathrm{Bi}(n, \theta_1)$ distribution (see Chapter 2).

The symmetry of the joint distribution implies that equivalent results hold for all combinations of the random variables. So, in general, (X_i, X_j) follows a $\mathrm{Mu}(n, \theta_i, \theta_j)$ or $\mathrm{Tri}(n, \theta_i, \theta_j)$ distribution $(i \neq j)$ and X_i follows a $\mathrm{Bi}(n, \theta_i)$ distribution.

Results previously derived for the binomial and trinomial distributions now allow us to write

$$\mathbb{E}(\mathbf{X}) = \begin{bmatrix} n\theta_1 \\ n\theta_2 \\ \vdots \\ n\theta_m \end{bmatrix}, \quad \mathrm{cov}(\mathbf{X}) = \begin{bmatrix} n\theta_1(1 - \theta_1) & -n\theta_1\theta_2 & \cdots & -n\theta_1\theta_m \\ -n\theta_1\theta_2 & n\theta_2(1 - \theta_2) & \cdots & -n\theta_2\theta_m \\ \vdots & \vdots & \ddots & \vdots \\ -n\theta_1\theta_m & -n\theta_2\theta_m & \cdots & n\theta_m(1 - \theta_m) \end{bmatrix}.$$

The m-dimensional random vector $\mathbf{X}$ is said to be a *(jointly) continuous random vector* if there exists a *joint probability density function* $f_{\mathbf{X}}(\mathbf{x})$ such that, for every $\mathbf{x} \in \mathbb{R}^m$,

$$\int_{-\infty}^{x_1} \int_{-\infty}^{x_2} \cdots \int_{-\infty}^{x_m} f_{\mathbf{X}}(\mathbf{t}) \, dt_m \cdots dt_2 \, dt_1 = F_{\mathbf{X}}(\mathbf{x}).$$

It follows that

$$\int_{-\infty}^{\infty} \int_{-\infty}^{\infty} \cdots \int_{-\infty}^{\infty} f_{\mathbf{X}}(\mathbf{x}) \, dx_m \cdots dx_2 \, dx_1 = F_{\mathbf{X}}(\infty, \infty, \ldots, \infty) = 1.$$

Also

$$f_{\mathbf{X}}(\mathbf{x}) = \frac{\partial^m}{\partial x_1 \, \partial x_2 \ldots \partial x_m} F_{\mathbf{X}}(\mathbf{x}).$$

It is clear that $F_{\mathbf{X}}(\mathbf{x})$ must be non-decreasing in every dimension, and so we require

$$f_{\mathbf{X}}(\mathbf{x}) \geq 0, \qquad \text{for all } \mathbf{x} \in \mathbb{R}^m,$$

since the relevant partial derivatives must all be non-negative.

Example 5.6

In a satellite survey of land use in part of Scotland, each photographic 'frame' captures a region of land with area $1 \, \text{km}^2$. The proportions of urban, farming and forestry land in a randomly selected 'frame' are denoted by the random variables X_1, X_2 and X_3, respectively. It is believed that the random vector $\mathbf{X} = [X_1 \, X_2 \, X_3]^{\mathrm{T}}$ is uniformly distributed on its range space:

$$R_{\mathbf{X}} = \{(x_1, x_2, x_3) : 0 \leq x_1, x_2, x_3 \leq 1; \; x_1 + x_2 + x_3 \leq 1\}.$$

This means that there is a real constant, k, such that

$$f_{\mathbf{X}}(\mathbf{x}) = \begin{cases} k, & \mathbf{x} \in R_{\mathbf{X}}, \\ 0, & \text{otherwise.} \end{cases}$$

In order to make the joint probability density function valid, k must be greater than 0. Also,

$$\int\int\int_{R_X} k\,d\mathbf{x} = 1$$

$$\iff \int_0^1 \int_0^{1-x_1} \int_0^{1-x_1-x_2} k\,dx_3\,dx_2\,dx_1 = 1$$

$$\iff k\int_0^1 \int_0^{1-x_1} [x_3]_{x_3=0}^{x_3=1-x_1-x_2}\,dx_2\,dx_1 = 1$$

$$\iff k\int_0^1 \int_0^{1-x_1} (1 - x_1 - x_2)\,dx_2\,dx_1 = 1$$

$$\iff k\int_0^1 \left[(1 - x_1)x_2 - \frac{1}{2}x_2^2\right]_{x_2=0}^{x_2=1-x_1}\,dx_1$$

$$\iff k\int_0^1 \frac{1}{2}(1 - x_1)^2\,dx_1 = 1 \qquad (5.1)$$

$$\iff \frac{k}{2}\left[-\frac{1}{3}(1 - x_1)^3\right]_{x_1=0}^{x_1=1} = 1$$

$$\iff \frac{k}{6} = 1$$

$$\iff k = 6.$$

So

$$f_X(\mathbf{x}) = 6, \qquad \mathbf{x} \in R_X.$$

We now find the probability that less than half the area in a randomly selected 'frame' is urban:

$$P(X_1 < 0.5) = \int_0^{0.5} \int_0^{1-x_1} \int_0^{1-x_1-x_2} 6\,dx_3\,dx_2\,dx_1$$

$$= 6\int_0^{0.5} \frac{1}{2}(1 - x_1)^2\,dx_1 \qquad \text{(using equation (5.1))}$$

$$= 3\left[x_1 - x_1^2 + \frac{1}{3}x_1^3\right]_0^{0.5}$$

$$= 3\left(\frac{1}{2} - \frac{1}{4} + \frac{1}{24}\right)$$

$$= 0.875.$$

We can also find the probability that, in total, less than half the area in a randomly selected 'frame' is devoted to these three uses.

$$P(X_1 + X_2 + X_3 < 0.5) = \int_0^{0.5} \int_0^{0.5-x_1} \int_0^{0.5-x_1-x_2} 6 \, dx_3 \, dx_2 \, dx_1$$

$$= 6 \int_0^{0.5} \int_0^{0.5-x_1} [0.5 - x_1 - x_2] \, dx_2 \, dx_1$$

$$= 6 \int_0^{0.5} \left[(0.5 - x_1)x_2 - \frac{1}{2} x_2^2 \right]_{x_2=0}^{x_2=0.5-x_1} \, dx_1$$

$$= 6 \int_0^{0.5} \frac{1}{2} (0.5 - x_1)^2 \, dx_1$$

$$= 3 \int_0^{0.5} \left(\frac{1}{4} - x_1 + x_1^2 \right) dx_1$$

$$= 3 \left[0.25x_1 - 0.5x_1^2 + \frac{1}{3} x_1^3 \right]_0^{0.5}$$

$$= 3 \left(\frac{1}{8} - \frac{1}{8} + \frac{1}{24} \right)$$

$$= 0.125.$$

Suppose that $\mathbf{X} = [X_1 \cdots X_m]^T$ is a continuous random vector. Then the *marginal distribution function* of X_i is the function

$$F_i(x_i) = P(X_i \le x_i) = F_{\mathbf{X}}(\infty, \ldots, \infty, x_i, \infty, \ldots, \infty), \qquad x_i \in \mathbb{R}.$$

When it exists, the *marginal probability density function* of X_i is the function

$$f_i(x_i) = \frac{d}{dx_i} F_i(x_i), \quad x_i \in \mathbb{R}.$$

Proposition 5.1

Let $\mathbf{X} = [X_1 \ldots X_m]^T$ be a continuous random vector with joint probability density function $f_X(\mathbf{x})$. Then the marginal probability density function of X_i is

$$f_i(x_i) = \int_{-\infty}^{\infty} \cdots \int_{-\infty}^{\infty} f_{\mathbf{X}}(x_1, \ldots, x_i, \ldots, x_m) \, dx_1 \cdots dx_{i-1} \, dx_{i+1} \cdots dx_m.$$

Proof.

$$f_i(x_i) = \frac{d}{dx_i} F_i(x_i)$$

$$= \frac{d}{dx_i} F(\infty, \ldots, x_i, \ldots \infty)$$

$$= \frac{d}{dx_i} \int_{-\infty}^{\infty} \cdots \int_{-\infty}^{x_i} \cdots \int_{-\infty}^{\infty} f_{\mathbf{X}}(\mathbf{t}) \, dt_1 \cdots dt_m$$

$$= \frac{d}{dx_i} \int_{-\infty}^{x_i} \left\{ \int_{-\infty}^{\infty} \cdots \int_{-\infty}^{\infty} f_{\mathbf{X}}(\mathbf{t}) \, dt_1 \cdots dt_{i-1} \, dt_{i+1} \cdots dt_m \right\} dt_i$$

$$= \int_{-\infty}^{\infty} \cdots \int_{-\infty}^{\infty} f_{\mathbf{X}}(\mathbf{t}) \, dx_1 \cdots dx_{i-1} \, dx_{i+1} \cdots dx_m$$

$$= \int_{-\infty}^{\infty} \cdots \int_{-\infty}^{\infty} f_{\mathbf{X}}(x_1, \ldots, x_i, \ldots, x_m) \, dx_1 \cdots dx_{i-1} \, dx_{i+1} \cdots dx_m. \quad \blacksquare$$

Example 5.6 – continued

We now find the marginal probability density function and distribution function of X_1, the proportion of urban land in a randomly selected 'frame'.

We have

$$f_{\mathbf{X}}(\mathbf{x}) = 6, \qquad \mathbf{x} \in R_{\mathbf{X}},$$

so

$$f_1(x_1) = \int_0^{1-x_1} \int_0^{1-x_1-x_2} 6 \, dx_3 \, dx_2$$

$$= 6 \int_0^{1-x_1} ((1-x_1) - x_2) \, dx_2$$

$$= 6 \left[(1-x_1)x_2 - \frac{x_2^2}{2} \right]_0^{1-x_1}$$

$$= 6 \left\{ (1-x_1)^2 - \frac{(1-x_1)^2}{2} \right\}$$

$$= 3(1-x_1)^2, \qquad 0 \le x_1 \le 1.$$

It follows that

$$F_1(x_1) = \int_0^{x_1} f_1(t_1) \, dt_1 = \int_0^{x_1} 3(1-t_1)^2 \, dt_1 = [-(1-t_1)^3]_0^{x_1} = 1 - (1-x_1)^3,$$

$$0 \le x_1 \le 1.$$

Often we want to know about the joint (marginal) distribution of a subset of variables. Partition the random vector $\mathbf{X}$ into k sub-vectors $\mathbf{X}^{(1)}, \ldots, \mathbf{X}^{(k)}$ as shown below, where $\mathbf{X}^{(j)}$ has dimension m_j, and let $\mathbf{x} \in \mathbb{R}^m$ be partitioned conformably:

$$\mathbf{X} = \begin{bmatrix} \mathbf{X}^{(1)} \\ \mathbf{X}^{(2)} \\ \vdots \\ \mathbf{X}^{(k)} \end{bmatrix}, \qquad \mathbf{x} = \begin{bmatrix} \mathbf{x}^{(1)} \\ \mathbf{x}^{(2)} \\ \vdots \\ \mathbf{x}^{(k)} \end{bmatrix}.$$

Then the *(joint) marginal distribution function of* $\mathbf{X}^{(j)}$ is the function

$$F^{(j)}(\mathbf{x}^{(j)}) = P(\mathbf{X}^{(j)} \leq \mathbf{x}^{(j)}), \qquad \mathbf{x}^{(j)} \in \mathbb{R}^{m_j}.$$

When it exists, the *marginal probability density function* of $\mathbf{X}^{(j)}$ is the function

$$f^{(j)}(\mathbf{x}^{(j)}) = \frac{\partial^{m_j}}{\partial \mathbf{x}^{(j)}} F^{(j)}(\mathbf{x}^{(j)}), \qquad \mathbf{x}^{(j)} \in \mathbb{R}^{m_j}.$$

In practice, a (joint) marginal probability density function can usually be obtained by a method analogous to that suggested for the one-dimensional case by Proposition 5.1. This is illustrated by the following example.

Example 5.6 – continued

The joint marginal probability density function of (X_1, X_2) is

$$f_{12}(x_1, x_2) = \int_0^{1-x_1-x_2} 6 \, dx_3 = 6[x_3]_{x_3=0}^{x_3=1-x_1-x_2} = 6(1 - x_1 - x_2),$$

$$0 \leq x_1, x_2 \leq 1; \; x_1 + x_2 \leq 1.$$

Hence the marginal probability density function of X_1 is

$$f_1(x_1) = \int_0^{1-x_1} f_{12}(x_1, x_2) \, dx_2 = 6\left[(1 - x_1)x_2 - \tfrac{1}{2}x_2^2\right]_0^{1-x_1} = 3(1 - x_1)^2,$$

$$0 \leq x_1 \leq 1.$$

This agrees with the answer obtained previously.

Let $\mathbf{X}$ be a continuous random vector with joint probability density function $f_{\mathbf{X}}(\mathbf{x})$. Let $g(\mathbf{X})$ be any real-valued function of $\mathbf{X}$, i.e. $g: \mathbb{R}^m \to \mathbb{R}$. Then the *expected value* of $g(\mathbf{X})$ is defined by

$$E\{g(\mathbf{X})\} = \int_{-\infty}^{\infty} \cdots \int_{-\infty}^{\infty} g(\mathbf{x}) f_{\mathbf{X}}(\mathbf{x}) \, d\mathbf{x}.$$

The expected value is only well defined when the above integral is absolutely convergent.

From the above definition, it follows that, when the expected value of X_i exists, it is given by

$$
\mathbb{E}(X_i) = \int_{-\infty}^{\infty} \cdots \int_{-\infty}^{\infty} x_i f_{\mathbf{X}}(\mathbf{x}) \, d\mathbf{x}
$$

$$
= \int_{-\infty}^{\infty} x_i \left\{ \int_{-\infty}^{\infty} \cdots \int_{-\infty}^{\infty} f_{\mathbf{X}}(\mathbf{x}) \, dx_1 \cdots dx_{i-1} \, dx_{i+1} \cdots dx_m \right\} dx_i
$$

$$
= \int_{-\infty}^{\infty} x_i f_i(x_i) \, dx_i,
$$

as we would expect.

We have already defined the expected value of a random vector $\mathbf{X}$. In a similar way, the *expected value* of the $n \times m$ random matrix X is defined to be the $n \times m$ matrix $\mathbb{E}(X)$ whose (i,j)th element is $\mathbb{E}(X_{ij}), 1 \leq i \leq n, 1 \leq j \leq m$.

It follows immediately from this definition that, if X and Y are $n \times m$ random matrices with expected values $\mathbb{E}(X)$ and $\mathbb{E}(Y)$, respectively, then

$$
\mathbb{E}(X + Y) = \mathbb{E}(X) + \mathbb{E}(Y),
$$

$$
\mathbb{E}(X^{\mathrm{T}}) = \{\mathbb{E}(X)\}^{\mathrm{T}}.
$$

The covariance (or variance–covariance) matrix of the m-dimensional random vector $\mathbf{X}$ has already been defined. It is now clear that the covariance matrix is the expected value of an $m \times m$ matrix:

$$
\mathrm{cov}(\mathbf{X}) = \mathbb{E}\{(\mathbf{X} - \mathbb{E}(\mathbf{X}))(\mathbf{X} - \mathbb{E}(\mathbf{X}))^{\mathrm{T}}\}.
$$

It can be shown (see Exercises) that

$$
\mathrm{cov}(\mathbf{X}) = \mathbb{E}(\mathbf{X}\mathbf{X}^{\mathrm{T}}) - \mathbb{E}(\mathbf{X})\{\mathbb{E}(\mathbf{X})\}^{\mathrm{T}}.
$$

Example 5.6 – continued

We have already seen (using the symmetry of the distribution) that

$$
f_i(x_i) = 3(1 - x_i)^2, \qquad 0 \leq x_i \leq 1,
$$

$$
f_{ij}(x_i, x_j) = 6(1 - x_i - x_j), \qquad 0 \leq x_i, x_j \leq 1 \quad \text{and} \quad 0 \leq x_i + x_j \leq 1 (i \neq j).
$$

Using the marginal probability density function of X_i,

$$\mathbb{E}(X_i) = \int_0^1 x_i\, p_i(x_i)\, dx_i = 3\int_0^1 x_i(1-x_i)^2\, dx_i = 3\cdot\mathbb{B}(2,3) = 3\cdot\frac{1!\,2!}{4!} = \frac{1}{4}$$

$$\mathbb{E}(X_i^2) = \int_0^1 x_i^2\, p_i(x_i)\, dx_i = 3\int_0^1 x_i^2(1-x_i)^2\, dx_i = 3\cdot\mathbb{B}(3,3) = 3\cdot\frac{2!\,2!}{5!}$$

$$= 3\cdot\frac{4}{120} = \frac{1}{10}$$

$$\mathrm{var}(X_i) = \frac{1}{10} - \frac{1}{16} = \frac{16-10}{160} = \frac{3}{80}.$$

Using the (joint) marginal probability density function of (X_i, X_j),

$$\mathbb{E}(X_i\, X_j) = \int_0^1\int_0^{1-x_i} 6(1-x_i-x_j)x_i\, x_j\, dx_j\, dx_i$$

$$= 6\int_0^1 x_i \int_0^{1-x_i}(1-x_i)x_i - x_j^2\, dx_j\, dx_i$$

$$= 6\int_0^1 x_i\left[\frac{(1-x_i)x_j^2}{2} - \frac{x_j^3}{3}\right]_{x_j=0}^{x_j=1-x_i} dx_i$$

$$= \int_0^1 x_i(1-x_i)^3\, dx_i$$

$$= \mathbb{B}(2,4)$$

$$= \frac{1!\,3!}{5!} = \frac{1}{20}$$

and

$$\mathrm{cov}(X_i, X_j) = \mathbb{E}(X_i\, X_j) - \mathbb{E}(X_i)\mathbb{E}(X_j) = \frac{1}{20} - \frac{1}{16} = \frac{8-10}{160} = -\frac{1}{80}.$$

So

$$\mathbb{E}(\mathbf{X}) = \frac{1}{4}\begin{bmatrix}1\\1\\1\end{bmatrix}, \qquad \mathrm{cov}(\mathbf{X}) = \frac{1}{80}\begin{bmatrix}3 & -1 & -1\\-1 & 3 & -1\\-1 & -1 & 3\end{bmatrix}.$$

Proposition 5.2

Suppose that $\mathbf{X}$ is an m-dimensional, jointly continuous random vector with finite expected value $\mathbb{E}(\mathbf{X})$. Then, for any vector of constants $\mathbf{a} \in \mathbb{R}^m$ and any constant $b \in \mathbb{R}$,

$$\mathbb{E}(\mathbf{a}^\mathsf{T}\mathbf{X} + b) = \mathbf{a}^\mathsf{T}\mathbb{E}(\mathbf{X}) + b.$$

Proof. Denote the probability density function of $\mathbf{X}$ by $f_{\mathbf{X}}(\mathbf{x})$. By definition,

$$\mathbb{E}(\mathbf{a}^{\mathrm{T}}\mathbf{X}+b) = \int_{-\infty}^{\infty}\cdots\int_{-\infty}^{\infty}(\mathbf{a}^{\mathrm{T}}\mathbf{x}+b)f_{\mathbf{X}}(\mathbf{x})\,\mathrm{d}\mathbf{x}$$

$$= \int_{-\infty}^{\infty}\cdots\int_{-\infty}^{\infty}(\mathbf{a}^{\mathrm{T}}\mathbf{x})f_{\mathbf{X}}(\mathbf{x})\,\mathrm{d}\mathbf{x} + b\int_{-\infty}^{\infty}f_{\mathbf{X}}(\mathbf{x})\,\mathrm{d}\mathbf{x}$$

$$= \int_{-\infty}^{\infty}\cdots\int_{-\infty}^{\infty}\left\{\sum_{i=1}^{m}a_i x_i\right\}f_{\mathbf{X}}(\mathbf{x})\,\mathrm{d}\mathbf{x} + b\cdot 1$$

$$= \sum_{i=1}^{m}\left\{\int_{-\infty}^{\infty}\cdots\int_{-\infty}^{\infty}a_i x_i f_{\mathbf{X}}(\mathbf{x})\,\mathrm{d}\mathbf{x}\right\} + b$$

$$= \sum_{i=1}^{m}\left\{a_i\int_{-\infty}^{\infty}\cdots\int_{-\infty}^{\infty}x_i f_{\mathbf{X}}(\mathbf{x})\,\mathrm{d}\mathbf{x}\right\} + b$$

$$= \sum_{i=1}^{m}a_i\mathbb{E}(X_i) + b$$

$$= \mathbf{a}^{\mathrm{T}}\mathbb{E}(\mathbf{X}) + b.\qquad\blacksquare$$

From now on, we shall denote the set of all $m \times q$ real matrices by the symbol M_{mq} (where m and q are both positive integers). The set of m-dimensional real vectors may therefore be denoted M_{m1} or $\mathbb{R}^m$ (depending on the context).

Proposition 5.3

Suppose that X is an $n \times m$ random matrix with finite expected value $\mathbb{E}(X)$. Let $A \in M_{rn}, B \in M_{mq}, C \in M_{rq}$ be matrices of constants, where $r, q \geq 1$. Then

$$\mathbb{E}(AXB + C) = A\cdot\mathbb{E}(X)\cdot B + C.$$

Proof. The (i,j)th element of $AXB + C(1 \leq i \leq r, 1 \leq j \leq q)$ is

$$[AXB + C]_{ij} = [AXB]_{ij} + [C]_{ij}$$

$$= \sum_{k=1}^{n}\sum_{\ell=1}^{m}[A]_{ik}[X]_{k\ell}[B]_{\ell j} + [C]_{ij}$$

$$= \sum_{k=1}^{n}\sum_{\ell=1}^{m}\{[A]_{ik}[B]_{\ell j}\}[X_{k\ell}] + [C]_{ij}$$

$$\therefore\quad \mathbb{E}\{[AXB + C]_{ij}\} = \sum_{k=1}^{n}\sum_{\ell=1}^{m}\{[A]_{ik}[B]_{\ell j}\}\cdot\mathbb{E}\{X_{k\ell}\} + [C]_{ij}$$

$$= [A\cdot\mathbb{E}(X)\cdot B + C]_{ij}.\qquad\blacksquare$$

Proposition 5.4

Suppose that $\mathbf{X}$ is an m-dimensional random vector with expected value $\mathbb{E}(\mathbf{X})$ and covariance matrix $\text{cov}(\mathbf{X})$. Let $A \in M_{qm}$ and $\mathbf{b} \in \mathbb{R}^q$ be a matrix and a vector of constants (where $q \geq 1$). Then:

(a) $\mathbb{E}(A\mathbf{X} + \mathbf{b}) = A \cdot \mathbb{E}(\mathbf{X}) + \mathbf{b}$,

(b) $\text{cov}(A\mathbf{X} + \mathbf{b}) = A \cdot [\text{cov}(\mathbf{X})] \cdot A^{\mathrm{T}}$.

Proof. (a) This is just a special case of Proposition 5.3.

$$
\begin{aligned}
\text{(b) } \text{cov}(A\mathbf{X} + \mathbf{b}) &= \mathbb{E}\left\{(A\mathbf{X} + \mathbf{b})(A\mathbf{X} + \mathbf{b})^{\mathrm{T}}\right\} - \mathbb{E}(A\mathbf{X} + \mathbf{b})\left\{\mathbb{E}(A\mathbf{X} + \mathbf{b})\right\}^{\mathrm{T}} \\
&= \mathbb{E}\left\{A\mathbf{X}\mathbf{X}^{\mathrm{T}}A^{\mathrm{T}} + \mathbf{b}\mathbf{X}^{\mathrm{T}}A^{\mathrm{T}} + A\mathbf{X}\mathbf{b}^{\mathrm{T}} + \mathbf{b}\mathbf{b}^{\mathrm{T}}\right\} \\
&\quad - [A \cdot \mathbb{E}(\mathbf{X}) + \mathbf{b}][A \cdot \mathbb{E}(\mathbf{X}) + \mathbf{b}]^{\mathrm{T}} \\
&= \mathbb{E}(A\mathbf{X}\mathbf{X}^{\mathrm{T}}A^{\mathrm{T}}) + \mathbb{E}(\mathbf{b}\mathbf{X}^{\mathrm{T}}A^{\mathrm{T}}) + \mathbb{E}(A\mathbf{X}\mathbf{b}^{\mathrm{T}}) + \mathbb{E}(\mathbf{b}\mathbf{b}^{\mathrm{T}}) \\
&\quad - A\,\mathbb{E}(\mathbf{X})[\mathbb{E}(\mathbf{X})]^{\mathrm{T}}A^{\mathrm{T}} - A\,\mathbb{E}(\mathbf{X})\mathbf{b}^{\mathrm{T}} - \mathbf{b}[\mathbb{E}(\mathbf{X})]^{\mathrm{T}}A^{\mathrm{T}} - \mathbf{b}\mathbf{b}^{\mathrm{T}} \\
&= A \cdot \mathbb{E}(\mathbf{X}\mathbf{X}^{\mathrm{T}}) \cdot A^{\mathrm{T}} + \mathbf{b}\,\mathbb{E}(\mathbf{X}^{\mathrm{T}})A^{\mathrm{T}} + A\,\mathbb{E}(\mathbf{X})\mathbf{b}^{\mathrm{T}} + \mathbf{b}\mathbf{b}^{\mathrm{T}} \\
&\quad - A\,\mathbb{E}(\mathbf{X})\left\{\mathbb{E}(\mathbf{X})\right\}^{\mathrm{T}}A^{\mathrm{T}} - A\,\mathbb{E}(\mathbf{X})\mathbf{b}^{\mathrm{T}} \\
&\quad - \mathbf{b}\left\{\mathbb{E}(\mathbf{X})\right\}^{\mathrm{T}}A^{\mathrm{T}} - \mathbf{b}\mathbf{b}^{\mathrm{T}} \\
&= A\left\{\mathbb{E}(\mathbf{X}\mathbf{X}^{\mathrm{T}}) - \mathbb{E}(\mathbf{X})[\mathbb{E}(\mathbf{X})]^{\mathrm{T}}\right\}A^{\mathrm{T}} \\
&= A[\text{cov}(\mathbf{X})]A^{\mathrm{T}}. \qquad\blacksquare
\end{aligned}
$$

The exercises at the end of this section list some further important properties of the expected value and covariance matrix.

Proposition 5.5

Suppose that $\mathbf{X}$ is an m-dimensional random vector with expected value $\mathbb{E}(\mathbf{X})$ and covariance matrix $\text{cov}(\mathbf{X})$. Then:

(a) $\text{cov}(\mathbf{X})$ is a symmetric, positive semi-definite matrix;

(b) $\text{cov}(\mathbf{X})$ is positive definite unless there are real constants $a_1, \ldots, a_m, a_{m+1}$ (not all zero) such that $\sum_{i=1}^m a_i X_i = a_{m+1}$ with probability 1.

Proof. We have already noted that $\text{cov}(\mathbf{X})$ is symmetric.

Let $\mathbf{a} \in \mathbb{R}^m$ be any constant vector. Then, by Proposition 5.4,

$$\mathbf{a}^{\mathrm{T}} \cdot \text{cov}(\mathbf{X}) \cdot \mathbf{a} = \text{var}(\mathbf{a}^{\mathrm{T}}\mathbf{X}) \geq 0,$$

and so $\text{cov}(\mathbf{X})$ is positive semi-definite [see Appendix A4].

For $\text{cov}(\mathbf{X})$ to be positive definite, we require

$$\mathbf{a}^{\mathrm{T}} \cdot \text{cov}(\mathbf{X}) \cdot \mathbf{a} > 0,$$

unless $\mathbf{a} = \mathbf{0}$. Therefore, $\text{cov}(\mathbf{X})$ is positive definite unless there exists an $\mathbf{a}(\neq \mathbf{0}) \in \mathbb{R}^m$ such that $\text{var}(\mathbf{a}^T \mathbf{X}) = 0$, i.e. there exists an $\mathbf{a}(\neq \mathbf{0}) \in \mathbb{R}^m$ and $a_{m+1} \in \mathbb{R}$ such that $\mathbf{a}^T \mathbf{X} = a_{m+1}$. ■

When $\text{cov}(\mathbf{X})$ exists and is positive definite, then it is non-singular (Appendix A4). When $\text{cov}(\mathbf{X})$ exists and is positive semi-definite, but not positive definite, then it is a singular matrix (i.e. $[\text{cov}(\mathbf{X})]^{-1}$ does not exist). This can cause some problems, as we shall see when discussing the multivariate normal distribution (Chapter 8). This case is not of much practical interest, though, for the following reason.

Consider first a bivariate random vector $[X_1\, X_2]^T$ such that $\text{cov}(\mathbf{X})$ is non-singular. Then there exist $a_1, a_2, a_3 \in \mathbb{R}$ such that $a_2 \neq 0$ and

$$a_1 X_1 + a_2 X_2 = a_3,$$

i.e

$$X_2 = \frac{a_3}{a_2} - \frac{a_1}{a_2} X_1.$$

So the probability density function of $[X_1\, X_2]$ is concentrated on the line $X_2 = (a_3/a_2) - (a_1/a_2) X_1$.

Another way of putting this is to say that, if we observe $X_1 = x_1$, then (with probability 1) $X_2 = (a_3/a_2) - (a_1/a_2) X_1$. So, the value of $\mathbf{X} = [X_1\, X_2]^T$ is determined merely by observing X_1 (or X_2). To observe (X_1, X_2) gives no further information; we have (so to speak) observed 'too many' variables. We need only observe X_1 or X_2.

Similarly, if $\mathbf{X}$ is m-dimensional and $\text{cov}(\mathbf{X})$ is singular, then $f_{\mathbf{X}}(\mathbf{x})$ is concentrated on (at most) an $(m-1)$-dimensional subspace of $\mathbb{R}^m$.

The *correlation matrix* between the m-dimensional random vector $\mathbf{X}$ and the p-dimensional random vector $\mathbf{Y}$ is the $m \times p$ matrix $\rho(\mathbf{X}, \mathbf{Y})$ whose (i, j)th element is

$$\rho(X_i, Y_j) = \frac{\text{cov}(X_i, Y_j)}{\sqrt{\text{var}(X_i)\text{var}(Y_j)}}.$$

In particular, the *correlation matrix* of the random vector $\mathbf{X}$ itself is the $m \times m$ matrix

$$\rho(\mathbf{X}) = \begin{bmatrix} 1 & \rho(X_1, X_2) & \cdots & \rho(X_1, X_m) \\ \rho(X_2, X_1) & 1 & \cdots & \rho(X_2, X_m) \\ \vdots & \vdots & & \vdots \\ \rho(X_m, X_1) & \rho(X_m, X_2) & \cdots & 1 \end{bmatrix}.$$

We shall now introduce the following matrix notation: if A is a $p \times p$ matrix with elements $[a_{ij}]$, then D_A will represent the diagonal matrix

$$D_A = \text{diag}(a_{11}, a_{22}, \ldots, a_{pp}).$$

Letting

$$\mathrm{cov}(\mathbf{X}) = \Sigma,$$

then

$$D_\Sigma = \mathrm{diag}(\mathrm{var}(X_1), \mathrm{var}(X_2), \ldots, \mathrm{var}(X_m))$$

and

$$\rho(\mathbf{X}) = D_\Sigma^{-1/2}\, \Sigma D_\Sigma^{-1/2}.$$

Example 5.6 – continued

When $i \neq j, \mathrm{cov}(X_i, X_j) = -\frac{1}{80}$, and so

$$\rho(X_i, X_j) = \frac{\mathrm{cov}(X_i, X_j)}{\sqrt{\mathrm{var}(X_i)\,\mathrm{var}(X_j)}} = -\frac{\frac{1}{80}}{\sqrt{\frac{3}{80} \times \frac{3}{80}}} = -\frac{1}{3}.$$

Therefore,

$$\rho(\mathbf{X}) = \begin{bmatrix} 1 & -\frac{1}{3} & -\frac{1}{3} \\ -\frac{1}{3} & 1 & -\frac{1}{3} \\ -\frac{1}{3} & -\frac{1}{3} & 1 \end{bmatrix}.$$

Exercises

1 Suppose that the discrete random vector $\mathbf{X} = [X_1\, X_2\, X_3]^\mathrm{T}$ follows the $\mathrm{Mu}(4, \frac{1}{4}, \frac{1}{4}, \frac{1}{4})$ distribution.

   (a) Write down the joint range space of $\mathbf{X}$ and its joint probability mass function.

   (b) Obtain the (joint) marginal probability mass function of $[X_1\, X_2]^\mathrm{T}$. Recognize this as a trinomial distribution and hence obtain $\mathrm{cov}(X_1, X_2)$.

   (c) Use the (joint) marginal probability mass function of $[X_1\, X_2]^\mathrm{T}$ to obtain the marginal probability mass function of X_1. Hence obtain the expected value and variance of X_1.

   (d) Write down $\mathbb{E}(\mathbf{X})$ and $\mathrm{cov}(\mathbf{X})$.

2 One hundred experienced drivers are to be divided at random into five groups, each group to road-test a different make of 1,400 cc car (labelled A, B, C, D, E respectively). Find

   (a) the probability that exactly 20 drivers are assigned to each make of car;

   (b) the probability that exactly 20 drivers are assigned to each of makes A, B and C.

3 $\mathbf{X}$ is a three-dimensional random vector with joint range space

$$R_{\mathbf{X}} = \{(x_1, x_2, x_3): 0 < x_1, x_2, x_3 < 1\}.$$

Its joint distribution function is

$$F_{\mathbf{X}}(\mathbf{x}) = k\, x_1\, x_2\, x_3\, (x_1 + x_2 + x_3), \qquad \mathbf{x} \in R_{\mathbf{X}},$$

for some real constant k.

(a) Write down k and hence find the (joint) probability density function of $\mathbf{X}$.

(b) Derive the (joint) marginal probability density function of $[X_1\ X_2]^{\mathrm{T}}$ and hence the marginal probability density function of X_1.

(c) Evaluate $P(X_3 > X_2 > X_1)$, $P(X_2 > X_1)$ and $P(X_1 > 0.5)$.

4 Suppose that $\mathbf{X}$ is a three-dimensional, jointly continuous random vector with joint probability density function

$$f_{\mathbf{X}}(\mathbf{x}) = k\, x_1\, x_2\, x_3, \qquad 0 < x_1 < x_2 < x_3 < 1,$$

for some real constant k.

(a) Find k.

(b) Derive the (joint) marginal probability density function of $[X_1\ X_2]^{\mathrm{T}}$. Hence find the marginal probability density function of X_1.

(c) Find $\mathbb{E}(\mathbf{X})$, $\mathrm{cov}(\mathbf{X})$ and $\rho(\mathbf{X})$.

5 Prove that, for any random vector $\mathbf{X}$,

$$\mathrm{cov}(\mathbf{X}) = \mathbb{E}(\mathbf{X}\mathbf{X}^{\mathrm{T}}) - \mathbb{E}(\mathbf{X})\left\{\mathbb{E}(\mathbf{X})\right\}^{\mathrm{T}}.$$

6 Suppose that $\mathbf{X}$ and $\mathbf{Y}$ are, respectively, a p-dimensional and a q-dimensional random vector, with finite expected values and finite covariance matrices. Let $\mathbf{a}$, $\mathbf{a}_1$ and $\mathbf{a}_2 \in \mathbb{R}^p$ and $\mathbf{b} \in \mathbb{R}^q$ be vectors of constants, and let $A, A_1, A_2 \in M_{mp}$ and $B \in M_{rq}$ be matrices of constants.

(a) Show that $\mathrm{cov}(\mathbf{Y}, \mathbf{X}) = [\mathrm{cov}(\mathbf{X}, \mathbf{Y})]^{\mathrm{T}}$.

(b) Prove that $\mathrm{cov}(\mathbf{X} - \mathbf{a}, \mathbf{Y} - \mathbf{b}) = \mathrm{cov}(\mathbf{X}, \mathbf{Y})$. Deduce that $\mathrm{cov}(\mathbf{X} - \mathbf{a}) = \mathrm{cov}(\mathbf{X})$.

(c) Prove that $\mathrm{cov}(\mathbf{a}^{\mathrm{T}}\mathbf{X}, \mathbf{b}^{\mathrm{T}}\mathbf{Y}) = \mathbf{a}^{\mathrm{T}}\mathrm{cov}(\mathbf{X}, \mathbf{Y})\mathbf{b}$. Deduce that $\mathrm{cov}(\mathbf{a}_1^{\mathrm{T}}\mathbf{X}, \mathbf{a}_2^{\mathrm{T}}\mathbf{X}) = \mathbf{a}_1^{\mathrm{T}}\mathrm{cov}(\mathbf{X})\mathbf{a}_2$.

(d) Prove that $\mathrm{cov}(A\mathbf{X}, B\mathbf{Y}) = A\mathrm{cov}(\mathbf{X}, \mathbf{Y})B^{\mathrm{T}}$. Deduce that $\mathrm{cov}(A_1\mathbf{X}, A_2\mathbf{X}) = A_1\mathrm{cov}(\mathbf{X})A_2^{\mathrm{T}}$.

(e) Show that, when $m = p$, $\mathbb{E}(\mathbf{X}^{\mathrm{T}}A\mathbf{X}) = \mathrm{tr}\{A\,\mathrm{cov}(\mathbf{X})\} + \{\mathbb{E}(\mathbf{X})\}^{\mathrm{T}}A\mathbb{E}(\mathbf{X})$.

5.2 Independence and conditional distributions

The random variables $X_1, \ldots, X_m$ are said to be *independent* if

$$F_{\mathbf{X}}(\mathbf{x}) = \prod_{i=1}^{m} F_i(x_i) \qquad \text{for all } \mathbf{x} = [x_1 \ldots x_m] \in \mathbb{R}^m.$$

An equivalent definition in the discrete case is to say that the random variables $X_1, \ldots, X_m$ are independent if

$$p_{\mathbf{X}}(\mathbf{x}) = \prod_{i=1}^{m} p_i(x_i) \qquad \text{for all } \mathbf{x} = [x_1 \ldots x_m] \in \mathbb{R}^m.$$

Example 5.4 – continued

Consider again the discrete random vector $\mathbf{X} = [X_1\, X_2\, X_3]$ with the probability mass function tabulated below.

$x_3 = 0$	x_1			$x_3 = 1$	x_1	
$p_{\mathbf{X}}(x_1, x_2, 0)$	0	1		$p_{\mathbf{X}}(x_1, x_2, 1)$	0	1
x_2 0	0.10	0.15		x_2 0	0.15	0.10
1	0.15	0.10		1	0.10	0.15

We can check from the marginal probability mass function of X_1 and X_2 that X_1 and X_2 are independent, since $p_{12}(x_1, x_2) = p_1(x_1)p_2(x_2)$ for all possible choices of x_1 and x_2. It is easily shown, by constructing tables of $p_{13}(x_1, x_3)$ and $p_{23}(x_2, x_3)$, that X_1 and X_3 are also independent, as are X_2 and X_3. We say that X_1, X_2 and X_3 are *pairwise independent* random variables.

This is not enough, however, to allow us to conclude that X_1, X_2 and X_3 are independent random variables. In this example, they are not. For instance,

$$p_{\mathbf{X}}(0, 0, 0) = 0.10$$

whilst

$$p_1(0)p_2(0)p_3(0) = 0.5 \times 0.5 \times 0.5 = 0.125.$$

This example shows that it is possible for a set of random variables to be pairwise independent but not independent. However, independence of random variables necessarily implies pairwise independence. This is easily proved in the discrete case. Suppose that

$$p_{\mathbf{X}}(\mathbf{x}) = p_1(x_1)p_2(x_2)\ldots p_m(x_m) \qquad \text{for all } (x_1, x_2, \ldots, x_m) \in \mathbb{R}^m.$$

Then, for all $(x_1, x_2) \in \mathbb{R}^2$,

$$p_{12}(x_1, x_2) = \sum_{x_3} \cdots \sum_{x_m} p_{\mathbf{X}}(\mathbf{x})$$

$$= \sum_{x_3} \cdots \sum_{x_m} p_1(x_1) p_2(x_2) \cdots p_m(x_m)$$

$$= p_1(x_1) p_2(x_2) \sum_{x_3} p_3(x_3) \cdots \sum_{x_m} p_m(x_m)$$

$$= p_1(x_1) p_2(x_2).$$

So, X_1 and X_2 are pairwise independent. So, too, are any other pair of the random variables, as can be shown by a simple modification of the above proof. An extension of this proof shows that, if $X_1, X_2, \ldots, X_m$ are independent discrete random variables, then any subset of them, of any size between 2 and m, must also be independent random variables.

With more than two discrete random variables, there is a richer collection of conditional probability mass functions. For example, in the case $m = 3$, the univariate *conditional* probability mass function of X_3 given that $X_1 = x_1, X_2 = x_2$ is defined as follows:

$$p_{3|12}(x_3|x_1, x_2) = P(X_3 = x_3 | X_1 = x_1, X_2 = x_2)$$

$$= \frac{P(X_1 = x_1, X_2 = x_2, X_3 = x_3)}{P(X_1 = x_1, X_2 = x_2)}$$

$$= \frac{p_{123}(x_1, x_2, x_3)}{p_{12}(x_1, x_2)}.$$

There are also bivariate conditional probability mass functions. For example, the bivariate *conditional* probability mass function of X_2 and X_3 given that $X_1 = x_1$ is defined as follows:

$$p_{23|1}(x_2, x_3|x_1) = P(X_2 = x_2, X_3 = x_3 | X_1 = x_1)$$

$$= \frac{P(X_1 = x_1, X_2 = x_2, X_3 = x_3)}{P(X_1 = x_1)}$$

$$= \frac{p_{123}(x_1, x_2, x_3)}{p_1(x_1)}.$$

Example 5.4 – continued

The *conditional* probability mass function of X_3 given that $X_1 = 1, X_2 = 0$ is

$$p_{3|12}(0 \mid 1, 0) = \frac{p_{123}(1,0,0)}{p_{12}(1,0)} = \frac{0.15}{0.25} = 0.6,$$

$$p_{3|12}(1 \mid 1, 0) = \frac{p_{123}(1,0,1)}{p_{12}(1,0)} = \frac{0.10}{0.25} = 0.4.$$

Each layer of the table of the joint probability mass function can be used to obtain a *conditional* probability mass function of X_1 and X_2 *given* X_3 by dividing through by the total probability of that layer, as shown below.

$p_{12|3}(x_1, x_2 \mid 0)$		x_1 0	1		$p_{12|3}(x_1, x_2 \mid 1)$		x_1 0	1
x_2	0	0.20	0.30		x_2	0	0.30	0.20
	1	0.30	0.20			1	0.20	0.30

Measures of *conditional association* can be obtained from a bivariate conditional distribution of this sort. For example,

$\mathbb{E}(X_1 \mid X_3 = 0) = 0.5,$ $\mathbb{E}(X_1 \mid X_3 = 1) = 0.5;$
$\mathbb{E}(X_2 \mid X_3 = 0) = 0.5,$ $\mathbb{E}(X_2 \mid X_3 = 1) = 0.5;$
$\mathbb{E}(X_1 X_2 \mid X_3 = 0) = 0.2,$ $\mathbb{E}(X_1 X_2 \mid X_3 = 1) = 0.3;$
$\text{cov}(X_1, X_2 \mid X_3 = 0)$ $\text{cov}(X_1, X_2 \mid X_3 = 1)$
$\quad = 0.2 - 0.5 \times 0.5 = -0.05,$ $\quad = 0.3 - 0.5 \times 0.5 = 0.05.$

Notice that, in this case, the conditional covariance of X_1 and X_2 is different, and actually of different sign, depending on whether we condition on $X_3 = 0$ or $X_3 = 1$. One consequence of independence of the random variables would be that the conditional covariance of any two given the other would be zero for all possible choices of conditioning event.

The conditional odds ratio of X_1 and X_2 given X_3 can also be calculated. Given $X_3 = 0$, the odds ratio is $(0.2 \times 0.2)/(0.3 \times 0.3) = \frac{4}{9} < 1$. Given $X_3 = 1$, the odds ratio is $(0.3 \times 0.3)/(0.2 \times 0.2) = \frac{9}{4} > 1$.

Given $X_3 = 0, \gamma = (0.04 - 0.09)/(0.04 + 0.09) = -0.385$. Given $X_3 = 1$, $\gamma = (0.09 - 0.04)/(0.04 + 0.09) = 0.385$.

All three of these measures of conditional association indicate that the relationship between X_1 and X_2 is different depending on the value of X_3. Notice that, in this example, the variables X_1 and X_2 are pairwise independent; this is only a marginal property.

Example 5.7

The table below shows the joint probability mass function of three binary random variables, X_1, X_2 and X_3.

$x_3 = 0$	x_1			$x_3 = 1$	x_1		
$p(x_1, x_2, 0)$	0	1		$p(x_1, x_2, 1)$	0	1	
x_2 0	0.125	0.125		x_2 0	0	0	
1	0.125	0.125		1	0	0.50	

The conditional probability mass functions of X_1 and X_2 given $X_3 = 0$ and given $X_3 = 1$ are shown below.

$p_{12|3}(x_1, x_2 \mid 0)$	x_1			$p_{12|3}(x_1, x_2 \mid 1)$	x_1		
	0	1			0	1	
x_2 0	0.25	0.25		x_2 0	0	0	
1	0.25	0.25		1	0	1	

In each of these tables, X_1 and X_2 are independent. When X_1 and X_2 are independent, given X_3, for every possible value x_3, we say that X_1 and X_2 are *conditionally independent* given X_3. We might assume that conditional independence implies marginal independence, but this is not necessarily the case. In this example, the (joint) marginal probability mass function of X_1 and X_2 is shown below. It is clear that these random variables are not independent.

$p_{12}(x_1, x_2)$	x_1	
	0	1
x_2 0	0.125	0.125
1	0.125	0.625

So it is possible for X_1 and X_2 to be conditionally independent given $X_3 = x_3$, whatever value of x_3 we choose, but not (marginally) independent.

An important consequence of the definitions of conditional probability mass functions is that

$$p_{12\ldots m}(x_1, x_2, \ldots, x_m) = p_{m|12\ldots(m-1)}(x_m \mid x_1, x_2, \ldots, x_{m-1}) \cdots p_{2|1}(x_2 \mid x_1) p_1(x_1).$$

In particular,

$$p_{123}(x_1, x_2, x_3) = p_{3|12}(x_3 \mid x_1, x_2) p_{12}(x_1, x_2) = p_{3|12}(x_3 \mid x_1, x_2) p_{2|1}(x_2 \mid x_1) p_1(x_1).$$

Suppose now that $\mathbf{X}$ is a continuous random vector and partition it into k sub-vectors $\mathbf{X}^{(1)}, \ldots, \mathbf{X}^{(k)}$ as shown below, where $\mathbf{X}^{(j)}$ has dimension m_j.

Let $\mathbf{x} \in \mathbb{R}^m$ be partitioned conformably:

$$\mathbf{X} = \begin{bmatrix} \mathbf{X}^{(1)} \\ \mathbf{X}^{(2)} \\ \vdots \\ \mathbf{X}^{(k)} \end{bmatrix}, \qquad \mathbf{x} = \begin{bmatrix} \mathbf{x}^{(1)} \\ \mathbf{x}^{(2)} \\ \vdots \\ \mathbf{x}^{(k)} \end{bmatrix}$$

The random sub-vectors $\mathbf{X}^{(1)}, \ldots, \mathbf{X}^{(k)}$ are said to be *independent* if

$$F_{\mathbf{X}}(\mathbf{x}) = \prod_{j=1}^{k} F^{(j)}(\mathbf{X}^{(j)}), \qquad \text{for all } \mathbf{x} \in \mathbb{R}^m.$$

Proposition 5.6

Let $\mathbf{X}$ be an m-dimensional jointly continuous random vector with joint probability density function $f_{\mathbf{X}}(\mathbf{x})$. Suppose $\mathbf{X}$ can be partitioned as above. Then, if $\mathbf{X}^{(1)}, \ldots, \mathbf{X}^{(k)}$ are independent, it follows that

$$f_{\mathbf{X}}(\mathbf{x}) = \prod_{j=1}^{k} f^{(j)}(\mathbf{x}^{(j)}), \qquad \text{for all } \mathbf{x} = \begin{bmatrix} \mathbf{x}^{(1)} \\ \vdots \\ \mathbf{x}^{(k)} \end{bmatrix} \in \mathbb{R}^m.$$

Proof. $\mathbf{X}^{(1)}, \ldots, \mathbf{X}^{(k)}$ are independent

$$\Rightarrow F_{\mathbf{X}}(\mathbf{x}) = \prod_{j=1}^{k} F^{(j)}(\mathbf{x}^{(j)}) \qquad \text{for all } \mathbf{x} \in \mathbb{R}^m$$

$$\Rightarrow f_{\mathbf{X}}(\mathbf{x}) = \frac{\partial^m}{\partial \mathbf{x}^{(1)} \partial \mathbf{x}^{(2)} \ldots \partial \mathbf{x}^{(k)}} \prod_{j=1}^{k} F^{(j)}(\mathbf{x}^{(j)}) \qquad \text{for all } \mathbf{x} \in \mathbb{R}^m$$

$$= \frac{\partial^{m-m_k}}{\partial \mathbf{x}^{(1)} \partial \mathbf{x}^{(2)} \ldots \partial \mathbf{x}^{(k-1)}} \left\{ \frac{\partial^{m_k}}{\partial \mathbf{x}^{(k)}} \prod_{j=1}^{k} F^{(j)}(\mathbf{x}^{(j)}) \right\} \qquad \text{for all } \mathbf{x} \in \mathbb{R}^m$$

$$= \frac{\partial^{m-m_k}}{\partial \mathbf{x}^{(1)} \partial \mathbf{x}^{(2)} \ldots \partial \mathbf{x}^{(k-1)}} \left\{ f^{(k)}(\mathbf{x}^{(k)}) \prod_{j=1}^{k} F^{(j)}(\mathbf{x}^{(j)}) \right\} \qquad \text{for all } \mathbf{x} \in \mathbb{R}^m$$

$$= f^{(k)}(\mathbf{x}^{(k)}) \frac{\partial^{m-m_k}}{\partial \mathbf{x}^{(1)} \ldots \partial \mathbf{x}^{(k-1)}} \prod_{j=1}^{k} F^{(j)}(\mathbf{x}^{(j)}) \qquad \text{for all } \mathbf{x} \in \mathbb{R}^m$$

$$= \ldots$$

$$= \prod_{j=1}^{k} f^{(j)}(\mathbf{x}^{(j)}) \qquad \text{for all } \mathbf{x} \in \mathbb{R}^m. \qquad \blacksquare$$

Proposition 5.7 – The factorization theorem for probability density functions

Let $\mathbf{X}$ be an m-dimensional random vector with joint probability density function $f_{\mathbf{X}}(\mathbf{x})$. Suppose $\mathbf{X}$ can be partitioned as before, and that

$$f_{\mathbf{X}}(\mathbf{x}) = \prod_{j=1}^{k} g^{(j)}(\mathbf{x}^{(j)}), \qquad \text{for all } \mathbf{x} = \begin{bmatrix} \mathbf{x}^{(1)} \\ \vdots \\ \mathbf{x}^{(k)} \end{bmatrix} \in \mathbb{R}^m,$$

where each $g^{(j)}(\mathbf{x}^{(j)})$ is a real-valued function of $\mathbf{x}^{(j)}$ alone. Then, $\mathbf{X}^{(1)}, \ldots, \mathbf{X}^{(k)}$ are independent random vectors and the marginal probability density function of $\mathbf{X}^{(j)}$ is proportional to $g^{(j)}$, i.e.

$$f^{(j)}(\mathbf{x}^{(j)}) \propto g^{(j)}(\mathbf{x}^{(j)}).$$

Proof.

$$f_{\mathbf{X}}(\mathbf{x}) = \prod_{j=1}^{k} g^{(j)}(\mathbf{x}^{(j)})$$

$$\Rightarrow f^{(1)}(\mathbf{x}^{(1)}) = \int_{-\infty}^{\infty} \cdots \int_{-\infty}^{\infty} f_{\mathbf{X}}(\mathbf{x}) \, d\mathbf{x}^{(2)} \ldots d\mathbf{x}^{(k)}$$

$$= g^{(1)}(\mathbf{x}^{(1)}) \cdot \int_{-\infty}^{\infty} \cdots \int_{-\infty}^{\infty} \left\{ \prod_{j=2}^{k} g^{(j)}(\mathbf{x}^{(j)}) \right\} d\mathbf{x}^{(2)} \ldots d\mathbf{x}^{(k)}$$

$$= g^{(1)}(\mathbf{x}^{(1)}) \cdot c_1,$$

where c_1 is a constant with respect to $\mathbf{x}^{(1)}$. Similarly,

$$f^{(j)}(\mathbf{x}^{(j)}) = g^{(j)}(\mathbf{x}^{(j)}) \cdot c_j, \qquad j = 2, \ldots, k.$$

But this means that

$$1 = \int_{-\infty}^{\infty} \cdots \int_{-\infty}^{\infty} f_{\mathbf{X}}(\mathbf{x}) \, d\mathbf{x}^{(1)} \cdots d\mathbf{x}^{(k)}$$

$$= \int_{-\infty}^{\infty} \cdots \int_{-\infty}^{\infty} \prod_{j=1}^{k} g^{(j)}(\mathbf{x}^{(j)}) \, d\mathbf{x}^{(1)} \cdots d\mathbf{x}^{(k)}$$

$$= \left\{ \int_{-\infty}^{\infty} \cdots \int_{-\infty}^{\infty} g^{(1)}(\mathbf{x}^{(1)}) \, d\mathbf{x}^{(1)} \right\} \cdots \left\{ \int_{-\infty}^{\infty} \cdots \int_{-\infty}^{\infty} g^{(k)}(\mathbf{x}^{(k)}) \, d\mathbf{x}^{(k)} \right\}$$

$$= \prod_{j=1}^{k} \left\{ \frac{1}{c_j} \int_{-\infty}^{\infty} \cdots \int_{-\infty}^{\infty} f^{(j)}(\mathbf{x}^{(j)}) \, d\mathbf{x}^{(j)} \right\}$$

$$= \prod_{j=1}^{k} \frac{1}{c_j}$$

and so, for all $\mathbf{x} \in \mathbb{R}^m$, we have

$$f_{\mathbf{X}}(\mathbf{x}) = \prod_{j=1}^{k} g^{(j)}(\mathbf{x}^{(j)})$$

$$= \prod_{j=1}^{k} c_j f^{(j)}(\mathbf{x}^{(j)})$$

$$= \left\{ \prod_{j=1}^{k} c_j \right\} \left\{ \prod_{j=1}^{k} f^{(j)}(\mathbf{x}^{(j)}) \right\}$$

$$= \prod_{j=1}^{k} f^{(j)}(\mathbf{x}^{(j)}).$$

Therefore, for all $\mathbf{x} \in \mathbb{R}^m$,

$$F_{\mathbf{X}}(\mathbf{x}) = \int_{-\infty}^{\infty} \cdots \int_{-\infty}^{\infty} f_{\mathbf{X}}(\mathbf{x}) \, d\mathbf{x}^{(1)} \cdots d\mathbf{x}^{(k)}$$

$$= \int_{-\infty}^{\infty} \cdots \int_{-\infty}^{\infty} \left\{ \prod_{j=1}^{k} f^{(j)}(\mathbf{x}^{(j)}) \right\} d\mathbf{x}^{(1)} \cdots d\mathbf{x}^{(k)}$$

$$= \prod_{j=1}^{k} \left\{ \int_{-\infty}^{\infty} \cdots \int_{-\infty}^{\infty} f^{(j)}(\mathbf{x}^{(j)}) \, d\mathbf{x}^{(j)} \right\}$$

$$= \prod_{j=1}^{k} F^{(j)}(\mathbf{x}^{(j)})$$

and so $\mathbf{X}^{(1)}, \ldots, \mathbf{X}^{(k)}$ are independent. ∎

The above two propositions are often applied when $k = m$ and each $\mathbf{X}^{(j)} = X_j (j = 1, \ldots, m)$.

Example 5.8

In each of the following cases, the joint probability density function of a three-dimensional random vector $\mathbf{X} = [X_1 \, X_2 \, X_3]^T$ is given:

(a) $f_{\mathbf{X}}(\mathbf{x}) = k e^{-(x_1 + x_2 + x_3)}$ $(x_1, x_2, x_3 \geq 0)$;

(b) $f_{\mathbf{X}}(\mathbf{x}) = k e^{-(x_1 + x_2 + x_3)}$ $(x_3 \geq x_2 \geq x_1 \geq 0)$.

In (a), X_1, X_2 and X_3 are independent whereas in (b) they are not. This is shown below.

(a) Since $R_{\mathbf{X}} = \{(x_1, x_2, x_3) : x_1, x_2, x_3 \geq 0\}$ is a 'rectangular' subspace of $\mathbb{R}^3$, it follows that

$$f_{\mathbf{X}}(\mathbf{x}) = \prod_{i=1}^{3} k_i e^{-x_i}, \qquad \text{for all } \mathbf{x} \in R_{\mathbf{X}},$$

where k_1, k_2, k_3, are constants such that $k_1 k_2 k_3 = k$. Therefore

$$f_i(x_i) = \begin{cases} k_i e^{-x_i}, & x_i \geq 0, \\ 0, & \text{otherwise.} \end{cases}$$

So $k_i = 1 (i = 1, 2, 3)$ and $X_i \sim \text{Ex}(1)$. Hence $k = 1$ and the factorization theorem tells us that the variables are independent.

(b) Because of the restriction on $R_{\mathbf{X}}$ it is not true that $f_{\mathbf{X}}(\mathbf{x})$ factorizes for all $\mathbf{x}$. For example, $f_{\mathbf{X}}(\frac{1}{2}, \frac{1}{4}, \frac{1}{8}) = 0$ but $f_1(\frac{1}{2})$, $f_2(\frac{1}{4})$, $f_3(\frac{1}{8}) > 0$. Hence, X_1, X_2 and X_3 are not independent.

Proposition 5.8

Suppose that $X_1, \ldots, X_m$ are independent random variables such that X_i has finite expected value μ_i and finite variance $\sigma_i^2 (i = 1, \ldots, m)$. Let $a_1, \ldots, a_m$ be real constants. Then

$$\mathbb{E}(a_1 X_1 + \cdots + a_m X_m) = a_1 \mu_1 + \cdots + a_m \mu_m$$

and

$$\text{var}(a_1 X_1 + \cdots + a_m X_m) = a_1^2 \sigma_1^2 + \cdots + a_m^2 \sigma_m^2.$$

Proof. Let $\mathbf{a} = [a_1 \, a_2 \cdots a_m]^T$ and $\mathbf{X} = [X_1 \, X_2 \cdots X_m]^T$. Then

$$a_1 X_1 + \cdots + a_m X_m = \mathbf{a}^T \mathbf{X}$$

and the first result follows immediately from Proposition 5.2.

Since $X_1, \ldots, X_m$ are independent,

$$\text{cov}(\mathbf{X}) = \text{diag}(\sigma_1^2, \sigma_2^2, \ldots, \sigma_m^2).$$

So Proposition 5.4 gives

$$\begin{aligned}
\mathrm{var}(a_1 X_1 + \cdots + a_m X_m) &= \mathrm{var}(\mathbf{a}^\mathrm{T} \mathbf{X}) \\
&= \mathbf{a}^\mathrm{T}[\mathrm{cov}(\mathbf{X})]\,\mathbf{a} \\
&= a_1^2 \sigma_1^2 + \cdots + a_m^2 \sigma_m^2.
\end{aligned}$$ ■

Independence of more than two random variables is a very strong property. As we have already seen, it is possible for all pairs of random variables to be independent (*pairwise independence*) while the random variables are not mutually independent.

Example 5.9

The random variables X_1, X_2 and X_3 have joint probability density function

$$f_\mathbf{X}(x_1, x_2, x_3) = \frac{1}{8}(1 + x_1 x_2 x_3), \qquad -1 \le x_1, x_2, x_3 \le 1.$$

So

$$\begin{aligned}
f_{12}(x_1, x_2) &= \int_{-1}^{1} \frac{1}{8}(1 + x_1 x_2 x_3)\,\mathrm{d}x_3 \\
&= \left[\frac{1}{8} x_3 + \frac{1}{16} x_1 x_2 x_3^2 \right]_{-1}^{1} \\
&= \left(\frac{1}{8} + \frac{1}{16} x_1 x_2 \right) - \left(-\frac{1}{8} + \frac{1}{16} x_1 x_2 \right) \\
&= \frac{1}{4}, \qquad -1 \le x_1, x_2 \le 1.
\end{aligned}$$

By symmetry,

$$f_{13}(x_1, x_3) = \frac{1}{4}, \qquad -1 \le x_1, x_3 \le 1,$$

$$f_{23}(x_2, x_3) = \frac{1}{4}, \qquad -1 \le x_2, x_3 \le 1.$$

Now, 'stepping down',

$$f_1(x_1) = \int_{-1}^{1} \frac{1}{4}\,\mathrm{d}x_2 = \frac{1}{2}, \qquad -1 \le x_1 \le 1.$$

By symmetry,

$$f_2(x_2) = \frac{1}{2}, \qquad -1 \le x_2 \le 1,$$

$$f_3(x_3) = \frac{1}{2}, \qquad -1 \le x_3 \le 1.$$

This means that the random variables are pairwise independent, since

$$f_{12}(x_1, x_2) = f_1(x_1) f_2(x_2),$$

$$f_{13}(x_1, x_3) = f_1(x_1) f_3(x_3),$$

$$f_{23}(x_2, x_3) = f_2(x_2) f_3(x_3).$$

On the other hand, the variables are not independent, since

there exists $x \in R_X$ such that $f(x_1, x_2, x_3) \neq f(x_1)f(x_2)f(x_3)$.

Sometimes we know something about the values of some of the components in $\mathbf{X}$, in which case we are interested in making *conditional* inferences about the other components.

Let the random vector $\mathbf{X}$ be partitioned as before: $\mathbf{X} = [\mathbf{X}^{(1)} \cdots \mathbf{X}^{(k)}]^\mathsf{T}$. Then the *conditional distribution function* of $\mathbf{X}^{(j)}$ given that $\mathbf{X}^{(k)} = \mathbf{x}^{(k)}$ $(j \neq k)$ is the function

$$F^{j|k}(\mathbf{x}^{(j)} \mid \mathbf{x}^{(k)}) = P(\mathbf{X}^{(j)} \leq \mathbf{x}^{(j)} \mid \mathbf{X}^{(k)} = \mathbf{x}^{(k)}), \qquad \mathbf{x}^{(j)} \in \mathbb{R}^{m_j}$$

(where the vector inequality applies elementwise). When it exists, the *conditional probability density function* of $\mathbf{X}^{(j)}$ given that $\mathbf{X}^{(k)} = \mathbf{x}^{(k)}$ $(j \neq k)$ is the function

$$f^{j|k}(\mathbf{x}^{(j)} \mid \mathbf{x}^{(k)}) = \frac{\partial^{m_j}}{d\mathbf{x}^{(j)}} F^{j|k}(\mathbf{x}^{(j)} \mid \mathbf{x}^{(k)}), \qquad \mathbf{x}^{(j)} \in \mathbb{R}^{m_j}.$$

It can be shown that $\mathbf{X}^{(j)}$ and $\mathbf{X}^{(k)}$ are independent if and only if

$$F^{j|k}(\mathbf{x}^{(j)} \mid \mathbf{x}^{(k)}) = F^{(j)}(\mathbf{x}^{(j)}), \qquad \forall \mathbf{x}^{(j)} \in \mathbb{R}^{p_j} \text{ and } \forall \mathbf{x}^{(k)} \in \mathbb{R}^{p_k}.$$

Also, if the probability density functions exist, then

$$f^{j|k}(\mathbf{x}^{(j)} \mid \mathbf{x}^{(k)}) = \frac{f(\mathbf{x}^{(j)}, \mathbf{x}^{(k)})}{f^{(k)}(\mathbf{x}^{(k)})}.$$

Example 5.6 – continued

Here

$$f_\mathbf{X}(x_1, x_2, x_3) = 6, \qquad 0 \leq x_1, x_2, x_3 \leq 1, \quad 0 \leq x_1 + x_2 + x_3 \leq 1.$$

We will now find the conditional probability density function of X_1 given $X_2 = x_2, X_3 = x_3$.

Write

$$\mathbf{X}^{(1)} = [X_1], \qquad \mathbf{X}^{(2)} = [X_2 \, X_3]^\mathsf{T}.$$

We have already seen that

$$f^{(2)}(\mathbf{x}^{(2)}) = f_{23}(x_2, x_3) = 6(1 - x_2 - x_3), \qquad 0 \leq x_2, x_3 \leq 1, \quad 0 \leq x_2 + x_3 \leq 1.$$

Therefore, for x_1 in the range $0 \leq x_1 \leq 1 - x_2 - x_3$,

$$f_{1|23}(x_1 \mid x_2, x_3) = \frac{f_\mathbf{X}(\mathbf{x})}{f_{23}(x_2, x_3)} = \frac{6}{6(1 - x_2 - x_3)} = \frac{1}{1 - x_2 - x_3}.$$

This is a uniform distribution on the range 0 to $1 - x_2 - x_3$. Conditional moments of X_1 *given* $X_2 = x_2$ and $X_3 = x_3$ may be found from this distribution in the usual way. For example, using properties of the uniform distribution introduced in Chapter 2, we may state that

$$\mathbb{E}(X_1 | x_2, x_3) = \frac{1}{2}(1 - x_2 - x_3).$$

This is the regression of X_1 on X_2 and X_3. Also,

$$\text{var}(X_1 | x_2, x_3) = \frac{1}{12}(1 - x_2 - x_3)^2.$$

Next, we will find the (joint) conditional probability density function of X_1 and X_2 given that $X_3 = x_3$. Write

$$\mathbf{X}^{(1)} = [X_1 \quad X_2]^T, \qquad \mathbf{X}^{(2)} = [X_3].$$

We have already seen that

$$f^{(2)}(\mathbf{x}^{(2)}) = f_3(x_3) = 3(1 - x_3)^2, \qquad 0 \le x_3 \le 1.$$

Therefore, for x_1 and x_2 such that $0 \le x_1, x_2 \le 1 - x_3$ and $0 \le x_1 + x_2 \le 1 - x_3$,

$$f_{12|3}(x_1, x_2 | x_3) = \frac{f_{\mathbf{X}}(\mathbf{x})}{f_3(x_3)} = \frac{6}{3(1 - x_3)^2} = \frac{2}{(1 - x_3)^2}.$$

Note that this is a bivariate uniform distribution.

Example 5.10

Suppose that $\mathbf{X} = [X_1 \, X_2 \, X_3 \, X_4]^T$ is a continuous random vector with the joint probability density function

$$f_{\mathbf{X}}(\mathbf{x}) = 384 \, x_1 \, x_2 \, x_3 \, x_4, \qquad 0 < x_1 < x_2 < x_3 < x_4 < 1.$$

We are aiming to find the conditional probability density function of X_1 and X_4 given (X_2, X_3). We must start by obtaining the (joint) marginal probability density function of X_2 and X_3, which we do by integrating out X_1 and X_4. We obtain

$$f_{23}(x_2, x_3) = \int_0^{x_2} \int_{x_3}^1 384 \, x_1 \, x_2 \, x_3 \, x_4 \, dx_4 \, dx_1$$

$$= 384 \, x_2 \, x_3 \int_0^{x_2} x_1 \, dx_1 \int_{x_3}^1 x_4 \, dx_4$$

$$= 384 \, x_2 \, x_3 \left[\frac{1}{2} x_1^2 \right]_0^{x_2} \left[\frac{1}{2} x_4^2 \right]_{x_3}^1$$

$$= 96 \, x_2^3 \, x_3 (1 - x_3^2).$$

Therefore

$$f_{14|23}(x_1, x_4 \mid x_2, x_3) = \frac{384\, x_1\, x_2\, x_3\, x_4}{96\, x_2^3\, x_3(1 - x_3^2)} = \frac{4\, x_1\, x_4}{x_2^2(1 - x_3^2)}, \qquad 0 < x_1 < x_2,\ x_3 < x_4 < 1.$$

This (joint) conditional density function is defined on a rectangular region. On this region, the joint density function factorizes into a function of x_1 multiplied by a function of x_4. This means that, given (X_2, X_3), X_1 and X_4 are independent. Using the factorization theorem, it follows that their (marginal) conditional density functions must be of the form

$$f_{1|23}(x_1 | x_2, x_3) \propto x_1 (0 < x_1 < x_2) \quad \text{and} \quad f_{4|23}(x_4 | x_2, x_3) \propto x_4 (x_3 < x_4 < 1)$$

so

$$f_{1|23}(x_1 | x_2, x_3) = \frac{2\, x_1}{x_2^2}(0 < x_1 < x_2) \quad \text{and}$$

$$f_{4|23}(x_4 | x_2, x_3) = \frac{2\, x_4}{1 - x_3^2}(x_3 < x_4 < 1)$$

In this case, X_1 and X_4 are conditionally independent given (X_2, X_3).

Notice that X_1 and X_4 are not independent in this example. This can be seen by noting that the joint marginal range space of X_1 and X_4 is the set $\{(x_1, x_4), 0 < x_1 < x_4 < 1\}$, which is not a rectangular region.

Proposition 5.9

Suppose that $X_1, \ldots, X_m$ are independent random variables. Then

$$\mathbb{E}\{X_1 X_2 \cdots X_m\} = \{\mathbb{E}(X_1)\}\{\mathbb{E}(X_2)\} \cdots \{\mathbb{E}(X_m)\}.$$

Proof. The random vector $\mathbf{X} = [X_1 X_2 \cdots X_m]^{\mathrm{T}}$ has probability density function

$$f_1(x_1) f_2(x_2) \cdots f_m(x_m).$$

By definition, then,

$$\mathbb{E}\{X_1 X_2 \cdots X_m\} = \int_{-\infty}^{\infty} \cdots \int_{-\infty}^{\infty} (x_1 \cdots x_p) f_1(x_1) \cdots f_m(x_m) \, dx_1 \cdots dx_m$$

$$= \left\{ \int_{-\infty}^{\infty} x_1 f_1(x_1) \, dx_1 \right\} \cdots \left\{ \int_{-\infty}^{\infty} x_m f_m(x_m) \, dx_m \right\}$$

$$= \mathbb{E}(X_1) \cdots \mathbb{E}(X_m).$$

The *covariance matrix* between the m-dimensional random vector $\mathbf{X}$ and the q-dimensional random vector $\mathbf{Y}$ is the $m \times q$ matrix $\mathrm{cov}(\mathbf{X}, \mathbf{Y})$ whose (i, j)th element is $\mathrm{cov}(X_i, Y_j)$. Since

$$\mathrm{cov}(X_i, Y_j) = \mathbb{E}\{[X_i - \mathbb{E}(X_i)][Y_j - \mathbb{E}(Y_j)]\} = \mathbb{E}(X_i Y_j) - \{\mathbb{E}(X_i)\}\{\mathbb{E}(Y_i)\},$$

it follows that

$$\text{cov}(\mathbf{X}, \mathbf{Y}) = \mathbb{E}\left\{[\mathbf{X} - \mathbb{E}(\mathbf{X})][\mathbf{Y} - \mathbb{E}(\mathbf{Y})]^{\mathrm{T}}\right\} = \mathbb{E}\left\{\mathbf{X}\mathbf{Y}^{\mathrm{T}}\right\} - \left\{\mathbb{E}(\mathbf{X})\right\}\left\{\mathbb{E}(\mathbf{Y})\right\}^{\mathrm{T}}.$$

∎

Proposition 5.10

Suppose that $\mathbf{X}$ and $\mathbf{Y}$ are independent random vectors. Then, $\text{cov}(\mathbf{X}, \mathbf{Y}) = 0$ (the zero matrix).

Proof. Since $\mathbf{X}$ and $\mathbf{Y}$ are independent, the random vector $\begin{bmatrix} \mathbf{X} \\ \mathbf{Y} \end{bmatrix}$ has joint probability density function

$$f_{\mathbf{X}}(\mathbf{x}) \cdot f_{\mathbf{Y}}(\mathbf{y}) \qquad \forall \mathbf{x} \in \mathbb{R}^m \quad \text{and} \quad \forall \mathbf{y} \in \mathbb{R}^q.$$

So, for any (i, j),

$$\mathbb{E}(X_i Y_j) = \int_{-\infty}^{\infty} \cdots \int_{-\infty}^{\infty} x_i y_j f_{\mathbf{X}}(\mathbf{x}) f_{\mathbf{Y}}(\mathbf{y}) \, d\mathbf{x} \, d\mathbf{y}$$

$$= \left\{\int_{-\infty}^{\infty} \cdots \int_{-\infty}^{\infty} x_i f_{\mathbf{X}}(\mathbf{x}) \, d\mathbf{x}\right\}\left\{\int_{-\infty}^{\infty} \cdots \int_{-\infty}^{\infty} y_i f_{\mathbf{Y}}(\mathbf{y}) \, d\mathbf{y}\right\}$$

$$= \mathbb{E}(X_i)\mathbb{E}(Y_j)$$

So

$$\text{cov}(X_i, Y_j) = \mathbb{E}(X_i Y_j) - \mathbb{E}(X_i)\mathbb{E}(Y_j) = 0.$$

∎

Exercises

1 Suppose that the m-dimensional random vector X follows a multinomial distribution. Prove that, for $q = 1, 2, \ldots, m-1$, the conditional distribution of $[X_1 X_2 \ldots X_q]$ given $[X_{q+1} \ldots X_m]$ is also multinomial.

2 Suppose that $\mathbf{X} = [X_1 X_2 X_3]^{\mathrm{T}} \sim \text{Mu}(3, \frac{1}{4}, \frac{1}{4}, \frac{1}{4})$. Tabulate the (joint) conditional probability mass function of X_1 and X_2 given $X_3 = x_3$ for all possible values x_3. Use the result of Exercise 1 to write down the conditional covariance and correlation of X_1 and X_2 in each case. Also evaluate the conditional value of γ between X_1 and X_2 given $X_3 = x_3$ for all possible values of x_3.

3 Suppose that $\mathbf{X} = [X_1 X_2 X_3]^{\mathrm{T}}$ is a continuous random vector with joint probability density function

$$f_{\mathbf{X}}(\mathbf{x}) = 48 x_1 x_2 x_3, \qquad 0 < x_3 < x_2 < x_1 < 1.$$

(a) Derive the (joint) marginal probability density function of (X_i, X_j) for all possible combinations of i and j ($i \neq j$).

(b) Hence derive the conditional probability density functions

$$f(x_i|X_j = x_j), \ f(x_i|X_j = x_j, \ X_k = x_k), \ f(x_i, x_j|X_k = x_k),$$

for all possible combinations of i, j and k.

(c) Confirm that X_1 and X_3 are conditionally independent given X_2, i.e.

$$f(x_1, x_3|X_2 = x_2) = f(x_1|X_2 = x_2)f(x_3|X_2 = x_2).$$

4 Consider again the random vector **X** introduced in Exercise 3 on Section 5.1. Find the conditional probability density function of X_3 given X_1 and X_2. Write down the conditional expected value of X_3 given X_1 and X_2.

5 Suppose that $\mathbf{X} = [X_1\, X_2\, X_3]^{\mathrm{T}}$ is a jointly continuous random vector such that the sub-vector $[X_1\, X_2]$ is independent of the sub-vector $[X_3]$. By considering the (joint) marginal probability density function of X_1 and X_3, show that X_1 is independent of X_3. Deduce that X_2 is also independent of X_3.

6 Suppose that $\mathbf{X} = [X_1\, X_2\, X_3]^{\mathrm{T}}$ is a jointly continuous random vector and that

$$f_{12|3}(x_1, x_2\,|\,x_3) = g(x_1, x_2), \qquad \text{for all possible } x_3.$$

In other words, the conditional probability density function of X_1 and X_2 given $X_3 = x_3$ does not depend on the value of x_3. Show that

$$f_{12}(x_1, x_2) = g(x_1, x_2), \qquad \text{for all } x_1 \text{ and } x_2.$$

Deduce that $[X_1\, X_2]$ is independent of $[X_3]$.

7 Suppose that $\mathbf{X} = [X_1\, X_2\, X_3]^{\mathrm{T}}$ is a jointly continuous random vector and that X_1 and X_2 are conditionally independent given X_3. Prove that, for all x_1, x_2 and x_3,

$$f_{1|23}(x_1\,|\,x_2, x_3) = f_{1|3}(x_1|x_3).$$

5.3 More about correlation

In this section, we will discuss three measures of the association between two random variables, say X_1 and X_2, within a jointly continuous random vector. The first of these is the correlation between X_1 and X_2, obtained (as in Chapter 4) from their (joint) marginal distribution. In order to distinguish this measure from others, it might be called the *marginal correlation*. The marginal correlation measures the strength of the (linear) relationship between X_1 and X_2 with no reference to the possible effects of the other random variables in the random vector. It is possible, though, for the marginal correlation between two random variables to be the result of the influence of a third variable on both the others. The alternative situation, where the presence of a third variable obscures the relationship between X_1 and X_2, may also occur.

In order to remove the influence of other random variables, a second measure of association is introduced; this is the *conditional correlation* between X_1 and X_2 given the value of some other random variable or variables. In general, this conditional correlation is different for different values of the conditioning variables. In the special case of the multivariate normal distribution, which is discussed at length in Chapter 8, and for some other distributions, the conditional correlation does not vary with the value of the conditioning variable. For this reason, much interest is paid to the form of the conditional correlation in the multivariate normal case, where it is usually termed the *partial correlation*. Partial correlations are often worked out for other distributions too; when this is done, the partial correlation, which does not depend on a particular value of the conditioning variable, represents an overall (in some sense, the average) value of the conditional correlation.

Example 5.8 – continued

Consider again the random vector $\mathbf{X} = [X_1\, X_2\, X_3]^T$ with joint probability density function

$$f_{\mathbf{X}}(x_1, x_2, x_3) = \frac{1}{8}(1 + x_1\, x_2\, x_3), \qquad -1 \le x_1, x_2, x_3 \le 1.$$

We have already found that the joint (marginal) distribution of X_i and X_j $(i \ne j)$ is

$$f_{ij}(x_i, x_j) = \frac{1}{4}, \qquad -1 \le x_i, x_j \le 1.$$

We have also seen that the marginal distribution of X_i $(i = 1, 2, 3)$ is

$$f_i(x_i) = \frac{1}{2}, \qquad -1 \le x_i \le 1.$$

This means that the random variables are pairwise independent, so the marginal correlation between X_i and X_j is $\rho_{12} = 0$ $(i \ne j)$.

We will now obtain the conditional correlation between X_1 and X_2 given x_3. First of all,

$$f_{12|3}(x_1, x_2 | x_3) = \frac{f_{\mathbf{X}}(x_1, x_2, x_3)}{f_3(x_3)} = \frac{\frac{1}{8}(1 + x_1\, x_2\, x_3)}{\frac{1}{2}} = \frac{1}{4}(1 + x_1\, x_2\, x_3),$$

$$-1 \le x_1, x_2 \le 1.$$

So

$$E(X_1 X_2 | x_3) = \frac{1}{4} \int_{-1}^{1} \int_{-1}^{1} x_1 x_2 (1 + x_1 x_2 x_3) \, dx_2 \, dx_1$$

$$= \frac{1}{4} \int_{-1}^{1} x_1 \left[\frac{1}{2} x_2^2 + \frac{1}{3} x_1 x_2^3 x_3 \right]_{x_2=-1}^{x_2=1} \, dx_1$$

$$= \frac{1}{6} \int_{-1}^{1} x_1^2 x_3 \, dx_1$$

$$= \frac{1}{9} x_3.$$

Since X_1 and X_3 are pairwise independent, then the conditional moments of X_1 given x_3 are just equal to the marginal moments of X_1. The marginal distribution of X_1 is $Un(-1, 1)$, which means that X_1 has expected value 0 and variance $\frac{1}{3}$. X_2 has the same marginal distribution as X_1. So

$$\rho_{12|3} = \frac{E(X_1 X_2 | x_3) - E(X_1 | x_3) E(X_2 | x_3)}{\sqrt{var(X_1 | x_3) var(X_2 | x_3)}} = \frac{x_3 / 9 - 0 \times 0}{\sqrt{\frac{1}{3} \times \frac{1}{3}}} = \frac{x_3}{3}.$$

Marginally, as we have seen, X_1 and X_2 are independent and therefore have zero correlation. Once account is taken of the value of X_3, the conditional correlation between X_1 and X_2 is not in general zero. The conditional correlation is negative when x_3 is negative and positive when x_3 is positive.

As Example 5.8 has shown, the conditional correlation $\rho_{12|3}$ generally varies with x_3. This would appear to be an important feature of the joint probability distribution and suggests that it is generally useful to examine the conditional correlations. Nevertheless, there is a very popular measure of the association between X_1 and X_2 that smoothes or averages out the effect of X_3 (taken across the whole range space of X_3). This measure is the *partial correlation* of X_1 and X_2 given X_3, which is defined by the following formula:

$$\rho_{12 \cdot 3} = \frac{\rho_{12} - \rho_{13} \rho_{23}}{\sqrt{\left(1 - \rho_{13}^2\right)\left(1 - \rho_{23}^2\right)}}.$$

Example 5.8 – continued

We have already noted that the random variables are pairwise independent. This means that all correlations ρ_{ij} are equal to 0. Therefore,

$$\rho_{12 \cdot 3} = \frac{\rho_{12} - \rho_{13} \rho_{23}}{\sqrt{\left(1 - \rho_{13}^2\right)\left(1 - \rho_{23}^2\right)}} = \frac{0 - 0 \times 0}{\sqrt{1 \times 1}} = 0.$$

This suggests that, overall, having allowed for the effect of X_3, there is zero correlation between X_1 and X_2. We have seen, though, that at all values of $x_3 \neq 0$, the *conditional* correlation of X_1 and X_2 *given* x_3 is non-zero, being negative for negative values of x_3 and positive for positive values of x_3. It is, therefore, true as indicated by the partial correlation that the conditional correlation is zero 'on average' over all the possible values of X_3, but simply quoting the partial correlation would obscure an important feature of the probability distribution. The conditional correlation (as a function of x_3) is much more informative.

Example 5.11

Consider the random vector $\mathbf{X} = [X_1 \; X_2 \; X_3]^{\mathsf{T}}$ with joint probability density function

$$f_{\mathbf{X}}(x_1, x_2, x_3) = e^{-x_3}, \qquad 0 < x_1 < x_2 < x_3.$$

The (joint) marginal distribution of X_1 and X_2 is

$$f_{12}(x_1, x_2) = \int_{x_2}^{\infty} f_{\mathbf{X}}(x_1, x_2, x_3)\, dx_3 = \int_{x_2}^{\infty} e^{-x_3}\, dx_3 = \left[-e^{-x_3}\right]_{x_3=x_2}^{x_3=\infty} = e^{-x_2},$$

$$0 < x_1 < x_2.$$

So the marginal distribution of X_1 is

$$f_1(x_1) = \int_{x_1}^{\infty} f_{12}(x_1, x_2)\, dx_2 = \int_{x_1}^{\infty} e^{-x_2}\, dx_2 = \left[-e^{-x_2}\right]_{x_2=x_1}^{x_2=\infty} = e^{-x_1},$$

$$0 < x_1.$$

This means that $X_1 \sim \text{Ex}(1)$, so in particular $\mathbb{E}(X_1) = 1$ and $\text{var}(X_1) = 1$. Also

$$f_2(x_2) = \int_{0}^{x_2} f_{12}(x_1, x_2)\, dx_1 = \int_{0}^{x_2} e^{-x_2}\, dx_1 = e^{-x_2}[x_1]_{x_1=0}^{x_1=x_2} = x_2 e^{-x_2},$$

$$0 < x_2.$$

This means that $X_2 \sim \text{Ga}(2, 1)$, so in particular $\mathbb{E}(X_2) = 2$ and $\text{var}(X_2) = 2$.

The joint (marginal) distribution of X_1 and X_3 is

$$f_{13}(x_1, x_3) = \int_{x_1}^{x_3} f_{\mathbf{X}}(x_1, x_2, x_3)\, dx_2 = \int_{x_1}^{x_3} e^{-x_3}\, dx_3 = e^{-x_3}[x_2]_{x_2=x_1}^{x_2=x_3}$$

$$= (x_3 - x_1)e^{-x_3}, \qquad 0 < x_1 < x_3.$$

Routine calculation now shows that $f_3(x_3) = \frac{1}{2} x_3^2 e^{-x_3}$. In other words, $X_3 \sim \text{Ga}(3, 1)$, so in particular $\mathbb{E}(X_3) = 3$ and $\text{var}(X_3) = 3$.

We can use the joint (marginal) distribution of X_1 and X_3 to obtain $\mathbb{E}(X_1 X_3)$ and hence the marginal covariance and correlation between these random variables.

$$\mathbb{E}(X_1 X_3) = \int_{0}^{\infty}\int_{0}^{x_3} x_1 x_3 (x_3 - x_1)e^{-x_3}\, dx_1\, dx_3 = \int_{0}^{\infty} \frac{1}{6} x_3^4 e^{-x_3}\, dx_3 = \frac{1}{6}\Gamma(5) = 4.$$

So

$$\text{cov}(X_1, X_3) = \mathbb{E}(X_1 X_3) - \mathbb{E}(X_1)\mathbb{E}(X_3) = 4 - 1 \times 3 = 1.$$

Therefore

$$\rho_{13} = \frac{\text{cov}(X_1, X_3)}{\sqrt{\text{var}(X_1)\text{var}(X_3)}} = \frac{1}{\sqrt{1 \times 3}} = \frac{1}{\sqrt{3}}.$$

This value suggests that, marginally, there is a moderately strong correlation between X_1 and X_3.

We will now look at the conditional correlation of X_1 and X_3 given X_2:

$$f_{13|2}(x_1, x_3 | x_2) = \frac{f_{\mathbf{X}}(x_1, x_2, x_3)}{f_2(x_2)} = \frac{e^{-x_3}}{x_2 \, e^{-x_2}} = \frac{1}{x_2} e^{-(x_3 - x_2)}, \qquad 0 < x_1 < x_2 < x_3.$$

Notice that the joint range space of X_1 and X_3 given $X_2 = x_2$ is a 'rectangular' region, namely $\{(x_1, x_3): 0 < x_1 < x_2, x_2 < x_3 < \infty\}$. The (joint) conditional probability density function of X_1 and X_3 factorizes on this (joint) conditional range space, which means that X_1 and X_3 are conditionally independent given X_2. Using the factorization theorem, we can show that

$$f_{1|2}(x_1 | x_2) = \frac{1}{x_2}, \qquad 0 < x_1 < x_2,$$

$$f_{3|2}(x_3 | x_2) = e^{-(x_3 - x_2)}, \qquad x_2 < x_3.$$

Anyway, the conditional independence of X_1 and X_3 means that $\rho_{13|2} = 0$ for all possible values of X_2. The marginal correlation between the random variables disappears once allowance is made for the effect of X_2 on both of the other variables.

In this example, unlike Example 5.8, the conditional correlation, $\rho_{13|2}$ is constant for all values of X_2. It would seem reasonable to expect, then, that the partial correlation $\rho_{13\cdot2}$ would be meaningful, and that is indeed the case. We have already obtained the result that $\rho_{13} = 1/\sqrt{3}$. It is routine to show in a similar way that $\rho_{12} = 1/\sqrt{2}$ and $\rho_{23} = \sqrt{2}/\sqrt{3}$. Hence,

$$\rho_{12\cdot3} = \frac{\rho_{12} - \rho_{13}\rho_{23}}{\sqrt{\left(1 - \rho_{13}^2\right)\left(1 - \rho_{23}^2\right)}} = \frac{\dfrac{1}{\sqrt{3}} - \dfrac{1}{\sqrt{2}}\dfrac{\sqrt{2}}{\sqrt{3}}}{\sqrt{\left(1 - \dfrac{1}{2}\right)\left(1 - \dfrac{2}{3}\right)}} = 0.$$

In this example, where the conditional correlation is constant for all possible values of the conditioning variable, the partial correlation is equal to the conditional correlation.

Exercises

1 Consider again the random vector discussed in Exercise 3 on Section 5.1. Find ρ_{12}, $\rho_{12|3}$ and $\rho_{12\cdot3}$.

2 Consider again the random vector discussed in Exercise 3 on Section 5.2. Find $\rho_{ij\cdot k}$ for all possible combinations of values of i, j and k. In each case, compare the partial correlation with the corresponding conditional correlation.

3 Suppose that $\mathbf{X} = [X_1 X_2 X_3]^T$ is a jointly continuous random vector such that the sub-vector $[X_1 X_2]^T$ is independent of the sub-vector $[X_3]$. Explain why $\rho_{13} = \rho_{23} = 0$. Show that $\rho_{12\cdot3} = \rho_{12}$. Show that, for all possible values x_3, the conditional correlation between X_1 and X_2 given $X_3 = x_3$ is also ρ_{12}.

5.4 The joint moment-generating function

As in the univariate case, we often find it useful to work with the moment-generating function of a random vector $\mathbf{X}$. When it exists, the *joint moment-generating function* of the m-dimensional random vector $\mathbf{X}$ is the real-valued function

$$M_{\mathbf{X}}(\mathbf{t}) = \mathbb{E}(e^{X_1 t_1 + \cdots + X_m t_m}) = \mathbb{E}(e^{\mathbf{X}^T \mathbf{t}}), \qquad \mathbf{t} = [t_1 \ldots t_m] \in \mathbb{R}^m.$$

Example 5.12

Suppose that the random vector $\mathbf{X} = [X_1 X_2 X_3]^T$ has (joint) probability density function

$$f_{\mathbf{X}}(\mathbf{x}) = 6 e^{-(x_1 + x_2 + x_3)}, \qquad x_3 > x_2 > x_1 > 0.$$

The joint moment-generating function of this random vector is found as follows:

$$M_{\mathbf{X}}(\mathbf{t}) = \mathbb{E}\left(e^{\mathbf{x}^T \mathbf{t}}\right)$$

$$= \int_0^\infty \int_{x_1}^\infty \int_{x_2}^\infty e^{\mathbf{x}^T \mathbf{t}} f_{\mathbf{X}}(\mathbf{x}) \, d\mathbf{x}$$

$$= \int_0^\infty \int_{x_1}^\infty \int_{x_2}^\infty e^{x_1 t_1 + x_2 t_2 + x_3 t_3} 6 e^{-(x_1 + x_2 + x_3)} \, dx_3 \, dx_2 \, dx_1$$

$$= 6 \int_0^\infty e^{-x_1(1 - t_1)} \int_{x_1}^\infty e^{-x_2(1 - t_2)} \int_{x_2}^\infty e^{-x_3(1 - t_3)} \, dx_3 \, dx_2 \, dx_1$$

$$= 6 \int_0^\infty e^{-x_1(1 - t_1)} \int_{x_1}^\infty e^{-x_2(1 - t_2)} \left[-\frac{1}{(1 - t_3)} e^{-x_3(1 - t_3)} \right]_{x_3 = x_2}^{x_3 = \infty} \, dx_2 \, dx_1$$

$$= \frac{6}{1 - t_3} \int_0^\infty e^{-x_1(1-t_1)} \int_{x_1}^\infty e^{-x_2(1-t_2)} e^{-x_2(1-t_3)} \, dx_2 \, dx_1$$

$$= \frac{6}{1 - t_3} \int_0^\infty e^{-x_1(1-t_1)} \int_{x_1}^\infty e^{-x_2(2-t_2-t_3)} \, dx_2 \, dx_1$$

$$= \frac{6}{1 - t_3} \int_0^\infty e^{-x_1(1-t_1)} \left[\frac{-1}{2 - t_2 - t_3} e^{-x_2(2-t_2-t_3)} \right]_{x_2=x_1}^{x_2=\infty} dx_1$$

$$= \frac{6}{(1 - t_3)(2 - t_2 - t_3)} \int_0^\infty e^{-x_1(3-t_1-t_2-t_3)} \, dx_1$$

$$= \frac{6}{(1 - t_3)(2 - t_2 - t_3)(3 - t_1 - t_2 - t_3)},$$

where the restrictions $t_3 < 1$, $t_2 + t_3 < 2$ and $t_1 + t_2 + t_3 < 3$ are required for convergence.

As its name suggests, the joint moment-generating function can be used to work out moments (e.g. $\mathbb{E}(X_3)$, $\text{cov}(X_1, X_4)$), and this can sometimes reduce the amount of work involved.

Proposition 5.11

Let **X** be an m-dimensional random vector with moment-generating function $M_X(t)$. When all the appropriate partial derivatives exist, then

$$\mathbb{E}(X_1^{k_1} \cdots X_m^{k_m}) = \frac{\partial^{k_1 + \cdots + k_m}}{\partial t_1^{k_1} \cdots \partial t_m^{k_m}} M_X(t) \Bigg|_{t=0}.$$

In practice, though, the joint moment-generating function is of greatest use for deriving theoretical results, just as univariate moment-generating functions are used to derive 'reproductive' properties of random variables (see Chapter 2). In this connection, the following uniqueness theorem (which is stated without proof) is important.

Proposition 5.12

Suppose that **X** and **Y** are m-dimensional random vectors. If their moment-generating functions exist and are equal for all **t** in an open region about the point $t = 0$, then **X** and **Y** follow the same joint probability distribution.

Some further important properties of the moment-generating function are given below.

Proposition 5.13

Suppose **X** is an m-dimensional random vector with moment-generating function $M_X(t)$.

(a) If $\mathbf{Y} = A\mathbf{X} + \mathbf{b}$, for some constant matrix $A \in M_{qm}$ and some constant vector $\mathbf{b} \in \mathbb{R}^q$, then $\mathbf{Y}$ has moment-generating function

$$M_{\mathbf{Y}}(\mathbf{t}) = e^{\mathbf{b}^{\mathrm{T}}\mathbf{t}} M_{\mathbf{X}}(A^{\mathrm{T}}\mathbf{t}), \qquad \mathbf{t} \in \mathbb{R}^q.$$

(b) If $\mathbf{Y} = c\mathbf{X}$, for some constant $c \in \mathbb{R}$, then $\mathbf{Y}$ has moment-generating function

$$M_{\mathbf{Y}}(\mathbf{t}) = M_{\mathbf{X}}(c\mathbf{t}), \qquad \mathbf{t} \in \mathbb{R}^m.$$

(c) If $\mathbf{X}$ is partitioned as

$$\mathbf{X} = \begin{bmatrix} \mathbf{X}^{(1)} \\ \mathbf{X}^{(2)} \end{bmatrix},$$

where $\mathbf{X}^{(1)}$ is a m_1-dimensional sub-vector, then:

(i) $\mathbf{X}^{(1)}$ has moment-generating function

$$M_{\mathbf{X}^{(1)}}(\mathbf{t}^{(1)}) = M_{\mathbf{X}} \left\{ \begin{bmatrix} \mathbf{t}^{(1)} \\ \mathbf{0} \end{bmatrix} \right\}, \qquad \mathbf{t}^{(1)} \in \mathbb{R}^{m_1}.$$

(ii) $\mathbf{X}^{(1)}$ and $\mathbf{X}^{(2)}$ are independent if and only if

$$M_{\mathbf{X}}(\mathbf{t}) = M_{\mathbf{X}} \left\{ \begin{bmatrix} \mathbf{t}^{(1)} \\ \mathbf{0} \end{bmatrix} \right\} \cdot M_{\mathbf{X}} \left\{ \begin{bmatrix} \mathbf{0} \\ \mathbf{t}^{(2)} \end{bmatrix} \right\},$$

$$\forall \, \mathbf{t}^{(1)} \in \mathbb{R}^{m_1}, \forall \mathbf{t}^{(2)} \in \mathbb{R}^{m-m_1}.$$

Proof. The proofs of (a) and (b) are left as an exercise. We shall prove (c) in the case where $\mathbf{X}$ is a continuous random vector whose probability density function exists. For part (i),

$$M_{\mathbf{X}^{(1)}}(\mathbf{t}^{(1)}) = \mathbb{E}\left\{ \exp\left(\mathbf{X}^{(1)\mathrm{T}} \cdot \mathbf{t}^{(1)} \right) \right\}$$

$$= \int_{-\infty}^{\infty} \cdots \int_{-\infty}^{\infty} \exp\left\{ \mathbf{x}^{(1)\mathrm{T}} \cdot \mathbf{t}^{(1)} \right\} f^{(1)}\left(\mathbf{x}^{(1)} \right) d\mathbf{x}^{(1)}$$

$$= \int_{-\infty}^{\infty} \cdots \int_{-\infty}^{\infty} \exp\left\{ \mathbf{x}^{(1)\mathrm{T}} \cdot \mathbf{t}^{(1)} \right\} \cdot \left\{ \int_{-\infty}^{\infty} \cdots \int_{-\infty}^{\infty} f(\mathbf{x}) d\mathbf{x}^{(2)} \right\} d\mathbf{x}^{(1)}$$

$$\text{[marginal p.d.f.]}$$

$$= \int_{-\infty}^{\infty} \cdots \int_{-\infty}^{\infty} \exp\left\{ \mathbf{x}^{(1)\mathrm{T}} \cdot \mathbf{t}^{(1)} + \mathbf{x}^{(2)\mathrm{T}} \cdot \mathbf{0} \right\} f_{\mathbf{X}}(\mathbf{x}) d\mathbf{x}$$

$$= M_{\mathbf{X}} \begin{bmatrix} \mathbf{t}^{(1)} \\ \mathbf{0} \end{bmatrix}$$

In order to prove part (ii), suppose first that $\mathbf{X}^{(1)}$ and $\mathbf{X}^{(2)}$ are independent. Then

$$f_{\mathbf{X}}^{(\mathbf{x})} = f^{(1)}(\mathbf{x}^{(1)}) \cdot f^{(2)}(\mathbf{x}^{(2)}) \qquad \text{for all } \mathbf{x} = \begin{bmatrix} \mathbf{x}^{(1)} & \mathbf{x}^{(2)} \end{bmatrix}^{\mathrm{T}} \in \mathbb{R}^m$$

So

$$M_{\mathbf{X}}(\mathbf{t}) = \int_{-\infty}^{\infty} \cdots \int_{-\infty}^{\infty} e^{\mathbf{x}^{\mathrm{T}}\mathbf{t}} f_{\mathbf{X}}(\mathbf{x}) \, d\mathbf{x}$$

$$= \int_{-\infty}^{\infty} \cdots \int_{-\infty}^{\infty} e^{\mathbf{x}^{(1)\mathrm{T}}\cdot\mathbf{t}^{(1)}} e^{\mathbf{x}^{(2)}\cdot\mathbf{t}^{(2)}} f^{(1)}(\mathbf{x}^{(1)}) f^{(2)}(\mathbf{x}^{(2)}) \, d\mathbf{x}^{(1)} d\mathbf{x}^{(2)}$$

$$= \left\{ \int_{-\infty}^{\infty} \cdots \int_{-\infty}^{\infty} e^{\mathbf{x}^{(1)\mathrm{T}}\cdot\mathbf{t}^{(1)}} f^{(1)}(\mathbf{x}^{(1)}) \, d\mathbf{x}^{(1)} \right\}$$

$$\times \left\{ \int_{-\infty}^{\infty} \cdots \int_{-\infty}^{\infty} e^{\mathbf{x}^{(2)\mathrm{T}}\cdot\mathbf{t}^{(2)}} f^{(2)}(\mathbf{x}^{(2)}) \, d\mathbf{x}^{(2)} \right\}$$

$$= M_{\mathbf{X}^{(1)}}(\mathbf{t}^{(1)}) \cdot M_{\mathbf{X}^{(2)}}(\mathbf{t}^{(2)})$$

$$= M_{\mathbf{X}}\begin{bmatrix} \mathbf{t}^{(1)} \\ \mathbf{0} \end{bmatrix} \cdot M_{\mathbf{X}}\begin{bmatrix} \mathbf{0} \\ \mathbf{t}^{(2)} \end{bmatrix} \qquad \text{(using part (i))}.$$

Now suppose that the moment-generating function of $\mathbf{X}$ factorizes as above. Let $\mathbf{Y}^{(1)}$ and $\mathbf{Y}^{(2)}$ be independent random vectors such that each $\mathbf{Y}^{(i)}$ has the same marginal distribution as $\mathbf{X}^{(i)}$. Then, using part (i) of (c),

$$M_{\mathbf{Y}^{(1)}}(\mathbf{t}^{(1)}) = M_{\mathbf{X}^{(1)}}(\mathbf{t}^{(1)}) = M_{\mathbf{X}}\begin{bmatrix} \mathbf{t}^{(1)} \\ \mathbf{0} \end{bmatrix},$$

$$M_{\mathbf{Y}^{(2)}}(\mathbf{t}^{(2)}) = M_{\mathbf{X}^{(2)}}(\mathbf{t}^{(2)}) = M_{\mathbf{X}}\begin{bmatrix} \mathbf{0} \\ \mathbf{t}^{(2)} \end{bmatrix}.$$

Since $\mathbf{Y}^{(1)}$ and $\mathbf{Y}^{(2)}$ are independent, then

$$M_{\mathbf{Y}}(\mathbf{t}) = M_{\mathbf{X}}\begin{bmatrix} \mathbf{t}^{(1)} \\ \mathbf{0} \end{bmatrix} \cdot M_{\mathbf{X}}\begin{bmatrix} \mathbf{0} \\ \mathbf{t}^{(2)} \end{bmatrix} = M_{\mathbf{X}}(\mathbf{t}) \qquad \text{(by assumption)}.$$

By Proposition 5.12, then, $\mathbf{X}$ and $\mathbf{Y}$ follow the same distribution. In particular, $\mathbf{X}^{(1)}$ and $\mathbf{X}^{(2)}$ are independent since $\mathbf{Y}^{(1)}$ and $\mathbf{Y}^{(2)}$ are independent. ∎

It is a consequence of Proposition 5.13(a) that the (marginal) moment-generating function of X_i is

$$M_i(t) = M_{\mathbf{X}}\begin{bmatrix} 0 & 0 & \cdots & t & 0 & \cdots & 0 \end{bmatrix}^{\mathrm{T}}$$

(where the t appears in the ith position in the vector). This is easily shown from the Proposition by replacing A with $\mathbf{A} = [0 \cdots 0 \; 1 \; 0 \cdots 0]^{\mathrm{T}}$ and $\mathbf{b} = [0 \cdots 0]^{\mathrm{T}}$. This can reduce the amount of work involved in finding the marginal moments of X_i, and also (sometimes) helps us to identify the (marginal) distribution of X_i.

Example 5.12 – continued

The marginal moment-generating function of X_1 is obtained from the joint moment-generating function as follows.

$$M_1(t) = M_X([\ t \quad 0 \quad 0\ ]^T)$$

$$= \frac{6}{(1-0)(2-0-0)(3-t-0-0)}, \qquad t+0+0 < 3,$$

$$= \frac{3}{3-t}, \qquad t < 3.$$

This is immediately recognized as the moment-generating function of an Ex(3) random vector. Using the uniqueness property of (univariate) moment-generating functions, this means that $X_1 \sim \text{Ex}(3)$. The expected value and variance of X_1 may be obtained from the marginal moment-generating function more easily than from the joint moment-generating function, as follows:

$$\mathbb{E}(X_1) = \left.\frac{d}{dt}M_1(t)\right|_{t=0} = \left.\frac{3}{(3-t)^2}\right|_{t=0} = \frac{1}{3}$$

$$\mathbb{E}(X_1^2) = \left.\frac{d^2}{dt^2}M_1(t)\right|_{t=0} = \left.\frac{6}{(3-t)^3}\right|_{t=0} = \frac{6}{27} = \frac{2}{9}$$

$$\therefore \qquad \text{var}(X_1) = \frac{2}{9} - \left(\frac{1}{3}\right)^2 = \frac{1}{9}.$$

Proposition 5.14 – The factorization theorem for moment-generating functions

Let $X = [X_1 \cdots X_m]^T$ be a random vector with moment-generating function $M_X(t)$. Suppose there are real-valued functions $g_1, \ldots, g_p$ such that

$$M_X(t) = \prod_{i=1}^{m} g_i(t_i).$$

Then $X_1, \ldots, X_m$ are independent and each X_i has moment-generating function

$$M_i(t) \propto g_i(t).$$

Example 5.12 – continued

We can use Proposition 5.13(a) in order to find the moment-generating function of the random vector $Y = [Y_1\ Y_2\ Y_3]^T$, where

$$Y_1 = X_1,$$

$$Y_2 = X_2 - X_1,$$

$$Y_3 = X_3 - X_2.$$

Since

$$Y = \begin{bmatrix} 1 & 0 & 0 \\ -1 & 1 & 0 \\ 0 & -1 & 1 \end{bmatrix} X = AX$$

(say), then

$$M_Y(t) = M_X(A^T t)$$

where

$$A^T t = \begin{bmatrix} 1 & -1 & 0 \\ 0 & 1 & -1 \\ 0 & 0 & 1 \end{bmatrix} t = \begin{bmatrix} t_1 - t_2 \\ t_2 - t_3 \\ t_3 \end{bmatrix},$$

therefore

$$M_Y(t) = M_X(A^T t) = \frac{6}{(1 - t_3)(2 - t_2)(3 - t_1)}.$$

By the factorization theorem, then, Y_1, Y_2 and Y_3 are independent and their marginal moment-generating functions are

$$M_1(t) = \frac{k_1}{3 - t}, \qquad M_2(t) = \frac{k_2}{2 - t}, \qquad M_3(t) = \frac{k_3}{1 - t}$$

$\therefore$ $k_1 = 3$ and $Y_1 \sim \text{Ex}(3)$, $k_2 = 2$ and $Y_2 \sim \text{Ex}(2)$, $k_3 = 1$ and $Y_3 \sim \text{Ex}(1)$

This means that

$$\mathbb{E}(Y) = \begin{bmatrix} \frac{1}{3} \\ \frac{1}{2} \\ 1 \end{bmatrix}, \qquad \text{cov}(Y) = \begin{bmatrix} \frac{1}{9} & 0 & 0 \\ 0 & \frac{1}{4} & 0 \\ 0 & 0 & 1 \end{bmatrix}.$$

We may now obtain the expected value and covariance matrix of X by inverse transformation, as follows.

$$\left. \begin{matrix} X_1 = Y_1 \\ X_2 = Y_1 + Y_2 \\ X_3 = Y_1 + Y_2 + Y_3 \end{matrix} \right\} \Rightarrow X = \underbrace{\begin{bmatrix} 1 & 0 & 0 \\ 1 & 1 & 0 \\ 1 & 1 & 1 \end{bmatrix}}_{=B(\text{say})} Y$$

therefore

$$\mathbb{E}(X) = B.\mathbb{E}(Y) = \begin{bmatrix} \frac{1}{3} \\ \frac{5}{6} \\ \frac{11}{6} \end{bmatrix},$$

$$\mathrm{cov}(\mathbf{X}) = B.\mathrm{cov}(\mathbf{X})B^{\mathrm{T}} = \begin{bmatrix} 1 & 0 & 0 \\ 1 & 1 & 0 \\ 1 & 1 & 1 \end{bmatrix} \begin{bmatrix} \frac{1}{9} & 0 & 0 \\ 0 & \frac{1}{4} & 0 \\ 0 & 0 & 1 \end{bmatrix} \begin{bmatrix} 1 & 1 & 1 \\ 0 & 1 & 1 \\ 0 & 0 & 1 \end{bmatrix}$$

$$= \begin{bmatrix} \frac{1}{9} & \frac{1}{9} & \frac{1}{9} \\ \frac{1}{9} & \frac{13}{36} & \frac{13}{36} \\ \frac{1}{9} & \frac{13}{36} & \frac{49}{36} \end{bmatrix}.$$

Proposition 5.15

Let $\mathbf{X}$ and $\mathbf{Y}$ be (respectively) an m-dimensional and a q-dimensional random vector. Define functions $\mathbf{a}\colon \mathbb{R}^m \to \mathbb{R}^n$ and $\mathbf{b}\colon \mathbb{R}^q \to \mathbb{R}^p$ such that

$$\mathbf{a}(\mathbf{X}) = \begin{bmatrix} a_1(\mathbf{X}) \\ a_2(\mathbf{X}) \\ \vdots \\ a_n(\mathbf{X}) \end{bmatrix}, \qquad \mathbf{b}(\mathbf{Y}) = \begin{bmatrix} b_1(\mathbf{Y}) \\ b_2(\mathbf{Y}) \\ \vdots \\ b_p(\mathbf{Y}) \end{bmatrix}.$$

If $\mathbf{X}$ and $\mathbf{Y}$ are independent, then so too are $\mathbf{a}(\mathbf{X})$ and $\mathbf{b}(\mathbf{Y})$.

Proof. We prove this proposition for the case where the required joint probability density functions exist.

Since $\mathbf{X}$ and $\mathbf{Y}$ are independent, the joint probability density function of $\mathbf{Z} = \begin{bmatrix} \mathbf{X} \\ \mathbf{Y} \end{bmatrix}$ is

$$f_{\mathbf{Z}}(\mathbf{z}) = f_{\mathbf{X}}(\mathbf{x}) \cdot f_{\mathbf{Y}}(\mathbf{y}) \qquad \text{for all } \mathbf{z} = \begin{bmatrix} \mathbf{x} \\ \mathbf{y} \end{bmatrix} \in \mathbb{R}^{m+q}$$

Writing

$$\mathbf{U} = \begin{bmatrix} \mathbf{a}(\mathbf{X}) \\ \mathbf{b}(\mathbf{Y}) \end{bmatrix},$$

then

$$M_{\mathbf{U}}(\mathbf{t}) = \mathbb{E}\left\{ \exp\left([\mathbf{a}(\mathbf{X})]^{\mathrm{T}} \cdot \mathbf{t}^{(1)} + [\mathbf{b}(\mathbf{Y})]^{\mathrm{T}} \cdot \mathbf{t}^{(2)} \right) \right\}$$

$$= \int_{-\infty}^{\infty} \cdots \int_{-\infty}^{\infty} \exp\left([\mathbf{a}(\mathbf{x})]^{\mathrm{T}} \cdot \mathbf{t}^{(1)} + [\mathbf{b}(\mathbf{y})]^{\mathrm{T}} \cdot \mathbf{t}^{(2)} \right) f_{\mathbf{Z}}(\mathbf{z}) \, d\mathbf{z}$$

(since $\mathbf{U}$ is a function of $\mathbf{Z}$)

$$= \left\{ \int_{-\infty}^{\infty} \cdots \int_{-\infty}^{\infty} \exp\left([\mathbf{a}(\mathbf{x})]^{\mathrm{T}} \cdot \mathbf{t}^{(1)}\right) \cdot f_{\mathbf{X}}(\mathbf{x}) \, d\mathbf{x} \right\}$$

$$\times \left\{ \int_{-\infty}^{\infty} \cdots \int_{-\infty}^{\infty} \exp\left([\mathbf{b}(\mathbf{y})]^{\mathrm{T}} \cdot \mathbf{t}^{(2)}\right) \cdot f_{\mathbf{Y}}(\mathbf{y}) \, d\mathbf{y} \right\}$$

(since $f_{\mathbf{Z}}(\mathbf{z})$ factorizes)

$$= M_{\mathbf{a}(\mathbf{X})}(\mathbf{t}^{(1)}) \cdot M_{\mathbf{b}(\mathbf{Y})}(\mathbf{t}^{(2)})$$

So $\mathbf{a}(\mathbf{X})$ and $\mathbf{b}(\mathbf{Y})$ are independent random vectors (Proposition 5.13(c)(ii)).   ∎

Setting $m = n = p = q = 1$ in Proposition 5.15, we deduce that, if X and Y are independent random variables and $a(\cdot)$ and $b(\cdot)$ are real functions, then $a(X)$ and $b(Y)$ are also independent.

Proposition 5.16

Suppose $\mathbf{X}_1, \ldots, \mathbf{X}_n$ are independent random vectors of dimension m, with moment-generating functions $M_{\mathbf{X}_1}(\mathbf{t}), \ldots, M_{\mathbf{X}_n}(\mathbf{t})$. If $\mathbf{Y} = \mathbf{X}_1 + \cdots + \mathbf{X}_n$, then $\mathbf{Y}$ has moment-generating function

$$M_{\mathbf{Y}}(\mathbf{t}) = \prod_{i=1}^{n} M_{\mathbf{X}_i}(\mathbf{t}).$$

Proof.

$$M_{\mathbf{Y}}(\mathbf{t}) = \mathbb{E}\{\exp(\mathbf{Y}^{\mathrm{T}}\mathbf{t})\} = \mathbb{E}\left\{ \exp\left(\sum_{i=1}^{n} \mathbf{X}_i^{\mathrm{T}}\mathbf{t} \right) \right\} = \mathbb{E}\left\{ \prod_{i=1}^{n} \exp(\mathbf{X}_i^{\mathrm{T}}\mathbf{t}) \right\}$$

But $\exp(\mathbf{X}_1^{\mathrm{T}}\mathbf{t}), \ldots, \exp(\mathbf{X}_n^{\mathrm{T}}\mathbf{t})$ are independent (Proposition 5.15) and so

$$M_{\mathbf{Y}}(\mathbf{t}) = \mathbb{E}\left\{ \prod_{i=1}^{n} \exp(\mathbf{X}_i^{\mathrm{T}}\mathbf{t}) \right\} = \prod_{i=1}^{n} \{\mathbb{E}\left(\exp(\mathbf{X}_i^{\mathrm{T}}\mathbf{t})\right)\} = \prod_{i=1}^{n} M_{\mathbf{X}_i}(\mathbf{t}).$$

This is the required result.   ∎

In Chapter 6, we make continual use of Proposition 5.16 for the special case of independent random variables ($m = 1$).

Exercises

1 Suppose that $X_1, \ldots, X_m$ are independent random variables such that each X_i follows an $\mathrm{Ex}(\theta_i)$ distribution. Find the distribution of

$$Y = \theta_1 X_1 + \theta_2 X_2 + \cdots + \theta_m X_m$$

using a multivariate moment-generating function.

2 Let **X** be a three-dimensional random vector with (joint) probability density function

$$f(\mathbf{x}) = \begin{cases} e^{-x_3}, & 0 < x_1 < x_2 < x_3, \\ 0, & \text{otherwise.} \end{cases}$$

(a) Check that this is a valid probability density function.

(b) Derive the (joint) moment-generating function of **X**. Hence find the marginal distributions of X_1, X_2 and X_3, and state their expected values and variances.

(c) Using moment-generating functions, show that $Y_1 = X_1$, $Y_2 = X_2 - X_1$ and $Y_3 = X_3 - X_2$ and independent and identically distributed Ex(1) random variables. Use this result to find $\mathbb{E}(\mathbf{X})$ and $\text{cov}(\mathbf{X})$.

3 The $(m+1)$-dimensional *discrete* random vector **X** has probability mass function

$$p_\mathbf{X}(x_1, \ldots, x_{m+1}) = \frac{n!}{x_1! \cdots x_m! x_{m+1}!} \theta_1^{x_1} \cdots \theta_m^{x_m} \theta_{m+1}^{x_{m+1}},$$

$$x_1, \ldots, x_{m+1} = 0, 1, \ldots, n, x_1 + \cdots + x_{m+1} = n.$$

This means that $[X_1\, X_2 \cdots X_m]^{\mathrm{T}}$ follows the $\mu(n, \theta_1, \cdots, \theta_m)$ distribution.

(a) Show that the moment-generating function of **X** is

$$M_\mathbf{X}(\mathbf{t}) = \left\{ e^{t_1}\theta_1 + \cdots + e^{t_{m+1}}\theta_{m+1} \right\}^n.$$

(b) Hence find $\mathbb{E}(\mathbf{X})$ and $\text{cov}(\mathbf{X})$.

(c) Write down an $(m+1)$-dimensional vector of constants, **a**, such that $\mathbf{a} \neq \mathbf{0}$ but $\mathbf{a}^{\mathrm{T}}\mathbf{X}$ is a constant (with probability 1). Show that $\mathbf{a}^{\mathrm{T}} \cdot \text{cov}(\mathbf{X}) \cdot \mathbf{a} = 0$, and deduce that $\text{cov}(\mathbf{X})$ is a singular matrix.

4 Prove parts (a) and (b) of Proposition 5.13.

Summary

This chapter has extended previous work on two random variables to the general case of a multi-dimensional random vector. Particular interest has focused on relationships among the random variables in a random vector. The concepts of independence, pairwise independence and conditional independence have been introduced and contrasted. Marginal correlation has been compared with conditional correlation and partial correlation. The joint moment-generating function has been introduced and several of its most important properties have been stated and proved.

6
Sequences of random variables

The previous two chapters have covered the basic theory of joint probability distributions for more than one random variable. In this chapter and the next, we will consider how to derive probability models for functions of sets of random variables. We will begin, in this chapter, with two of the most important functions – the sum and average of a collection of random variables. We will restrict ourselves here to the case of independent random variables. These are often obtained by running a basic experiment on a number of occasions and recording the value of the same random variable each time. In some very important cases, when we are dealing with independent random variables that follow the same probability distribution, the sum or average is found to follow the same kind of distribution as the original variables. In general, the sum or average of a sufficiently large number of random variables converges to a normal distribution.

6.1 Sums and averages of random variables

Example 6.1

Cocoa beans are used to make chocolate. The beans are found inside cocoa pods; we shall assume that the number of beans inside a randomly selected pod is a $Po(\theta)$ random variable. In particular, then, $\mathbb{E}(X_i) = \theta$ and $var(X_i) = \theta$.

Assume that n pods grow on a particular cocoa tree, and let $X_1, X_2, \ldots, X_n$ be the numbers of beans inside the n pods. The total number of beans that may be harvested from this tree is

$$S_n = X_1 + X_2 + \cdots + X_n.$$

Assuming that the numbers of beans inside different pods are independent, then Proposition 5.8 shows that

$$\mathbb{E}(S_n) = \mathbb{E}(X_1) + \mathbb{E}(X_2) + \cdots + \mathbb{E}(X_n) = n\theta$$

and

$$var(S_n) = var(X_1) + var(X_2) + \cdots + var(X_n) = n\theta.$$

Noting that the expected value and variance of S_n are equal, we might suspect that S_n follows a Poisson distribution. We shall show later that this is correct.

Now suppose that N, the number of pods that develop on a randomly selected cocoa tree in the course of a year, is a $Po(\lambda)$ random variable. Then, $\mathbb{E}(N) = \lambda$ and $var(N) = \lambda$. The total number of beans that may be harvested from a randomly selected cocoa tree is

$$S_N = X_1 + X_2 + \cdots + X_N$$

where N is a random variable. We can find the expected value and variance of S_N using the formulae for iterated expectation and variance (Proposition 4.5):

$$\mathbb{E}(S_N) = \mathbb{E}\{\mathbb{E}(S_N|N)\} = \mathbb{E}(N\theta) = \theta\mathbb{E}(N) = \theta\lambda$$

and

$$var(S_N) = \mathbb{E}\{var(S_N|N)\} + var\{\mathbb{E}(S_N|N\} = \mathbb{E}(N\theta) + var(N\theta) = \theta\lambda + \theta^2\lambda.$$

The expected value and variance of S_N are no longer equal, so S_N does not follow a Poisson distribution when the number of pods is allowed to vary.

Example 6.2

Your favourite breakfast cereal is running a new promotional offer. In future, each pack will contain one figure of a twentieth-century British prime minister. There are six different figures in the set and one is chosen at random for inclusion in each pack. How many packets of cereal will you have to buy, on average, in order to collect the whole set of six different figures?

Let the random variable X_1 denote the number of packets you must buy until you obtain the first figurine you require for your set. Clearly X_1 takes the value 1 with probability 1 (since you have none of the figurines at the start).

Now let X_i ($i = 2, \ldots, 6$) denote the number of *further* packets you must buy, *after* obtaining $i - 1$ different figurines for your set, in order to obtain the next different figurine. You now have probability $(7 - i)/6$ of obtaining a new type of figurine in any packet of cereal you buy, and this probability is unchanged from packet to packet until you obtain the next figurine for your set. This means that X_i is a geometric random variable with success probability $(7 - i)/6$, and so

$$\mathbb{E}(X_i) = \frac{1}{(7-i)/6} = \frac{6}{7-i},$$

$$var(X_i) = \frac{1 - (7-i)/6}{((7-i)/6)^2} = 6\frac{i-1}{(7-i)^2}.$$

The total number of packets you must buy in order to obtain a full set of six figurines is

$$S = X_1 + X_2 + \cdots + X_6.$$

Since all the X_i are independent,

$$\mathbb{E}(S) = \mathbb{E}(X_1) + \cdots + \mathbb{E}(X_6) = 1 + \frac{6}{5} + \frac{6}{4} + \frac{6}{3} + \frac{6}{2} + \frac{6}{1} = 14.7$$

and

$$\text{var}(S) = \text{var}(X_1) + \cdots + \text{var}(X_6) = 0 + 6\frac{1}{25} + 6\frac{2}{16} + 6\frac{3}{9} + 6\frac{4}{4} + 6\frac{5}{1} = 38.99.$$

This means that, on average, you will have to buy 14.7 packets of cereal in order to collect your set of British prime ministers. No problem, you say! Note, though, that the variance is very large relative to the expected value, so there is a good chance that you will have to buy a lot more than this average number. We are not in a position to obtain such probabilities algebraically, but this context is relatively easy to simulate and a large simulation (of, say, 10,000 attempts to collect a full set of figures) would give a reasonable approximation to almost any probability of interest.

Example 6.3

Suppose you have designed a lift that can safely carry loads up to 1,000 kg. You wish to sell lifts of this kind for carrying people, so you need to be able to indicate how many people can safely get into the lift. After some research, you discover that the distribution of weights of adults in your population has expected value 70 kg and standard deviation 6 kg. Suppose that a random group of n adults gets on the lift, and let $X_1, X_2, \ldots, X_n$ be their individual weights. Then their total weight is

$$S_n = X_1 + X_2 + \cdots + X_n.$$

S_n has expected value $70n$ kg and standard deviation $6\sqrt{n}$ kg (since the variance of S_n is $36n$). On average, then, 14 adults could safely be carried in the lift, since $70 \times 14 = 980 < 1,000$, but 15 adults could not be carried safely, since $70 \times 15 = 1,050 > 1,000$.

However, the variance in the total weight of 14 adults is substantial – $14 \times 36 = 504$ – with the result that there must be a fairly high probability that a group of 14 adults will weigh more than 1,000 kg. In calculating the average weight, such groups are compensated for by groups that weigh less than 1,000 kg, but that is not going to be of any comfort to a particular group that happens to exceed the maximum safe load for the lift. We really need to be able to calculate the probability that this will happen.

In order to calculate an exact probability associated with the random variable S_n, the form of its probability distribution must be known. Without the probability density function, we can obtain limited approximations to the probability of interest. The following result sometimes allows us to calculate a lower bound on such a probability.

Proposition 6.1 – Chebyshev's inequality

Let X be a random variable with finite expected value μ. If c is a real constant such that $\mathbb{E}\left\{[X-c]^2\right\}$ is finite, then

$$P\{|X-c| < \varepsilon\} \geq 1 - \frac{1}{\varepsilon^2}\mathbb{E}\left\{[X-c]^2\right\}, \qquad \forall \varepsilon > 0.$$

In particular, if X has finite variance $\sigma^2 = \mathbb{E}\left\{[X-\mu]^2\right\}$, then

$$P\{|X-\mu| < \varepsilon\} \geq 1 - \frac{\sigma^2}{\varepsilon^2}, \qquad \forall \varepsilon > 0.$$

Proof. This is easily proved when X is a continuous random variable with probability density function $f_X(x)$. In this case,

$$
\begin{aligned}
\mathbb{E}\left\{[X-c]^2\right\} &= \int_{-\infty}^{\infty} (x-c)^2 f_X(x)\,\mathrm{d}x \\
&\geq \int_{-\infty}^{c-\varepsilon} (x-c)^2 f_X(x)\,\mathrm{d}x + \int_{c+\varepsilon}^{\infty} (x-c)^2 f_X(x)\,\mathrm{d}x \\
&\geq \int_{-\infty}^{c-\varepsilon} \varepsilon^2 f_X(x)\,\mathrm{d}x + \int_{c+\varepsilon}^{\infty} \varepsilon^2 f_X(x)\,\mathrm{d}x \\
&= \varepsilon^2 \left\{ \int_{-\infty}^{c-\varepsilon} f_X(x)\,\mathrm{d}x + \int_{c+\varepsilon}^{\infty} f_X(x)\,\mathrm{d}x \right\} \\
&= \varepsilon^2 \left[1 - P\{|X-c| < \varepsilon\}\right] \\
\therefore P\{|X-c| < \varepsilon\} &\geq 1 - \frac{1}{\varepsilon^2}\mathbb{E}\left\{[X-c]^2\right\}.
\end{aligned}
$$

The second part of the theorem follows immediately from the first, putting $c = \mu$. ∎

Example 6.3 – continued

We would like to find an approximation to $P(S_{14} < 1,000)$, knowing only that $\mu = \mathbb{E}(S_{14}) = 14 \times 70 = 980$ and $\sigma^2 = \mathrm{var}(S_{14}) = 14 \times 36 = 504$. The probability of interest can be written in the form

$$P\left\{|S_{14} - 500| < 500\right\}.$$

Using Chebyshev's inequality with $c = 500$ and $\varepsilon = 500$,

$$P\left\{|S_{14} - 500| < 500\right\} \geq 1 - \frac{1}{500^2}\mathbb{E}\left\{[S_{14} - 500]^2\right\}.$$

Now

$$\mathbb{E}\left\{[S_{14} - 500]^2\right\} = \mathbb{E}\left\{[(S_{14} - 980) + 480]^2\right\}$$

$$= \mathbb{E}\left\{[(S_{14} - 980)^2 + 960(S_{14} - 980) + 480^2]\right\}$$

$$= \mathbb{E}\left\{[S_{14} - 980]^2\right\} + 960\mathbb{E}\left\{[S_{14} - 980]\right\} + 480^2$$

$$= \text{var}(S_{14}) + 0 + 480^2 = 230,904.$$

So

$$P\left\{|S_{14} - 500| < 500\right\} \geq 1 - \frac{1}{500^2} 230,904 = 0.076.$$

This is not a very helpful result, since the lower bound is too low to be useful in practice. This example illustrates how little information we often have about probabilities when the distribution function of a random variable is not fully known. The central limit theorem, which is a much more powerful tool for obtaining approximate probabilities in these circumstances, is discussed in the next section. In the meantime, we shall look at some important special cases where the distribution of a sum of random variables may be determined exactly.

Although it is useful to be able to determine the expected value and variance of a sum of random variables and calculate lower bounds on some probabilities associated with the sum, we would have a much better understanding if we could obtain the probability mass function or probability density function of the sum. There are many important examples for which we may do this, using moment-generating functions.

Proposition 5.16 tells us that, when $X_1, X_2, ..., X_n$ are independent random variables, then the moment-generating function of $S = X_1 + X_2 + \cdots + X_n$ is

$$M_S(t) = \prod_{i=1}^{n} M_i(t).$$

This fundamental result allows us to obtain a variety of important results, known as reproductive properties, concerning the sum of independent random variables drawn from the same family of distributions.

Example 6.4

Suppose that $X_1, X_2, ..., X_n$ are independent random variables and that $X_i \sim \text{Po}(\theta_i)$. The moment-generating function of the Poisson distribution was obtained in Chapter 2; the moment-generating function of X_i is

$$M_i(t) = \exp\left\{\theta_i(e^t - 1)\right\}.$$

If $S = X_1 + X_2 + \cdots + X_n$, then the independence of the sequence of random variables means that the moment-generating function of S may be found as follows:

$$M_S(t) = \prod_{i=1}^{n} M_i(t) = \prod_{i=1}^{n} \exp\{\theta_i (e^t - 1)\} = \exp\left\{(e^t - 1) \sum_{i=1}^{n} \theta_i\right\}.$$

This is the moment-generating function of a Poisson distribution with expected value $\sum_{i=1}^{n} \theta_i$. Using the uniqueness property of moment-generating functions, Proposition 2.5, we conclude that S follows a Poisson distribution with expected value $\sum_{i=1}^{n} \theta_i$.

Notice that the first part of Example 6.1 was a particular example of this result when all the θ_i were equal.

Example 6.5

Suppose that $X_1, X_2, \ldots, X_n$ are independent random variables and that $X_i \sim N(\mu_i, \sigma_i^2)$. The moment-generating function of the normal distribution was obtained in Chapter 2; using this result, we find that the moment-generating function of X_i is

$$M_i(t) = \exp\left\{\mu_i t + \tfrac{1}{2}\sigma_i^2 t^2\right\}.$$

Let $S = a_0 + a_1 X_1 + a_2 X_2 + \cdots + a_n X_n$, where $a_0, a_1, a_2, \ldots, a_n$ are real constants and not all of $a_1, a_2, \ldots, a_n$ are zero. Then (using Proposition 2.6 along with Proposition 5.16) the independence of the sequence of random variables means that the moment-generating function of S is

$$M_S(t) = e^{a_0 t} \prod_{i=1}^{n} M_i(a_i t)$$

$$= e^{a_0 t} \prod_{i=1}^{n} \exp\left\{\mu_i a_i t + \tfrac{1}{2}\sigma_i^2 a_i^2 t^2\right\}$$

$$= \exp\left\{(a_0 + a_1\mu_1 + \cdots + a_n\mu_n) t + \tfrac{1}{2}\left(a_1^2 \sigma_1^2 + \cdots + a_n^2 \sigma_n^2\right) t^2\right\}.$$

This is the moment-generating function of the normal distribution with expected value $a_0 + a_1\mu_1 + \cdots + a_n\mu_n$ and variance $a_1^2 \sigma_1^2 + \cdots + a_n^2 \sigma_n^2$. Using the uniqueness property of moment-generating functions, Proposition 2.5, $S \sim N(a_0 + a_1\mu_1 + \cdots + a_n\mu_n, a_1^2 \sigma_1^2 + \cdots + a_n^2 \sigma_n^2)$.

Now suppose that $X_1, X_2, \ldots, X_n$ are identically distributed, as well as independent. So, $\mu_1 = \mu_2 = \cdots = \mu_n = \mu$ and $\sigma_1 = \sigma_2 = \cdots = \sigma_n = \sigma$, for some real value μ and $\sigma > 0$. Consider the sum of the random variables, $S = X_1 + X_2 + \cdots + X_n$. Setting $a_0 = 0$ and $a_1 = a_2 = \cdots = a_n = 1$, then the general result we have just proved shows that $S \sim N(n\mu, n\sigma^2)$.

Consider next the sample mean (or average) of the random variables,

$$\overline{X} = \frac{1}{n} (X_1 + \cdots + X_n).$$

Setting $a_0 = 0$ once more, but $a_1 = a_2 = \cdots = a_n = 1/n$, then the above general result shows that $\overline{X} \sim \mathrm{N}\left(\mu, \sigma^2\right)$.

If the random vector $\mathbf{X} = [X_1 X_2 \ldots X_n]^{\mathrm{T}}$ consists of random variables that are independent and all normally distributed, then $\mathbf{X}$ itself follows a probability distribution of the multivariate normal type. The above result is just one example of the general result, to be proved in Chapter 8, that any linear function of a multivariate normal random vector follows a multivariate normal distribution.

Exercises

1 Suppose that $X_1, X_2, \ldots, X_n$ are independent random variables, with each $X_i \sim \mathrm{Po}(\theta t_i)$, for some $t_i > 0$ and $\theta > 0$. Find the expected value and variance of

(a) $\dfrac{1}{n} \displaystyle\sum_{i=1}^{n} \dfrac{X_i}{t_i}$ 　　　(b) $\dfrac{\sum_{i=1}^{n} X_i}{\sum_{i=1}^{n} t_i}$.

2 A sequence of independent trials is to be conducted. On each trial the probability of a success will be θ. Let the random variable X_1 be the number of trials that have to be carried out until the first success is recorded, so that $X_1 \sim \mathrm{Ge}(\theta)$. For $i = 2, 3, \ldots$, let the random variable X_i be the number of trials that have to be carried out after the $(i-1)$th success until the ith success is recorded. Explain why $X_i \sim \mathrm{Ge}(\theta)$ for $i = 2, 3, \ldots$, and why all the random variables $X_1, X_2, \ldots$ are independent.

The random variable $S_k = X_1 + X_2 + \cdots + X_k$ is the total number of trials that have to be carried out until the kth success is recorded. Find the expected value and variance of S_k. Using your background knowledge of the experiment, decide what standard probability distribution S_k must follow, and hence check your results.

3 Suppose that $S = X_1 + X_2 + \cdots + X_N$, where N is a discrete random variable and the random variables $X_1, X_2, \ldots, X_N$ are independent and identically distributed with expected value $\mathbb{E}(X)$ and variance $\mathrm{var}(X)$. Show that

$$\mathbb{E}(S) = \mathbb{E}(X)\mathbb{E}(N),$$

$$\mathrm{var}(S) = \mathrm{var}(X)\mathbb{E}(N) + \left\{\mathbb{E}(X)\right\}^2 \mathrm{var}(N).$$

4 Suppose that $X_1, X_2, \ldots, X_n$ are independent random variables and that $X_i \sim \mathrm{Po}(\theta_i)$. In Example 6.4, it was shown that $S = X_1 + X_2 + \cdots + X_n$ also follows a Poisson distribution. Show that the conditional distribution of $X_1, X_2, \ldots, X_{n-1}$ given $S = s$ is a multinomial distribution and write down its parameters.

5 Suppose that the number of messages that arrives at a communications channel in a time interval of one minute is a random variable, $N \sim \mathrm{Po}(\lambda)$.

The number of characters in a randomly selected message is a $Ge(\theta)$ random variable and different messages have sizes that are stochastically independent. Find the expected value and variance of the total number of characters in all the messages that arrive in one minute.

6 The binary random variable, X, has the range space $\{-1, 1\}$ and is equally likely to take these two values. Find the expected value and variance of X. Find $P\{|X| \geq k\}$, for $k > 0$, and plot this value as a function of k. Now use Chebyshev's inequality to establish an upper bound for the same probability. Superimpose this upper bound on the same plot, and comment.

7 On a certain stretch of road, the number of road accidents involving a pedestrian that occur during a randomly selected day is a $Po(\lambda)$ random variable. Suppose that, in any of these accidents, independently of whatever happens in any other accident, the pedestrian dies with probability θ. (It may be assumed that a single road accident involves no more than one pedestrian.)

Let the random variable N be the total number of road accidents involving pedestrians and S the total number of pedestrian deaths in these accidents. For $s = 0, 1, \ldots$ and $n = 0, 1, \ldots$, find $P(S = s | N = n)$. Hence find $P(S = s)$ and deduce that S follows a $Po(\lambda\theta)$ distribution.

8 The radius (cm) of a randomly selected wheel produced by a certain manufacturing process is an $N(1, 0.0001)$ random variable. These wheels are produced independently then paired as they come off the production line. A pair is satisfactory if the radii of the two wheels differ by less than 0.03 cm.

(a) Find the probability that a pair of wheels is not satisfactory.

(b) On average, how many pairs of wheels must be manufactured in order to obtain four satisfactory pairs?

9 Suppose that $X_1, X_2, \ldots, X_n$ are independent random variables with each $X_i \sim Bi(m_i, \theta)$ for $0 < \theta < 1$. Show that $S_n = X_1 + X_2 + \cdots + X_n$ also follows a binomial distribution.

10 Suppose that $X_1, X_2, \ldots, X_n$ are independent and identically distributed random variables, each following the $Ex(\theta)$ distribution for some $\theta > 0$.

(a) Show that $\sum_{i=1}^{n} X_i \sim Ga(n, \theta)$.

(b) What is the distribution of $Y = 2n\theta\bar{X}$?

11 Suppose that $X_1, \ldots, X_n$ are independent random variables, with each $X_i \sim Ga(\alpha_i, \theta)$ for $\alpha_i > 0$ and $\theta > 0$. Find the distribution of $\sum_{i=1}^{n} X_i$ and of $\bar{X}$.

12 The binary random variables, $X_1, X_2, \ldots, X_n$ are independent and identically distributed, each with probability distribution:

$$P(X = -1) = 1 - \theta, \qquad P(X = +1) = \theta$$

(where $0 < \theta < 1$). Let S_n be the sum of these n random variables. Derive the moment-generating function of S_n. Check your answer by considering the relationship between S_n and a Bi(n, θ) random variable.

6.2 Limit theorems and convergence

In the previous section, we saw how to determine the probability distribution of the sum of some sequences of random variables. We were then able to calculate exact probabilities associated with the sum or average. Limit theorems allow us to find approximations to probabilities of interest, even when we cannot write down the probability distribution of the sum explicitly.

Proposition 6.2 – The weak law of large numbers

Let $X_1, X_2, \ldots,$ be a sequence of independent and identically distributed random variables, each with finite expected value μ. Let

$$\overline{X}_n = \frac{1}{n} \sum_{i=1}^{n} X_i.$$

Then, for every real constant $\varepsilon > 0$,

$$P\left\{ \left| \overline{X}_n - \mu \right| < \varepsilon \right\} \to 1 \qquad \text{as } n \to \infty.$$

Proof. This result is easily proved for the case in which the random variables have a finite common variance, σ^2 (though this restriction is not required for the weak law of large numbers to hold).

For all $n \geq 1$, $\overline{X}_n$ has expected value μ and variance σ^2/n. By Chebyshev's inequality (Proposition 6.1), for any $\varepsilon > 0$,

$$P\left\{ \left| \overline{X}_n - \mu \right| < \varepsilon \right\} \geq 1 - \frac{\sigma^2/n}{\varepsilon^2} \to 1 \qquad \text{as } n \to \infty. \qquad \blacksquare$$

The weak law describes convergence in probability, sometimes written

$$\operatorname{p\,lim}\left(\overline{X}_n \right) = \mu.$$

Informally, it tells us that the probability distribution of $\overline{X}_n$ is ultimately very highly concentrated around the value μ, but it does not tell us anything about the process which leads to that concentration.

Proposition 6.3 – The strong law of large numbers

Let $X_1, X_2, \ldots,$ be a sequence of independent and identically distributed random variables, each with finite expected value μ. Let $\overline{X}_n = \frac{1}{n} \sum_{i=1}^{n} X_i$. Then, for all real constants $\varepsilon > 0$ and $\delta > 0$, there is an integer N such that

$$P\left\{ \left| \overline{X}_n - \mu \right| < \varepsilon \right\} > 1 - \delta, \qquad n = N, N+1, \ldots.$$

The strong law describes what is known as 'convergence with probability 1'. For a proof of it, see Feller (1968). In contrast with the weak law, we now know more about the process by which the distribution of the mean becomes concentrated at the value μ. We know that, for all $n \geq N$, there is a very high probability that $\overline{X}_n$ is within a distance ε of μ.

The laws of large numbers are very interesting theoretical results, but they do not help us calculate probabilities of interest associated with $\overline{X}_n$ since they do not tell us how close $\overline{X}_n$ is to μ for a given value of n. We shall next consider the central limit theorem, which does just that, although it is restricted to the case of distributions with finite variance.

Proposition 6.4 – The central limit theorem

Suppose that $X_1, \ldots, X_n$ is a sequence of independent and identically distributed random variables, each with finite expected value μ and finite variance σ^2. For sufficiently large values of n, the random variable

$$Z_n = \frac{\sum_{i=1}^{n} X_i - n\mu}{\sqrt{n\sigma^2}}$$

approximately follows the $N(0, 1)$ distribution, in the sense that

$$\lim_{n \to \infty} P\{Z_n \leq z\} \approx \Phi(z), \quad \forall z.$$

Proof. We will prove this result when the moment-generating function of the random variables exists. This is not a requirement for the validity of the central limit theorem, but makes the proof easier.

Since $X_1, \ldots, X_n$ follow the same probability distribution, then they have a common moment-generating function $M_X(t)$. Expanding $M_X(t)$ in a Maclaurin series around the value $t = 0$ gives

$$M_X(t) = M_X(0) + \frac{M_X'(0)}{1!} t + \frac{M_X''(0)}{2!} t^2 + \cdots$$

$$= 1 + \frac{\mathbb{E}(X)}{1!} t + \frac{\mathbb{E}(X^2)}{2!} t^2 + \cdots$$

$$= 1 + \mu t + \frac{(\mu^2 + \sigma^2)}{2} t^2 + \cdots.$$

Now

$$Z_n = \sum_{i=1}^{n} \frac{1}{\sqrt{n}\sigma} X_i - \frac{\sqrt{n}\mu}{\sigma}.$$

Using Proposition 2.6, this implies that Z_n has moment-generating function

$$M_Z(t) = \left\{ M_X \left(\frac{t}{\sqrt{n}\sigma} \right) \right\}^n \cdot \exp\left(-\frac{\sqrt{n}\mu}{\sigma} t \right).$$

Taking the logarithm of both sides gives

$$\log_e [M_Z(t)] = n \log_e \left\{ M_X \left(\frac{t}{\sqrt{n}\sigma} \right) \right\} - \frac{\sqrt{n}\mu}{\sigma} t$$

$$= n \log_e \left\{ 1 + \mu \frac{t}{\sqrt{n}\sigma} + (\mu^2 + \sigma^2) \frac{t^2}{2n\sigma^2} + \cdots \right\} - \frac{\sqrt{n}\mu}{\sigma} t.$$

The Maclaurin expansion of $\log_e(1 + u)$, when $-1 < u < 1$, is

$$\log_e (1 + u) = u - \frac{u^2}{2} + \frac{u^3}{3} - \cdots.$$

For sufficiently large values of n,

$$u = \mu \frac{t}{\sqrt{n}\sigma} + (\mu^2 + \sigma^2) \frac{t^2}{2n\sigma^2} + \cdots < 1,$$

so

$$\log_e [M_Z(t)] = n \left[\mu \frac{t}{\sqrt{n}\sigma} + (\mu^2 + \sigma^2) \frac{t^2}{2n\sigma^2} + \cdots \right]$$

$$- n \frac{1}{2} \left[\mu \frac{t}{\sqrt{n}\sigma} + (\mu^2 + \sigma^2) \frac{t^2}{2n\sigma^2} + \cdots \right]^2 + \cdots - \frac{\sqrt{n}\mu}{\sigma} t$$

$$= n\mu \frac{t}{\sqrt{n}\sigma} + n (\mu^2 + \sigma^2) \frac{t^2}{2n\sigma^2} - \frac{1}{2} n \left(\mu \frac{t}{\sqrt{n}\sigma} \right)^2 + \cdots - \frac{\sqrt{n}\mu}{\sigma} t$$

$$= \frac{1}{2} t^2 + \text{(terms in negative powers of } n)$$

$$\rightarrow \frac{1}{2} t^2 \qquad \text{as} \quad n \rightarrow \infty,$$

i.e.

$$\lim_{n \to \infty} \log_e [M_Z(t)] = \frac{1}{2} t^2$$

$$\therefore \lim_{n \to \infty} [M_Z(t)] = \exp\left(\frac{1}{2}t^2\right).$$

This is the moment-generating function of the $N(0,1)$ distribution. Using the uniqueness property of moment-generating functions, this means that Z_n converges in distribution to the $N(0,1)$ distribution as $n \to \infty$. ∎

The central limit theorem is often used in one of two equivalent forms:

1. $\sum_{i=1}^{n} X_i$ approximately follows the $N(n\mu, n\sigma^2)$ distribution for 'large' n.

2. $\overline{X}$ approximately follows the $N(\mu, \sigma^2/n)$ distribution for 'large' n.

Note that we have already derived these expected values and variances for the sum and average of independent and identically distributed random variables. The central limit theorem tells us about the approximate normality of their distribution for 'large' n. Again, we have already shown that, in the special case where $X_1, X_2, \ldots, X_n$ are normally distributed, then these are exact (not approximate) results (Example 6.3 above).

The central limit theorem is an extremely powerful result, since it requires so few assumptions to be made about the probability distribution of the X_i. What values of n are 'sufficiently large' to make the normal approximation useful is a question that can be investigated in any particular case using simulation. Generally speaking, the more symmetric the distribution, the smaller the value of n required in order to obtain a good normal approximation.

Example 6.3 – continued

We are considering a new design of lift that can safely carry loads up to 1,000 kg. The distribution of weights of adults in the population has expected value 70 kg and standard deviation 6 kg. A random group of n adults, with individual weights $X_1, X_2, \ldots, X_n$, has total weight

$$S_n = X_1 + X_2 + \cdots + X_n.$$

S_n has expected value 70n kg and standard deviation $6\sqrt{n}$ kg (since the variance of S is 36n). For example, 14 is the largest number of adults whose total weight is less than 1,000 kg on average, since S_{14} has expected value 980. But S_{14} has variance 504, so there is a good chance that a randomly selected group of 14 adults would exceed the capacity of the lift. Using the central limit theorem, we can obtain an approximation to this probability:

$$P(S_{14} > 1,000) = P\left(\frac{S_{14} - 980}{\sqrt{504}} > \frac{1,000 - 980}{\sqrt{504}}\right)$$
$$= P(Z > 0.89) \qquad (\text{where } Z \sim N(0,1))$$

$$= 1 - \Phi(0.89)$$
$$= 1 - 0.8133$$
$$= 0.1867$$

This means that a random group of 14 adults would exceed the safe capacity of the lift almost 20% of the time. Clearly, the lift could not be advertised as safe for carrying 14 adults.

However, we might be able to use the central limit theorem in order to find a value n such that S_n is very unlikely to exceed 1,000. This value of n will be less than 14, as we have already seen, so the central limit theorem might not give a very good approximation to the probability of interest unless the parent distribution of adult weights is reasonably symmetric. Continuing anyway, we might wish to find n such that

$$P(S_n > 1,000) < 0.0001$$

$$\Longleftrightarrow \quad P\left(\frac{S_n - 70n}{6\sqrt{n}} > \frac{1,000 - 70n}{6\sqrt{n}}\right) < 0.0001$$

$$\Longleftrightarrow \quad P\left(Z > \frac{1,000 - 70n}{6\sqrt{n}}\right) < 0.0001$$

$$\Longleftrightarrow \quad 1 - \Phi\left(\frac{1,000 - 70n}{6\sqrt{n}}\right) < 0.0001$$

$$\Longleftrightarrow \quad \frac{1,000 - 70n}{6\sqrt{n}} > \Phi^{-1}(0.9999) = 3.62$$

$$\Longleftrightarrow \quad n + 0.31\sqrt{n} - 14.29 < 0 \qquad \text{(after some tedious arithmetic)}.$$

This is a quadratic equation in $\sqrt{n}$. We require $\sqrt{n}$ to be less than the larger of the two roots of the equation, i.e.

$$\sqrt{n} < \frac{-0.31 + \sqrt{0.31^2 + 4 \times 14.29}}{2} = 3.62$$

$$\therefore n < 13.16.$$

Therefore, if we want to ensure that the capacity of the lift is only exceeded once in 10,000 times it is used by the largest number of adults we allow, we should advertise that it is safe for use by $n = 13$ adults.

Example 6.6

The general proof of the central limit theorem can be adapted for specific cases when the moment-generating function exists. For example, suppose that $X_1, X_2, \ldots, X_n$ are independently distributed, each following the $\text{Ex}(\theta)$

distribution. Then the standardized random variable

$$Z = \frac{\theta}{\sqrt{n}} (X_1 + \cdots + X_n) - \sqrt{n}$$

has moment-generating function

$$M_Z(t) = e^{-t\sqrt{n}} \left\{ M_X \left(\frac{\theta t}{\sqrt{n}} \right) \right\}^n$$

$$= e^{-t\sqrt{n}} \left\{ \frac{1}{1 - t/\sqrt{n}} \right\}^n$$

$$= e^{-t\sqrt{n}} \left\{ 1 + (-t/\sqrt{n}) \right\}^n$$

Taking logarithms,

$$\log_e M_Z(t) = -t\sqrt{n} - n \log_e \left\{ 1 + (-t/\sqrt{n}) \right\}$$

$$= -t\sqrt{n} - n \left\{ -\frac{t}{\sqrt{n}} - \frac{1}{2} \left(\frac{t}{\sqrt{n}} \right)^2 - \frac{1}{3} \left(\frac{t}{\sqrt{n}} \right)^3 - \cdots \right\}$$

$$= \frac{1}{2} (t^2) + \frac{1}{3} \left(\frac{t^3}{\sqrt{n}} \right) + \cdots$$

$$\rightarrow \frac{1}{2} t^2 \quad \text{as} \quad n \rightarrow \infty.$$

Therefore,

$$M_Z \rightarrow \exp \left\{ \frac{1}{2} t^2 \right\} \quad \text{as} \quad n \rightarrow \infty.$$

Since this limit is the moment-generating function of a standard normal random variable, it follows that, in the limit, Z follows a standard normal distribution.

The central limit theorem only applies when the parent distribution has a finite expected value and variance. A well-known example where the sum or average never becomes approximately normal is the Cauchy distribution, which has probability density function $f_X(x) = 1/\{\pi(1 + x^2)\}$. This distribution does not have a finite expected value. The average of independent Cauchy random variables is always a Cauchy random variable.

The normal distribution is commonly used to evaluate approximate 'tail' probabilities for the binomial distribution when the sample size, n, is large. This approximation is justified by the central limit theorem, using the following argument.

Suppose $X \sim \text{Bi}(n, \theta)$. Then X is the number of 'successes' in n independent trials, where each trial has 'success' probability θ. Let X_i be the number of 'successes' in the ith trial $(i = 1, \ldots, n)$. Then X_i takes the value 0 or 1, with $P(X_i = 0) = 1 - \theta$ and $P(X_i = 1) = \theta$. This means that $X_1, \ldots, X_n$ are independent and identically distributed random variables, each with expected value θ

and variance $\theta(1-\theta)$. By definition, $X = \sum_{i=1}^{n} X_i$. So, applying the central limit theorem leads to the following approximation.

Proposition 6.5

Suppose that the random variable X follows the $\text{Bi}(n, \theta)$ distribution, where $n \geq 20, n\theta \geq 5$ and $n(1-\theta) \geq 5$. Then the random variable $(X - n\theta)/\sqrt{n\theta(1-\theta)}$ approximately follows the standard normal distribution. X itself approximately follows the $\text{N}(n\theta, n\theta(1-\theta))$ distribution.

A similar argument justifies the following normal approximation to the Poisson distribution.

Proposition 6.6

Suppose that the random variable X follows the $\text{Po}(\theta)$ distribution, where $\theta \geq 30$. Then the random variable $(X - \theta)/\sqrt{\theta}$ approximately follows the standard normal distribution. X itself approximately follows the $\text{N}(\theta, \theta)$ distribution.

Both these results require a discrete distribution to be approximated by a continuous one. This introduces some inconsistencies, since the probability that a normal random variable equals any particular value is 0. Using a *continuity correction* improves the approximation. For integer values, m, correcting for continuity means evaluating:

$$P(m - \tfrac{1}{2} < X < m + \tfrac{1}{2}) \qquad \text{rather than } P(X = m);$$
$$P(X < m + \tfrac{1}{2}) \qquad \text{rather than } P(X \leq m) \text{ or } P(X < m+1);$$
$$P(X > m - \tfrac{1}{2}) \qquad \text{rather than } P(X \geq m) \text{ or } P(X > m - 1).$$

Example 6.7

A random digit generator is to be used to obtain $100k$ random digits, for some positive integer k. Let the random variable X be the number of zeros generated. Then $X \sim \text{Bi}(100k, 0.1)$. Roughly one-tenth of all the digits generated should be zeros, i.e. X should take a value close to $10k$ on average. Let us find the probability that $9k < X < 11k$, i.e. that the proportion of zeros lies between 0.09 and 0.11.

Using Proposition 6.5, $(X - 10k)/3\sqrt{k}$ is approximately distributed as $\text{N}(0, 1)$. So,

$$P\left\{9k + \frac{1}{2} < X < 11k - \frac{1}{2}\right\} = P\left\{\frac{1-2k}{6\sqrt{k}} < \frac{X - 10k}{3\sqrt{k}} < \frac{2k-1}{6\sqrt{k}}\right\}$$
$$= 2\Phi\left(\frac{2k-1}{6\sqrt{k}}\right) - 1.$$

As the sample size increases, i.e. as $k \to \infty$, $(2k - 1)/6\sqrt{k} \to \infty$, the probability that the proportion of zeros lies between 0.09 and 0.11 tends to 1. This probability is strictly increasing with k. This is what we would expect from the laws of large numbers. Use of the central limit theorem provides further insight into the rate of convergence. For example, the probability exceeds 0.9 for sample sizes of at least 26,000 ($k = 26$), exceeds 0.99 for sample sizes of at least 61,000 and exceeds 0.999 for sample sizes of at least 99,000 random digits.

Exercises

1 Suppose that the discrete random variable, X, follows the Poisson distribution with mean $\mu (\mu > 0)$. Find the moment-generating function of

$$W = \frac{X - \mu}{\sqrt{\mu}}.$$

Find the limiting form of this moment-generating function as $\mu \to \infty$. [*Hint*: consider taking the limit of the logarithm of the moment-generating function.] By recognizing the limiting moment-generating function, name the limiting distribution of W.

2 Using a continuity correction, find the approximate value of each of the following probabilities.

(a) $P(T > 95)$ where $T \sim \text{Bi}(100, 0.9)$;

(b) $P(30 \leq X < 40)$ where $X \sim \text{Bi}(50, 0.8)$;

(c) $P(Y \geq 50)$ where $Y \sim \text{Po}(40)$;

(d) $P(Z_1 + Z_2 \geq 100)$ where Z_1, Z_2 are independent $\text{Po}(40)$ random variables.

3 A certain car manufacturer's research and development team has designed a new gearbox. The team must now test ten prototype gearboxes on the firm's accelerated test facility. These gearboxes are to be tested consecutively, one at a time, until failure. Once a gearbox fails, its place on the test-bed is taken immediately by another gearbox. It is believed that, for this new design of gearbox, time to failure on this test-bed is an exponential distribution with expected value 10 hours.

The team wishes to know the shortest period for which they should book the test facility while still ensuring with probability at least 0.99 that the ten tests will be completed within the allotted time. Find an approximate answer using the central limit theorem.

4 A research worker intends to draw a sample from a large population and estimate the population mean by the sample mean. She wishes to guarantee a probability of at least 0.99 that the sample mean differs from the population

mean by no more than 10% of the population standard deviation. Use the central limit theorem to find (approximately) the smallest sample size she should use.

5 (a) An electronic system consists of 100 identical components which all function independently. The probability that any given component will fail between scheduled maintenance periods is 0.10. The system fails unless at least 90 components continue to function. Find the probability that the system fails between scheduled maintenance periods.

(b) The system may be modified so that it consists of a larger number, N, of the same components. Again the system fails unless at least 90 components continue to function. Approximately what value of N is required to reduce the probability of a system failure between maintenance periods to 0.05?

6 About 10% of bookings on scheduled airline flights are cancelled too late for airlines to reallocate the seats to other intending passengers. To reduce the revenue loss incurred in this way, airlines slightly overbook flights in advance. They hope that, after late cancellations, the number of passengers checking in for a flight will not exceed the number of available seats.

There are 200 seats available on a certain flight. Determine the largest number of tickets that the airline may issue in advance while still ensuring with probability at least 0.95 that 200 or fewer passengers check in for this flight. You may assume that all passengers booking to travel on the flight are making independent travel plans.

7 Explain how the central limit theorem justifies the use of a normal approximation to the $\text{NeBi}(k,\theta)$ distribution, for large enough values of k. (You might find it helpful to refer back to Exercise 2 on Section 6.1.) Write down the parameters of the approximate normal distribution.

8 Consider a general binary random variable, X, that takes the value c_1 with probability θ and $c_2 (>c_1)$ with probability $1 - \theta$. Find a normal approximation to the distribution of $S = X_1 + X_2 + \cdots + X_n$, where the X_i are independent and identically distributed random variables following this binary distribution. By considering the relationship between S and a binomial random variable, suggest conditions on n and θ that would ensure a good approximation by the appropriate normal distribution.

9 (a) Suppose that $X_1, X_2, \ldots, X_n$ are independent random variables, with each $X_i \sim \chi_1^2$. Use the moment-generating functions to prove that $\sum_{i=1}^{n} X_i \sim \chi_n^2$. Hence, justify a normal approximation to the χ_n^2 distribution for large n.

(b) If $Z_1, Z_2, \ldots, Z_n$ are independent random variables, with each $Z_i \sim N(0,1)$, write down the exact distribution of $\sum_{i=1}^{n} Z_i^2$ and an approximation to this distribution for large n.

Summary

In this chapter, we have looked at properties of sums and averages of independent random variables. We have used the results obtained for general multivariate distributions in order to find expressions for the expected value and variance, and the moment-generating function of sums and averages. For many cases that are important in practice, this has allowed us to obtain the actual probability distribution of the sum or average. In general, obtaining probabilities relating to sums and averages requires use of the central limit theorem which provides a means of finding a normal distribution that approximates the distribution of the sum. Generalizing the use of the central limit theorem has allowed us to write down and use normal approximations even to discrete distributions, particularly the binomial and Poisson. The use of the continuity correction in such cases has been discussed.

7
Functions of a random vector

It is often of interest to investigate a new random variable that is defined by a function of several base random variables whose joint distribution we know. We have already seen (Chapter 5) how to use the joint moment-generating function to do this for a linear function of random variables in some simple cases. In Chapter 6, we looked at sums and averages of independent random variables. Now we will consider the general case; we will not require the original variables to be independent or the functions to be linear. For example, if X_1, X_2 and X_3 are random variables measuring the speed of a particle in each of three orthogonal directions, then there might be particular interest in probabilities associated with the total speed, $Y = \sqrt{X_1^2 + X_2^2 + X_3^2}$. We will often find it useful to obtain the joint probability distribution of several new random variables. The general methods we introduce will also allow further methods of simulation to be explored.

7.1 Functions of discrete random variables

Suppose that $X_1, \ldots, X_m$ are discrete random variables, and that a new random variable is defined by $Y = g(X_1, \ldots, X_m)$, where $g(\cdot)$ is a single-valued real function of $X_1, \ldots, X_m$. For example, $g(X_1, X_2)$ could be $X_1 - X_2$ or $X_1 X_2$. Then Y must also be a discrete random variable (though its range space need not be a subset of the set of integers). In order to find the probability mass function of Y, working from the joint probability mass function of $(X_1, \ldots, X_m)$, we need to identify:

- the range space of Y, R_Y;
- the event $E_y = \{(x_1, \ldots, x_m): g(x_1, \ldots, x_m) = y\}$, for every $y \in R_Y$;
- the probability $p_Y(y) = P(Y = y) = P((x_1, \ldots, x_m) \in E_y) = \sum_{E_y} p_X(x_1, \ldots, x_m)$.

Example 7.1

Suppose that X_1 and X_2 have the following joint probability mass function:

		x_1		
$p_X(x_1, x_2)$		-1	0	1
x_2	-1	$\frac{1}{9}$	$\frac{1}{9}$	$\frac{1}{9}$
	0	$\frac{1}{9}$	$\frac{1}{9}$	$\frac{1}{9}$
	1	$\frac{1}{9}$	$\frac{1}{9}$	$\frac{1}{9}$

Define the new discrete random variable, Y, by $Y = X_1 X_2$. Then $R_Y = \{-1, 0, 1\}$.

$$E_{-1} = \{(-1, 1), (1, -1)\},$$

$$\therefore \quad p_Y(-1) = p_X(-1, 1) + p_X(1, -1) = \frac{2}{9};$$

$$E_0 = \{(-1, 0), (0, -1), (0, 0), (0, 1), (1, 0)\},$$

$$\therefore \quad p_Y(0) = p_X(-1, 0) + p_X(0, -1) + p_X(0, 0) + p_X(0, 1) + p_X(1, 0) = \frac{5}{9};$$

$$E_1 = \{(-1, -1), (1, 1)\},$$

$$\therefore \quad p_Y(-1) = p_X(-1, -1) + p_X(1, 1) = \frac{2}{9}.$$

The probability mass function of the discrete random variable Y is

y	−1	0	1
$p_Y(y)$	$\frac{2}{9}$	$\frac{5}{9}$	$\frac{2}{9}$

This method can be extended to find the joint probability mass function of a random vector $\mathbf{Y}$ that consists of several new random variables, all defined as single-valued, real functions of the base variables: $Y_1 = g_1(X_1, \ldots, X_m), \ldots, Y_q = g_q(X_1, \ldots, X_m)$. It is now necessary to determine, in turn,

- the joint range space of $\mathbf{Y}$, $R_{\mathbf{Y}}$;
- the event $E_{(y_1, \ldots, y_q)} = \{(x_1, \ldots, x_m) : g_1(x_1, \ldots, x_m) = y_1, \ldots, g_q(x_1, \ldots, x_m) = y_q\}$, for every point $\mathbf{y} = (y_1, \ldots, y_q) \in R_{\mathbf{Y}}$;
- the probability $p_{\mathbf{Y}}(\mathbf{y}) = P(Y_1 = y_1, \ldots, Y_q = y_q) = P((x_1, \ldots, x_m) \in E_{\mathbf{y}}) = \sum_{E_{\mathbf{y}}} p_{\mathbf{X}}(x_1, \ldots, x_m)$.

Example 7.1 – continued

Suppose we want to find the joint probability mass function of (Y_1, Y_2), where $Y_1 = X_1 X_2$ as before and $Y_2 = X_2^2$. In order to determine the events $E_{\mathbf{y}}$, it is helpful to draw up a table of the values of y_1 and y_2 for each pair (x_1, x_2) as follows:

		x_1		
(y_1, y_2)		−1	0	1
x_2	−1	$(1, 1)$	$(0, 1)$	$(-1, 1)$
	0	$(0, 0)$	$(0, 0)$	$(0, 0)$
	1	$(-1, 1)$	$(0, 1)$	$(1, 1)$

Therefore,

$$R_{\mathbf{Y}} = \{(-1, 1), (0, 0), (0, 1), (1, 1)\}$$

and

$$p_Y(-1,1) = p_X(-1,1) + p_X(1,-1) = \frac{2}{9},$$

$$p_Y(0,0) = p_X(-1,0) + p_X(0,0) + p_X(1,0) = \frac{3}{9},$$

$$p_Y(0,1) = p_X(0,-1) + p_X(0,1) = \frac{2}{9},$$

$$p_Y(1,1) = p_X(-1,-1) + p_X(1,1) = \frac{2}{9}.$$

The joint probability mass function of Y_1 and Y_2 is

		y_1		
$p_Y(y_1, y_2)$		-1	0	1
y_2	0	0	$3/9$	0
	1	$2/9$	$2/9$	$2/9$

Example 7.2

Suppose that (X_1, X_2) follow the trinomial distribution with joint probability mass function

$$p_X(x_1, x_2) = \frac{n!}{x_1! x_2! (n - x_1 - x_2)!} \theta_1^{x_1} \theta_2^{x_2} (1 - \theta_1 - \theta_2)^{n - x_1 - x_2},$$

$$x_1 = 0, 1, \ldots, n, x_2 = 0, 1, \ldots, n, x_1 + x_2 \le n$$

(where n is a positive integer and $0 \le \theta_1 \le 1, 0 \le \theta_2 \le 1, 0 \le \theta_1 + \theta_2 \le 1$). In Chapter 5, it was shown that X_1 and X_2 both marginally follow binomial distributions, with 'success' probabilities θ_1 and θ_2 respectively. We can now show that $Y = X_1 + X_2$ also follows a binomial distribution, with 'success' probability $\theta_1 + \theta_2$.

As a result of the restrictions on R_X, it is clear that $R_Y = \{0, 1, \ldots, n\}$. Let $y \in R_Y$. For Y to equal y, then (X_1, X_2) must take the values $(0, y)$ or $(1, y - 1)$ or ... or $(y, 0)$, i.e.

$$E_y = \{(0, y), (1, y - 1), \ldots, (y, 0)\}.$$

The diagram below shows E_2. Now

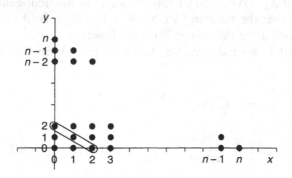

$$p_Y(y) = \sum_{(x_1, x_2) \in E_y} p_\mathbf{X}(x_1, x_2)$$

$$= \sum_{x_1=0}^{y} p_\mathbf{X}(x_1, y - x_1)$$

$$= \sum_{x_1=0}^{y} \frac{n!}{x_1!(y - x_1)!(n - x_1 - (y - x_1))!} \theta_1^{x_1} \theta_2^{y-x_1} (1 - \theta_1 - \theta_2)^{n - x_1 - (y - x_1)}$$

$$= \sum_{x_1=0}^{y} \frac{n!}{x_1!(y - x_1)!(n - y)!} \theta_1^{x_1} \theta_2^{y-x_1} (1 - \theta_1 - \theta_2)^{n-y}$$

$$= \frac{n!}{y!(n - y)!} (1 - \theta_1 - \theta_2)^{n-y} \sum_{x_1=0}^{y} \binom{y}{x} \theta_1^{x_1} \theta_2^{y-x_1}$$

$$= \binom{n}{y} (\theta_1 + \theta_2)^y (1 - \theta_1 - \theta_2)^{n-y} \qquad \text{(Result A1.8).}$$

In other words, $Y = X_1 + X_2 \sim \text{Bi}(n, \theta_1 + \theta_2)$.

This makes intuitive sense. X_1 is the number of type 1 items in the sample and X_2 is the number of type 2 items in the sample of size n. Each item sampled is type 1 or type 2 with probability $\theta_1 + \theta_2$. Hence the number of type 1 or type 2 items in the sample is a binomial random variable with 'success' probability $\theta_1 + \theta_2$.

When X_1 and X_2 are binomial random variables, $X_1 + X_2$ is not always a binomial random variable. This result does hold, however, in the special case when X_1 and X_2 are independent binomials with the same 'success' probability, θ, as Exercise 4 shows.

Example 7.3

Suppose that X_1 and X_2 are independent Poisson random variables, with $X_1 \sim \text{Po}(\theta)$ and $X_2 \sim \text{Po}(\lambda)$. In Chapter 6, we used moment-generating functions to prove that the random variable $Y = X_1 + X_2 \sim \text{Po}(\theta + \lambda)$. Here is an alternative proof, using the probability mass function.

Since X_1 and X_2 are independent, their joint probability mass function is

$$p_\mathbf{X}(x_1, x_2) = p(x_1)p(x_2) = \frac{e^{-\theta} \theta^{x_1}}{x_1!} \cdot \frac{e^{-\lambda} \lambda^{x_2}}{x_2!}$$

$$= \frac{e^{-(\theta + \lambda)} \theta^{x_1} \lambda^{x_2}}{x_1! x_2!}, \qquad x_1 = 0, 1, \ldots; \; x_2 = 0, 1, \ldots.$$

Letting $Y = X_1 + X_2$, then $R_Y = \{0, 1, \dots\}$. For $y \in R_Y$, $E_y = \{(0, y), (1, y - 1), \dots, (y, 0)\}$ and so

$$p_Y(y) = \sum_{x_1=0}^{y} p_{\mathbf{X}}(x_1, y - x_1)$$

$$= \sum_{x_1=0}^{y} \frac{e^{-(\theta+\lambda)} \theta^{x_1} \lambda^{y-x_1}}{x_1!(y-x_1)!}$$

$$= \frac{e^{-(\theta+\lambda)}}{y!} \sum_{x_1=0}^{y} \frac{y!}{x_1!(y-x_1)!} \theta^{x_1} \lambda^{y-x_1}$$

$$= \frac{e^{-(\theta+\lambda)}}{y!} \sum_{x_1=0}^{y} \binom{y}{x_1} \theta^{x_1} \lambda^{y-x_1}$$

$$= \frac{e^{-(\theta+\lambda)}}{y!} (\theta + \lambda)^y$$

by the binomial theorem (Result A1.8). So $Y = X_1 + X_2 \sim \mathrm{Po}(\theta + \lambda)$. This makes sense – if X_1 is the number of 'events' in one time interval and X_2 is the number of 'events' in a further, independent, time interval, then Y is the total number of 'events' in the two time intervals.

Suppose that $Y = X_1 + X_2$ is discovered to have the value y, for $y = 1, 2, \dots$. We can show that the conditional distribution of X_1, given that $Y = y$, is binomial.

Suppose that $Y = y$, for some positive integer y. Then X_1 must take values in the range $0, 1, \dots, y$. For any x_1 in this range,

$$P(X_1 = x_1 | Y = y)$$

$$= \frac{P(X_1 = x_1, Y = y)}{P(Y = y)}$$

$$= \frac{P(X_1 = x_1, X_2 = y - x_1)}{P(Y = y)}$$

$$= \frac{e^{-\theta} \theta^{x_1}}{x_1!} \cdot \frac{e^{-\lambda} \lambda^{y-x_1}}{(y-x_1)!} \Big/ \frac{e^{-(\theta+\lambda)} (\theta+\lambda)^y}{y!}$$

$$= \frac{y!}{x_1!(y-x_1)!} \cdot \frac{\theta^{x_1} \lambda^{y-x_1}}{(\theta+\lambda)^y}$$

$$= \binom{y}{x_1} \left(\frac{\theta}{\theta+\lambda}\right)^{x_1} \left(\frac{\lambda}{\theta+\lambda}\right)^{y-x_1}$$

$$= \binom{y}{x_1} \phi^{x_1} (1 - \phi)^{y-x_1},$$

where $\phi = \theta/(\theta + \lambda)$. So, conditional on the value of $Y = X_1 + X_2$, X_1 has a binomial distribution.

Exercises

1 The discrete random variables X_1 and X_2 have the joint probability mass function shown below.

| | | x_1 | | |
		-1	0	1
x_2	-1	0.05	0.10	0.05
	0	0.10	0.40	0.10
	1	0.05	0.10	0.05

For positive integers m and n, define $Y_{mn} = X_1^m X_2^n$.

(a) Find the probability mass function of (i) Y_{11}, (ii) Y_{21}, (iii) Y_{22}.

(b) Show that
$$E(Y_{mn}) = \begin{cases} 0.20, & \text{when } m \text{ and } n \text{ are both even,} \\ 0, & \text{otherwise.} \end{cases}$$

2 Suppose that X_1 and X_2 are independent binary random variables, each with probability θ of taking the value 1 and $1 - \theta$ of taking the value 0.

(a) Write down the joint probability mass function of (X_1, X_2).

(b) Determine the probability mass function of (i) $S = X_1 + X_2$, (ii) $T = X_1 - X_2$, (iii) $U = \min(X_1, X_2)$, (iv) $V = \max(X_1, X_2)$.

(c) Determine the joint probability mass function of U and V defined in part (b).

3 X_1 and X_2 are independent random digits.

(a) Write down the joint probability mass function of (X_1, X_2).

(b) Find the probability mass function of $S = X_1 + X_2$.

(c) Find the probability mass function of the mean, $M = \frac{1}{2}(X_1 + X_2)$.

(d) Compare the expected value and variance of M with those of the original random variables X_1 and X_2.

4 Suppose that X_1 and X_2 are independent random variables, with $X_1 \sim \text{Bi}(n, \theta)$ and $X_2 \sim \text{Bi}(m, \theta)$.

(a) Show that $Y = X_1 + X_2$ is a $\text{Bi}(n+m, \theta)$ random variable. Give a heuristic justification of this result.

(b) Suppose that Y is found to equal some fixed value y. Show that the conditional distribution of X_1 given Y is hypergeometric.

5 The discrete random variables X_1 and X_2 are independent and both follow the geometric distribution with 'success' probability θ $(0 < \theta < 1)$.

    (a) Show that $Y = X_1 + X_2$ is a negative binomial random variable. Give a heuristic justification of this result.

    (b) Suppose that Y is found to equal some fixed value y. Show that the conditional distribution of X_1 is a discrete uniform distribution.

6 The discrete random variables X_1 and X_2 independently follow $Ge(\theta)$ and $Ge(\phi)$ distributions. Show that $Y = \min(X_1, X_2)$ also follows a geometric distribution, with 'success' probability $1 - (1 - \theta)(1 - \phi)$. Give a heuristic justification of this result.

7.2 Functions of a continuous random vector

Sometimes we are interested in just one random variable that is defined by a continuous function of two continuous random variables. In this case, it is often best to start by deriving the distribution function of the new random variable, as in the following example.

Example 7.4

Copas and Heydari (1997) investigated patterns of reoffending by prisoners in England and Wales. They concluded that only a sub-population of prisoners will ever reoffend. In this population, the time to reconviction is the sum of two random variables: X_1, the time from the prisoner's release until he or she reoffends, and X_2, the time between the occurrence of the new offence and the offender's conviction (at the end of a criminal trial). According to Copas and Heydari, $X_1 \sim Ex(\theta)$ independently of $X_2 \sim Ex(\lambda)$.

We will now derive the probability density function of $T = X_1 + X_2$, which is the total time from release to reconviction. Assuming that $\theta > \lambda$, we shall find the distribution function of T.

Since X_1 and X_2 are independent, it follows that

$$f_X(x_1, x_2) = \theta\lambda \exp(-\theta x_1 - \lambda x_2), \qquad x_1 > 0, x_2 > 0,$$

and so (for $t > 0$)

$$F_T(t) = P(T \le t) = P(X_1 + X_2 \le t)$$

$$= \int_0^t \int_0^{t-x_1} \theta e^{-\theta x_1} \lambda e^{-\lambda x_2} \, dx_2 \, dx_1$$

$$= \int_0^t \theta e^{-\theta x_1} \int_0^{t-x_1} \lambda e^{-\lambda x_2} \, dx_2 \, dx_1$$

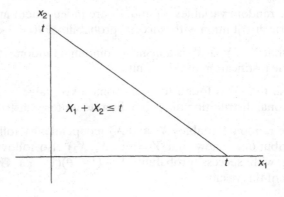

$$= \int_0^t \theta e^{-\theta x_1} \left(1 - e^{-\lambda(t-x_1)}\right) dx_1$$

$$= \int_0^t \theta e^{-\theta x_1} dx_1 - \theta e^{-\lambda t} \int_0^t e^{-(\theta-\lambda)x_1} dx_1$$

$$= 1 + \frac{\lambda}{\theta - \lambda} e^{-\theta t} - \frac{\theta}{\theta - \lambda} e^{-\lambda t}.$$

Differentiating once with respect to t, we obtain the probability density function of T:

$$f_T(t) = \frac{\lambda \theta}{\theta - \lambda} (e^{-\lambda t} - e^{-\theta t}), \qquad t > 0.$$

Notice that if $\theta = \lambda$, so that X and Y are identically distributed, then the above proof has to be amended (see Exercises) and T can be shown to follow a gamma distribution.

Example 7.5

In a certain post office there are two tellers serving a single queue of customers. As you join this queue, the tellers call the only two customers ahead of you, and begin to serve them. Assuming that the service time for a randomly selected customer in this post office is an $Ex(\theta)$ random variable, and that the service times for different customers are independent, find the distribution of Y, the time you will have to wait before being served.

Let X_1 and X_2 (respectively) denote the service times of the two customers who have just begun to be served. Then $X_1 \sim Ex(\theta)$ independently of $X_2 \sim Ex(\theta)$, and

$$f_1(x_1) = e^{-\theta x_1}, \qquad F_X(x) = 1 - e^{-\theta x_1},$$
$$f_2(x_2) = e^{-\theta x_2}, \qquad F_X(x) = 1 - e^{-\theta x_2}.$$

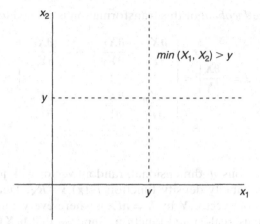

Now, $Y = \min(X_1, X_2)$. So

$$
\begin{aligned}
F_Y(y) &= P(Y \le y) \\
&= P(\min(X_1, X_2) \le y) \\
&= 1 - P(\min(X_1, X_2) > y) \\
&= 1 - P(X_1 > y, X_2 > y) \\
&= 1 - P(X_1 > y)P(X_2 > y) \\
&= 1 - \{1 - P(X_1 \le y)\}\{1 - P(X_2 \le y)\} \\
&= 1 - \{e^{-\theta y}\}\{e^{-\theta y}\} \\
&= 1 - e^{-2\theta y}
\end{aligned}
$$

and

$$
f_Y(y) = 2\theta e^{-2\theta y}, \qquad y \ge 0.
$$

This proves that $Y \sim \text{Ex}(2\theta)$. This means that the expected value of Y is $1/2\theta$, which is just one-half of the expected value of the service time of one customer.

Now suppose that we wish to find the joint probability density function of $Y_1, Y_2, \ldots, Y_m$, where each Y_i is a real-valued function of $\mathbf{X}$, i.e. $Y_i = h_i(\mathbf{X})$. We can write

$$
\mathbf{Y} = \begin{bmatrix} Y_1 \\ Y_2 \\ \vdots \\ Y_m \end{bmatrix} = \begin{bmatrix} h_1(\mathbf{X}) \\ h_2(\mathbf{X}) \\ \vdots \\ h_m(\mathbf{X}) \end{bmatrix} = \mathbf{h}(\mathbf{X}),
$$

where the vector function $\mathbf{h}$ maps $R_{\mathbf{X}}(\subseteq \mathbb{R}^m) \to \mathbb{R}^m$. We shall assume that $\mathbf{h}$ is a one-to-one transformation, i.e. for every $\mathbf{y} \in \mathbb{R}^m$ there is a unique $\mathbf{x} \in \mathbb{R}^m$ such that $\mathbf{h}(\mathbf{x}) = \mathbf{y}$. The *Jacobian* of this transformation is defined to be

$$J = \left| \frac{\partial \mathbf{X}}{\partial \mathbf{Y}} \right| = \begin{vmatrix} \dfrac{\partial X_1}{\partial Y_1} & \dfrac{\partial X_1}{\partial Y_2} & \cdots & \dfrac{\partial X_1}{\partial Y_m} \\ \vdots & \vdots & & \vdots \\ \dfrac{\partial X_m}{\partial Y_1} & \dfrac{\partial X_m}{\partial Y_2} & \cdots & \dfrac{\partial X_m}{\partial Y_m} \end{vmatrix}.$$

Proposition 7.1

Let $\mathbf{X}$ be a continuous, m-dimensional, random vector with joint range space $R_{\mathbf{X}}$ and joint probability density function $f_{\mathbf{X}}(\mathbf{x}), \mathbf{x} \in R_{\mathbf{X}}$. Define the new m-dimensional random vector $\mathbf{Y}$ by $\mathbf{Y} = \mathbf{h}(\mathbf{X})$, where every component $h_i(\cdot)$ of $\mathbf{h}(\cdot)$ is a continuous, real-valued function. Suppose that $\mathbf{h}(\mathbf{X})$ is a one-to-one transformation with inverse $\mathbf{X} = \mathbf{h}^{-1}(\mathbf{Y})$, that all the partial derivatives $\partial X_i / \partial Y_j$ exist and are continuous and that the Jacobian, J, of this transformation exists and is non-zero at all points in $R_{\mathbf{X}}$. Then the joint probability density function of the random vector $\mathbf{Y}$ is

$$f_{\mathbf{Y}}(\mathbf{y}) = \text{abs}(J) f_{\mathbf{X}}(\mathbf{h}^{-1}(\mathbf{y})), \qquad \mathbf{y} \in R_{\mathbf{Y}}.$$

The proof of this result is omitted. It follows immediately from rules for the change of variables in integral calculus.

Example 7.6

Suppose that X_1 and X_2 are independent random variables, with $X_i \sim \text{Ga}(\alpha_i, \theta)$, $i = 1, 2$. Notice that these two distributions have a common scale parameter, θ. Define new random variables Y_1 and Y_2 as follows:

$$Y_1 = \frac{X_1}{X_1 + X_2},$$

$$Y_2 = X_1 + X_2.$$

We shall now show that Y_2 has a gamma distribution, also with scale parameter θ, that Y_1 has a beta distribution and that Y_1 and Y_2 are independent.

Since X_1 and X_2 are independent, their joint p.d.f. is

$$f_{\mathbf{X}}(x_1, x_2) = \frac{\theta^{\alpha_1 + \alpha_2} x_1^{\alpha_1 - 1} x_2^{\alpha_2 - 1} e^{-\theta(x_1 + x_2)}}{\Gamma(\alpha_1) \Gamma(\alpha_2)}, \qquad x_1 > 0, \quad x_2 > 0.$$

The transformation is one-to-one and can be inverted to give:

$$X_1 = Y_1 Y_2,$$

$$X_2 = (1 - Y_1) Y_2,$$

so

$$J = \begin{vmatrix} \dfrac{\partial X_1}{\partial Y_1} & \dfrac{\partial X_1}{\partial Y_2} \\[2mm] \dfrac{\partial X_2}{\partial Y_1} & \dfrac{\partial X_2}{\partial Y_2} \end{vmatrix} = \begin{vmatrix} Y_2 & Y_1 \\ -Y_2 & 1 - Y_1 \end{vmatrix} = Y_2.$$

All the conditions for applying Proposition 7.1 are met, so Y_1 and Y_2 have joint p.d.f.

$$f_{\mathbf{Y}}(y_1, y_2) = \frac{y_2 \theta^{\alpha_1 + \alpha_2} (y_1 y_2)^{\alpha_1 - 1} \{(1 - y_1) y_2\}^{\alpha_2 - 1} e^{-\theta y_2}}{\Gamma(\alpha_1)\Gamma(\alpha_2)}, \qquad 0 < y_1 < 1, \ 0 < y_2$$

$$= \frac{\theta^{\alpha_1 + \alpha_2}}{\Gamma(\alpha_1)\Gamma(\alpha_2)} \left\{ y_1^{\alpha_1 - 1}(1 - y_1)^{\alpha_2 - 1} \right\} \left\{ y_2^{\alpha_1 + \alpha_2 - 1} e^{-\theta y_2} \right\},$$

$$0 < y_1 < 1, \quad 0 < y_2.$$

Since the range space of (Y_1, Y_2) is a rectangular region, and since $f_{\mathbf{Y}}(y_1, y_2)$ may be factorized into a function of Y_1 alone and a function of Y_2 alone, it follows that Y_1 and Y_2 are independent. Also, by the factorization theorem,

$$f_1(y_1) \propto y_1^{\alpha_1 - 1}(1 - y_1)^{\alpha_2 - 1}, \qquad 0 < y_1 < 1$$

and

$$f_2(y_2) \propto y_2^{\alpha_1 + \alpha_2 - 1} e^{-\theta y_2}, \qquad 0 < y_2.$$

Therefore, Y_1 follows the $\mathrm{Be}(\alpha_1, \alpha_2)$ distribution and Y_2 follows the $\mathrm{Ga}(\alpha_1 + \alpha_2, \theta)$ distribution.

Suppose now that X_1 and X_2 are independent and identically distributed exponential random variables. This is the special case where $\alpha_1 = \alpha_2 = 1$. Then $Y_1 = X_1/(X_1 + X_2)$ follows the $\mathrm{Be}(1, 1)$ or $\mathrm{Un}(0, 1)$ distribution, while $X_1 + X_2$ independently follows the $\mathrm{Ga}(2, \theta)$ distribution.

Example 7.7

The continuous random variable X follows a standard normal distribution. Independently of X, the continuous random variable Y follows the χ_k^2 distribution (i.e. the chi-squared distribution with k degrees of freedom):

$$f(y) = \frac{y^{k/2 - 1} \exp(-y/2)}{2^{k/2} \Gamma(k/2)}, \qquad y > 0$$

(where k is a positive integer). We shall now derive the joint probability density function of the random variables

$$U = \frac{X}{\sqrt{Y/k}} \quad \text{and} \quad V = \sqrt{Y/k},$$

and find the marginal density function of U.

We have

$$f_{XY}(x,y) = \frac{1}{\sqrt{2\pi}}\exp\left(-\frac{x^2}{2}\right) \cdot \frac{y^{k/2-1}\exp(-y/2)}{2^{k/2}\Gamma(k/2)}, \quad -\infty < x < \infty, \ 0 < y.$$

The joint range space of (U,V) is $R_{UV} = \{(u,v): -\infty < u < \infty, 0 < v\}$. Also

$$X = UV \quad \text{and} \quad Y = kV^2$$

so

$$J = \begin{vmatrix} \dfrac{\partial X}{\partial U} & \dfrac{\partial X}{\partial V} \\[2mm] \dfrac{\partial Y}{\partial U} & \dfrac{\partial Y}{\partial V} \end{vmatrix} = \begin{vmatrix} V & U \\ 0 & 2kV \end{vmatrix} = 2kV^2.$$

Therefore, for $(u,v) \in R_{UV}$,

$$f_{UV}(u,v) = \frac{1}{\sqrt{2\pi}}\exp\left(-\frac{u^2 v^2}{2}\right) \cdot \frac{k^{k/2-1}v^{k-2}\exp(-kv^2/2)}{2^{k/2}\Gamma(k/2)} \cdot 2kv^2$$

$$= 2^{(1-k)/2}k^{k/2}\frac{1}{\sqrt{\pi}\cdot\Gamma(k/2)}v^k\exp\left(-\frac{1}{2}v^2(u^2+k)\right)$$

and

$$f_U(u) = 2^{(1-k)/2}k^{k/2}\frac{1}{\sqrt{\pi}\cdot\Gamma(k/2)}\int_0^\infty v^k\exp\left(-\frac{1}{2}v^2(u^2+k)\right)dv$$

$$= 2^{(1-k)/2}k^{k/2}\frac{1}{\sqrt{\pi}\cdot\Gamma(k/2)}2^{(k-1)/2}(u^2+k)^{-(k+1)/2}\Gamma\left(\frac{k+1}{2}\right)$$

$$= \frac{\Gamma(k+1)}{\sqrt{k}(1+u^2/k)^{(k+1)/2}}\Gamma(1/2)\Gamma(k/2) \quad \left(\text{since}\sqrt{\pi} = \Gamma\left(\tfrac{1}{2}\right)\right)$$

$$= \frac{1}{\sqrt{k}(1+u^2/k)^{(k+1)/2}\mathbb{B}(k/2,1/2)} \quad \text{(Result A2.5)}$$

This marginal distribution of U is the *Student's t* distribution with k degrees of freedom, or t_k distribution (see Table 2.1(b)). This result has important applications in statistical inference (see Chapter 9).

Example 7.8

Consider a machine that has just been serviced and brought back into use. The machine will break down intermittently in the future and require further attention. It is believed that the time until the first breakdown is a $Ga(\alpha,\theta)$ random variable, and that the times between consecutive breakdowns thereafter are independent $Ga(\beta,\theta)$ random variables. This means that the times from now until each of the next m breakdowns are random variables $Y_1, \ldots, Y_m$ given by

$$Y_1 = X_1,$$
$$Y_2 = X_1 + X_2,$$
$$\vdots$$
$$Y_m = X_1 + X_2 + \cdots + X_m.$$

From Chapter 6, we know that each Y_i follows a gamma distribution. We will now obtain the joint probability density function of $Y_1, \ldots, Y_m$.

Since $X_1, X_2, \ldots, X_m$ are independent random variables, it follows that

$$f_{\mathbf{X}}(\mathbf{x}) = \frac{\theta^{\alpha} x_1^{\alpha-1} e^{-\theta x_1}}{\Gamma(\alpha)} \prod_{i=2}^{m} \frac{\theta^{\beta} x_i^{\beta-1} e^{-\theta x_i}}{\Gamma(\beta)}$$

$$= \frac{\theta^{\alpha+(m-1)\beta}}{\Gamma(\alpha)\{\Gamma(\beta)\}^{m-1}} x_1^{\alpha-1} \left\{ \prod_{i=2}^{m} x_i \right\}^{\beta-1} \exp\left(-\theta \sum_{i=1}^{m} x_i\right).$$

Inverting the transformation gives

$$\left. \begin{array}{l} Y_1 = X_1 \\ Y_2 = X_1 + X_2 \\ \vdots \\ Y_3 = X_1 + X_2 + \cdots + X_m \end{array} \right\} \Rightarrow \left. \begin{array}{l} X_1 = Y_1 \\ X_2 = Y_2 - Y_1 \\ \vdots \\ X_m = Y_m - Y_{m-1} \end{array} \right\}.$$

So

$$J = \left| \frac{\partial \mathbf{X}}{\partial \mathbf{Y}} \right| = \begin{vmatrix} 1 & 0 & \cdots & 0 \\ -1 & 1 & \cdots & 0 \\ \vdots & \vdots & \ddots & \vdots \\ 0 & 0 & \cdots & 1 \end{vmatrix} = 1$$

(since the determinant of a triangular matrix is the product of the elements on the leading diagonal). Therefore

$$f_{\mathbf{Y}}(\mathbf{y}) = \frac{\theta^{\alpha+(m-1)\beta}}{\Gamma(\alpha)\{\Gamma(\beta)\}^{m-1}} y_1^{\alpha-1} \left\{ \prod_{i=2}^{m} (y_i - y_{i-1}) \right\}^{\beta-1} \exp(-\theta y_m),$$

$$y_m > \cdots > y_2 > y_1 > 0.$$

Exercises

1 Suppose that the random variables X_1 and X_2 are independent, each following the $\mathrm{Un}(0, \theta)$ distribution for some $\theta > 0$. Find the distribution function and probability density function of $Y = \min(X_1, X_2)$. Hence find $\mathbb{E}(Y)$ and $\text{var}(Y)$.

2 Suppose that X_1 and X_2 are independent exponential random variables, as in Example 7.4, but that they both follow the same exponential distribution, $\text{Ex}(\theta)$. Amend the proof in Example 7.4 to show that $T = X_1 + X_2$ follows a gamma distribution.

3 Suppose that X_1 and X_2 are independent $\text{Ex}(\theta)$ random variables.

(a) Derive the distribution function, and hence the probability density function, of $Y = \max(X_1, X_2)$.

(b) Derive the distribution function, and hence the probability density function, of $Y = X_1 - X_2$. [*Hint*: deal separately with the cases $y < 0$ and $y \geq 0$.]

4 Suppose that a gas under constant pressure, equal in all directions, is constrained to flow in one plane. X_1 and X_2, the components of its velocity in the horizontal and vertical directions, are independent random variables which each follow the $N(0, \sigma^2)$ distribution. The speed of the gas is $U = \sqrt{X_1^2 + X_2^2}$. Derive the joint probability density function of the random variables U and ϕ defined implicitly by:

$$X_1 = U \cos(\phi), \qquad X_2 = U \sin(\phi).$$

Now find the marginal probability density function of U.

U is said to follow a *Rayleigh* distribution. The usual parameter of this distribution is α, where $\mathbb{E}(U) = \alpha$. Show that the marginal p.d.f. of U can be written in the form

$$f_U(u) = \frac{\pi}{2\alpha^2} \exp\left\{ -\frac{\pi}{4}\left(\frac{u}{\alpha}\right)^2 \right\}$$

Hence show that

$$\text{var}(U) = \alpha^2 \left(\frac{4}{\pi} - 1\right).$$

5 Suppose that X_1 and X_2 are independent random variables, each following the standard normal distribution. Define new random variables, Y_1 and Y_2, as follows:

$$Y_1 = \sqrt{\tfrac{1}{2}(1 + \rho)}X_1 + \sqrt{\tfrac{1}{2}(1 - \rho)}X_2,$$

$$Y_2 = \sqrt{\tfrac{1}{2}(1 + \rho)}X_1 - \sqrt{\tfrac{1}{2}(1 - \rho)}X_2.$$

Show that Y_1 and Y_2 are also standard normal random variables, but with correlation ρ. Hence derive one method for simulating from an arbitrary bivariate normal distribution, starting with a sequence of pseudo-random variates from the standard normal distribution.

6 The independent random variables, X and Y, follow chi-squared distributions with degrees of freedom m and n, respectively. Find the joint probability density function of

$$U = \frac{X/m}{Y/n} = \frac{nX}{mY} \quad \text{and} \quad V = \frac{Y}{n}.$$

Hence show that U follows the $F_{m,n}$ distribution.

7 Consider two random variables, X_1 and X_2, such that $X_1 \sim Ex(\theta)$ independently of $X_2 \sim Ex(\phi)$. By constructing a suitable one-to-one transformation, show that

$$Y_1 = \frac{\theta X_1}{\phi X_2} \sim F_{2,2}.$$

By considering the relationship between the $Ex(1)$ and χ_2^2 distributions, show that this result is just a special case of the result proved in Exercise 7 above.

8 X_1, X_2 and X_3 are independent random variables with each $X_i \sim Ga(\alpha_i, \theta)$. Define the random variables

$$Y_1 = X_1 + X_2 + X_3, \quad Y_2 = \frac{X_2 + X_3}{X_1 + X_2 + X_3}, \quad Y_3 = \frac{X_3}{X_2 + X_3}.$$

Derive the joint probability density function of Y_1, Y_2 and Y_3. Hence show that these random variables are all independent. Write down the marginal probability density functions of Y_1, Y_2 and Y_3, and identify these special distributions.

9 Suppose that $X_1, X_2, \ldots, X_n$ are independent, continuous random variables, each following the same distribution which has distribution function $F(x)$ and probability density function $f(x)$. Let Y be the minimum of $X_1, X_2, \ldots, X_n$.

(a) Show that

$$F_Y(y) = 1 - \{1 - F(y)\}^n,$$

$$f_Y(y) = nf(y)\{1 - F(y)\}^{n-1}.$$

(b) When each $X_i \sim We(\alpha, \theta)$, show that Y also follows a Weibull distribution.

10 Suppose that X_1 and X_2 are jointly continuous random variables with joint probability density function $f_\mathbf{X}(x_1, x_2)$. Let $Y_2 = g(X_2)$ be a continuous, one-to-one function of X_2. Show that the joint probability density function of X_1 and Y_2 is

$$f(x_1, y_2) = \left| \frac{dx_2}{dy_2} \right| f_\mathbf{X}(x_1, g^{-1}(y_2)).$$

Find the conditional probability density function of X_1 given Y_2. Show that information about X_2 and Y_2 is equivalent when it comes to calculating probabilities associated with X_1.

11 Suppose that $\mathbf{X}^{(1)}, \ldots, \mathbf{X}^{(k)}$ are independent random vectors. Define new random vectors $\mathbf{Y}^{(1)}, \ldots, \mathbf{Y}^{(k)}$ by $\mathbf{Y}^{(j)} = g_j\{\mathbf{X}^{(j)}\}$, where g_j is an arbitrary function of $\mathbf{X}^{(j)}$. Show that $\mathbf{Y}^{(1)}, \ldots, \mathbf{Y}^{(k)}$ are also independent random vectors. [*Hint*: obtain the joint distribution function of $\mathbf{Y}^{(1)}, \ldots, \mathbf{Y}^{(k)}$ in terms of the joint distribution function of $\mathbf{X}^{(1)}, \ldots, \mathbf{X}^{(k)}$, but do not assume that the conditions of Proposition 7.1 hold.]

7.3 Order statistics

Example 7.9

Suppose that three identical electronic components are assembled in series. This means that the assembly will fail as soon as any one of the components fails. If X_i is the time to failure of the ith component ($i = 1, 2, 3$), then the time to failure of the whole assembly is $Y = \min(X_1, X_2, X_3)$.

Example 7.10

Suppose that a flood defence is to be built to protect a flood plain. It seems sensible to build the defence to a height that is likely only to be exceeded by a flood every 50 years or so. The maximum height of flood water in any year is a random variable, so let $X_1, X_2, \ldots, X_{50}$ be the heights in 50 consecutive years. The random variable of greatest concern to the designer of the flood defence is $Y = \max(X_1, \ldots, X_{50})$.

Example 7.11

An electronic assembly is designed with built-in redundancy. It consists of 40 identical components, but will continue to operate as long as at least 30 of the components operate. The time to failure of the whole system is the 11th smallest of the individual failure times.

Example 7.12

A random sample, $X_1, \ldots, X_n$, is drawn from a particular population. Assuming that n is an odd number, the sample median is the $\frac{1}{2}(n + 1)$th smallest of the sample values. This is a natural estimator of the population median.

In general, suppose that $X_1, X_2, \ldots, X_m$ are independent and identically distributed random variables. Let $Y_j (j = 1, 2, \ldots, m)$ be the jth smallest of these values. So, for example, Y_1 is the smallest of $X_1, X_2, \ldots, X_m$ and Y_m is the largest of $X_1, X_2, \ldots, X_m$. In general,

$$Y_1 \leq Y_2 \leq \cdots \leq Y_m.$$

Then Y_j is known as the jth *order statistic* of the values $X_1, X_2, \ldots, X_m$.

In principle, order statistics might be of interest whether random variables follow a discrete or continuous distribution, but in practice it is the continuous case that has been found to be of greatest relevance. So, throughout this section, we will assume that $X_1, X_2, \ldots, X_m$ are continuous random variables with common distribution function $F(x)$ and common probability density function $f(x)$. Since the random variables are continuous, there is zero probability that any two of them are equal, and so we may assume that

$$Y_1 < Y_2 < \cdots < Y_m.$$

Proposition 7.2

The joint probability density function of the order statistics $(Y_1, \ldots, Y_m)$ is

$$g_{\mathbf{Y}}(y_1, y_2, \ldots, y_m) = m! f(y_1) f(y_2) \cdots f(y_m), \qquad y_1 < y_2 < \cdots < y_m.$$

Proof. Suppose that $(y_1, y_2, \ldots, y_m)$ is any point in $R_{\mathbf{Y}}$. We start by obtaining the value of the distribution function of Y at this point, i.e.

$$G_{\mathbf{Y}}(y_1, y_2, \ldots, y_m) = P(Y_1 \le y_1, Y_1 \le Y_2 \le y_2, \ldots, Y_{m-1} \le Y_m \le y_m).$$

Let E be the event $\{(y_1, y_2, \ldots, y_m): Y_1 \le y_1, Y_1 \le Y_2 \le y_2, \ldots, Y_{m-1} \le Y_m \le y_m\}$. The equivalent event in the range space $R_{\mathbf{X}}$ is made up of a disjoint union of sets, one of which is the following:

$$E_{123\ldots m} = \{(y_1, y_2, \ldots, y_m): X_1 \le y_1, X_1 \le X_2 \le y_2, \ldots, X_{m-1} \le X_m \le y_m\}.$$

This is the particular situation in which it so happens that $X_1 < X_2 < \cdots < X_m$ and, therefore, each $Y_j = X_j$. The probability that $\mathbf{X}$ lies in $E_{123\ldots m}$ is

$$\iint \cdots \int_{E_{123\ldots m}} f(x_1) f(x_2) \cdots f(x_m)\, dx_m \cdots dx_2\, dx_1.$$

Another set in E is

$$E_{213\ldots m} = \{(y_1, y_2, \ldots, y_m): X_2 \le y_1, X_2 \le X_1 \le y_2, \ldots, X_{m-1} \le X_m \le y_m\}.$$

Altogether, the event E is the disjoint union of $m!$ such sets (found by considering every possible permutation of the random variables $X_1, \ldots, X_m$). Because of the symmetry of the joint probability density function of $\mathbf{X}$, all these sets have the same probability, so

$$G_{\mathbf{Y}}(y_1, y_2, \ldots, y_m) = m! \iint \cdots \int_{E_{123\ldots m}} f(x_1) f(x_2) \cdots f(x_m)\, dx_m \cdots dx_2\, dx_1.$$

Therefore, the joint probability density function of the order statistics $(Y_1, \ldots, Y_m)$ is

$$g_{\mathbf{Y}}(y_1, y_2, \ldots, y_m) = m! f(y_1) f(y_2) \cdots f(y_m), \qquad y_1 < y_2 < \cdots < y_m.$$

∎

Example 7.13

Suppose that $X_1, X_2, \ldots, X_m$ are independent random variables each following the $\mathrm{Ex}(\theta)$ distribution for $\theta > 0$. Then the order statistics $Y_1, Y_2, \ldots, Y_m$ have the joint probability density function

$$g_\mathbf{Y}(y_1, y_2, \ldots, y_m) = m!\theta^m e^{-\theta(y_1 + y_2 + \cdots y_m)}, \qquad 0 < y_1 < y_2 < \cdots < y_m.$$

Proposition 7.3

The (marginal) probability density function of the order statistic $Y_j (j = 1, 2, \ldots, m)$ is

$$g_j(y_j) = \frac{m!}{(j-1)!(m-j)!} \{F(y_j)\}^{j-1} \{1 - F(y_j)\}^{m-j} f(y_j).$$

Proof. In order to find $g_j(y_j)$, we require to integrate the joint probability density function of $\mathbf{Y}$ with respect to all the other random variables. The integration has to be carried out over the region

$$Y_1 < \cdots < Y_{j-1} < y_j < Y_{j+1} < \cdots < Y_m.$$

The following sets of inequalities are best considered separately:

$$Y_1 < \cdots < Y_{j-1} < y_j \quad \text{and} \quad y_j < Y_{j+1} < \cdots < Y_m.$$

The required integral is

$$g_j(y_j) = \int_{-\infty}^{y_j} \cdots \int_{-\infty}^{y_2} \int_{y_j}^{\infty} \cdots \int_{y_{m-1}}^{\infty} m! f(y_1) \cdots f(y_m) \, dy_m \cdots dy_{j+1} dy_1 \cdots dy_{j-1}$$

$$= m! f(y_j) \left\{ \int_{-\infty}^{y_j} \cdots \int_{-\infty}^{y_2} f(y_1) \cdots f(y_{j-1}) \, dy_1 \cdots dy_{j-1} \right\}$$

$$\times \left\{ \int_{y_j}^{\infty} \cdots \int_{y_{m-1}}^{\infty} f(y_{j+1}) \cdots f(y_m) \, dy_m \cdots dy_{j+1} \right\}.$$

But

$$\int_{-\infty}^{y_j} \cdots \int_{-\infty}^{y_2} f(y_1) \cdots f(y_{j-1}) \, dy_1 \cdots dy_{j-1}$$

$$= \int_{-\infty}^{y_j} f(y_{j-1}) \cdots \int_{-\infty}^{y_3} f(y_2) \int_{-\infty}^{y_2} f(y_1) \, dy_1 \cdots dy_{j-1}$$

$$= \int_{-\infty}^{y_j} f(y_{j-1}) \cdots \int_{-\infty}^{y_3} f(y_2) F(y_2) \, dy_2 \cdots dy_{j-1}$$

$$= \int_{-\infty}^{y_j} f(y_{j-1}) \cdots \int_{-\infty}^{y_4} f(y_3) \frac{1}{2} \{F(y_3)\}^2 dy_3 \cdots dy_{j-1}$$

$$= \cdots$$

$$= \frac{1}{(j-1)!} \{F(y_j)\}^{j-1}$$

and

$$\int_{y_j}^{\infty} \cdots \int_{y_{m-1}}^{\infty} f(y_{j+1}) \cdots f(y_m) \, dy_m \cdots dy_{j+1}$$

$$= \int_{y_j}^{\infty} f(y_{j+1}) \cdots \int_{y_{m-2}}^{\infty} f(y_{m-1}) \int_{y_{m-1}}^{\infty} f(y_m) \, dy_m \cdots dy_{j+1}$$

$$= \int_{y_j}^{\infty} f(y_{j+1}) \cdots \int_{y_{m-2}}^{\infty} f(y_{m-1})\{1 - F(y_{m-1})\} \, dy_{m-1} \cdots dy_{j+1}$$

$$= \int_{y_j}^{\infty} f(y_{j+1}) \cdots \int_{y_{m-3}}^{\infty} f(y_{m-2}) \frac{1}{2}\{1 - F(y_{m-2})\}^2 \, dy_{m-2} \cdots dy_{j+1}$$

$$= \cdots$$

$$= \frac{1}{(m-j)!} \{1 - F(y_j)\}^{m-j}.$$

This proves the result. ∎

The form of the probability density function for Y_j can be remembered more easily from the following informal argument. In order for the jth largest of the original random variables, Y_j, to equal y_j, one of the m original random variables must be equal to y_j; the probability of this is proportional to $mf(y_j)$. Any one of the remaining $m - 1$ random variables is either less than y_j, with probability $F(y_j)$, or greater than y_j, with probability $1 - F(y_j)$.

Example 7.13 – continued

$X_1, \ldots, X_m$ all follow the Ex(θ) distribution with

$$f(x) = \theta e^{-\theta x}, \qquad F(x) = 1 - e^{-\theta x}.$$

Using Proposition 7.3, the minimum, Y_1, has probability density function

$$g_1(y_1) = m\{1 - F(y_1)\}^{m-1} f(y_1) = m\{e^{-\theta y_1}\}^{m-1} \theta e^{-\theta y_1} = m\theta e^{-m\theta y_1}, \qquad 0 < y_1.$$

This means that Y_1 follows the Ex($m\theta$) distribution. The maximum, Y_m, has probability density function

$$g_j(y_j) = m\{F(y_m)\}^{m-1} f(y_m) = m\theta\{1 - e^{-\theta y_m}\}^{m-1} e^{-\theta y_m}, \qquad 0 < y_m.$$

Proposition 7.4

The joint probability density function of the order statistics Y_j and $Y_k (j < k)$ is

$$g_{jk}(y_j, y_k) = \frac{m!}{(j-1)!(k-j-1)!(m-k)!} \{F(y_j)\}^{j-1} \{F(y_k) - F(y_j)\}^{k-j-1}$$

$$\times \{1 - F(y_k)\}^{m-k} f(y_j) f(y_k).$$

This proof is omitted, but is similar to that of the previous proposition.

Example 7.13 – continued

The joint probability density function of the minimum and maximum, Y_1 and Y_m, is (for $0 < y_1 < y_m$)

$$g_{1m}(y_1, y_m) = \frac{m!}{(m-2)!} \{F(y_m) - F(y_1)\}^{m-2} f(y_1) f(y_m)$$

$$= m(m-1)\theta^2 e^{-\theta y_1} e^{-\theta y_m} (e^{-\theta y_1} - e^{-\theta y_m})^{m-2}.$$

An important value is the *sample range*, $R = Y_m - Y_1$. In order to obtain the probability density function of the range in this example, consider the following transformation:

$$U_1 = Y_1, \qquad U_2 = Y_m - Y_1.$$

This is a one-to-one transformation with inverse

$$Y_1 = U_1, \qquad Y_m = U_1 + U_2.$$

The Jacobian is

$$J = \begin{vmatrix} \dfrac{\partial Y_1}{\partial U_1} & \dfrac{\partial Y_1}{\partial U_2} \\[2mm] \dfrac{\partial Y_m}{\partial U_1} & \dfrac{\partial Y_m}{\partial U_2} \end{vmatrix} = \begin{vmatrix} 1 & 0 \\ 1 & 1 \end{vmatrix} = 1.$$

So the joint probability density function of U_1 and U_2 is (for $0 < u_1, 0 < u_2$)

$$m(m-1)\theta^2 e^{-\theta u_1} e^{-\theta(u_1 + u_2)} (e^{-\theta u_1} - e^{-\theta(u_1+u_2)})^{m-2}$$

$$= m(m-1)\theta^2 (e^{-m\theta u_1}) e^{-\theta u_2} (1 - e^{-\theta u_2})^{m-2}.$$

Since this range space is a rectangular region and the joint density factorizes, U_1 and U_2 must be independent random variables. We already know that U_1, which is just Y_1, follows an $\text{Ex}(m\theta)$ distribution, so the probability density function of U_2 (or R) must be

$$(m-1)\theta e^{-\theta u_2}(1 - e^{-\theta u_2})^{m-2}, \qquad 0 < u_2.$$

In other words, the probability density function of R is

$$(m-1)\theta e^{-\theta r}(1 - e^{-\theta r})^{m-2}, \qquad 0 < r$$

It is easy to see from this that $e^{-\theta R}$ follows the $\text{Be}(2, m-1)$ distribution.

A much more extensive discussion of order statistics may be found in Stuart and Ord (1994, Chapter 14).

Exercises

1 Suppose that $X_1, X_2, \ldots, X_m$ are independent random variables, each of which follows the $\text{Un}(0,1)$ distribution. Find the marginal distribution of the *mid-range*, $\frac{1}{2}(Y_1 + Y_m)$, where Y_1 and Y_m are order statistics.

2 Suppose that $X_1, X_2, \ldots, X_m$ are independent random variables, each of which follows the $\text{Un}(0,1)$ distribution.

  (a) Obtain the joint probability density function of the order statistics.

  (b) Show that, marginally, each order statistic follows a beta distribution. Write down the expected value and variance of the order statistics.

  (c) Obtain the joint (marginal) probability density function of the order statistics Y_j and $Y_k (j < k)$.

  (d) Show that the sample range, $R = Y_m - Y_1$, has probability density function

$$f_R(r) = m(m-1)r^{m-2}(1-r), \qquad 0 < r < 1.$$

  Write down $\mathbb{E}(R)$ and $\text{var}(R)$, and show what happens to these values as the sample size $m \to \infty$.

3 Suppose that a random sample of size m is drawn from a population of continuous random variables with common distribution function $F(x)$ and probability density function $f(x)$. Let Y_1 and Y_m, respectively, denote the minimum and maximum values in the sample. Write down $P(Y_m < y_m)$ and $P(y_1 < Y_1, Y_m < y_m)$ in terms of F and f, and so derive the joint probability density function of Y_1 and Y_m in terms of F and f.

4 Suppose that $X_1, X_2, \ldots, X_m$ are independent random variables, each following the $\text{Ex}(\theta)$ distribution. Write down the joint probability density function of the order statistics $Y_1, Y_2, \ldots, Y_m$. Use the Jacobian method to obtain the joint probability density function of $Y_1, Y_2 - Y_1, \ldots, Y_m - Y_{m-1}$. Hence show

that these random variables are independent. Obtain their expected values and hence find the expected value of $Y_i(i = 1, 2, \ldots, m)$.

5 A triple modular redundant (TMR) system consists of three identical components, but the system will continue to function properly as long as any two of the components continue to function. Suppose that the lifetimes of the components in a TMR system are independent $Ex(\theta)$ random variables.

(a) Let the random variable Y_2 be the lifetime of the TMR system. Write down the probability density function, g_2, and the distribution function, G_2, of Y_2.

(b) The *reliability function* of the TMR system is the probability that the system is still functioning at time t, i.e.

$$R_{\text{TMR}}(t) = 1 - G_2(t) \qquad t > 0.$$

The reliability function of a single component is defined, in a similar way, as the probability that the component is still functioning at time t. Compare these two reliability functions.

7.4 Some further methods of simulation

The inverse distribution function method of simulation was introduced in Chapter 3. Its use is restricted to random variables whose distribution function can be inverted, usually algebraically. It is now possible to introduce further methods of simulation, which can be used even when this is not the case. We begin with *rejection methods*.

The (original) rejection method is used to generate pseudo-random variates from a distribution whose p.d.f., $f(x)$, is only non-zero on a finite interval $[a, a + b]$ and for which $f(x)$ has a finite maximum on the range $[a, a + b]$.

Let

$$\max_{a \leq x \leq a+b} [f(x)] = c.$$

Begin by generating two pseudo-random $Un(0, 1)$ variates, u_1 and u_2. The value $a + bu_2$ is a 'random' number in the range $[a, a + b]$. The value cu_1, on the

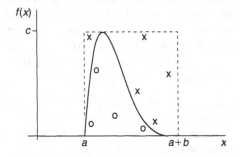

other hand, is a 'random' number between 0 and c. This means that the point $(a + bu_2, cu_1)$ is a 'random' point in the rectangle indicated on the figure.

The simulation proceeds as follows. If

$$cu_1 \leq f(a + bu_2),$$

then $x = a + bu_2$ is taken as a pseudo-random variate from the distribution $f(\cdot)$. Otherwise, the simulated value is rejected; i.e. no random variate is generated.

In the diagram above, several points $(a + bu_2, cu_1)$ are shown. Those that lie below the curve $f(x)$ correspond to simulated values. Those that lie above the p.d.f are rejected.

The point $(a + bU_2, cU_1)$ is uniformly distributed in the rectangle shown. Hence, the probability of rejection is

$$1 - \frac{\text{area under } f(x)}{\text{area of rectangle}} = 1 - \frac{1}{bc}.$$

Furthermore, the rejection method requires two uniform variates to be generated for every simulation. This means that it is much more expensive to generate simulations using the rejection method than using the inverse distribution function method, so the latter method should be used where possible.

Example 7.14

The continuous random variable X has the probability density function $f(x) = 2x$ $(0 \leq x \leq 1)$. Starting with the following sequence of random digits, obtain three variates from this distribution.

$$094 \quad 278 \quad 201 \quad 486 \quad 636 \quad 123 \quad 749 \quad 914 \quad 660 \quad 254$$

Here, $a = 0$, $b = 1$, $c = 2$.

First attempt:

$$u_1 = 0.094, \qquad \text{so } cu_1 = 0.188;$$
$$u_2 = 0.278, \qquad \text{so } a + bu_2 = 0.278, \quad \text{and} \quad f(a + bu_2) = 0.556.$$

since $cu_1 \leq a + bu_2$, accept simulation:

$$x = a + bu_2 = 0.278.$$

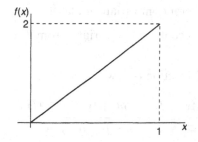

Second attempt:

$$u_1 = 0.201, \qquad \text{so } cu_1 = 0.402;$$
$$u_2 = 0.486, \qquad \text{so } a + bu_2 = 0.486, \quad \text{and} \quad f(a + bu_2) = 0.972.$$

Since $cu_1 \leq a + bu_2$, accept simulation:

$$x = a + bu_2 = 0.486.$$

Third attempt:

$$u_1 = 0.636, \qquad \text{so } cu_1 = 1.272;$$
$$u_2 = 0.123, \qquad \text{so } a + bu_2 = 0.123, \quad \text{and} \quad f(a + bu_2) = 0.246.$$

Since $cu_1 > a + bu_2$, reject simulation.
Fourth attempt:

$$u_1 = 0.749, \qquad \text{so } cu_1 = 1.498;$$
$$u_2 = 0.914, \qquad \text{so } a + bu_2 = 0.914, \quad \text{and} \quad f(a + bu_2) = 1.828.$$

Since $cu_1 \leq a + bu_2$, accept simulation:

$$x = a + bu_2 = 0.914.$$

In this example, the probability of rejection is $1 - 1/bc = 1 - 1/2 = 1/2$. Fairly high probabilities of rejection are not unusual with the original rejection method.

It is possible to reduce the probability of rejection substantially, using the *improved rejection method*. This method has the additional advantage that it is not restricted to use with probability density functions that are non-zero on a finite interval. In order to use this method, it must be possible to find a function $g(x)$ which is always (slightly) greater than $f(x)$, and also to simulate values from a proper p.d.f., $h(x)$, where $g(x) = K \cdot h(x)$, for some constant K. Proceed as follows:

1. Generate a random variate, x, from $h(x)$.

2. Generate a uniform random variate, u (say).

3. If $uKh(x) \leq f(x)$, accept x as a variate from $f(x)$, otherwise, reject the simulation.

The probability of rejection is now

$$\frac{\int_{-\infty}^{\infty} (g(x) - f(x)) \, dx}{\int_{-\infty}^{\infty} g(x) \, dx} = \frac{K \int_{-\infty}^{\infty} h(x) \, dx - \int_{-\infty}^{\infty} f(x) \, dx}{K \int_{-\infty}^{\infty} h(x) \, dx} = \frac{K - 1}{K} = 1 - \frac{1}{K}.$$

Example 7.15

The $Ga(\alpha, 1)$ distribution, for $\alpha > 1$, has probability density function

$$f(x) = \frac{x^{\alpha-1}e^{-x}}{\Gamma(\alpha)}, \qquad x > 0.$$

It is not possible to use the inverse distribution function method with this distribution, since $F(x)$ cannot be written (or inverted) in closed algebraic form. We use the improved rejection method instead.

In this case, a suitable choice of function $h(x)$ is the probability density function of the $Ex(1/\alpha)$ distribution:

$$h(x) = \frac{1}{\alpha}e^{-x/\alpha}, \qquad x > 0.$$

We must find a constant K such that $g(x) = Kh(x) \geq f(x)$ for all $x > 0$. A suitable choice of K is

$$K = \max_{0<x} \frac{f(x)}{h(x)} = \max_{0<x} \frac{\alpha}{\Gamma(\alpha)} x^{\alpha-1}e^{-x}e^{x/\alpha} = \max_{0<x} \frac{\alpha}{\Gamma(\alpha)} x^{\alpha-1}e^{-x(\alpha-1)/\alpha}.$$

Now

$$\frac{d}{dx} x^{\alpha-1}e^{-x(\alpha-1)/\alpha} = (\alpha-1)x^{\alpha-2}e^{-x(\alpha-1)/\alpha} - \frac{\alpha-1}{\alpha}x^{\alpha-1}e^{-x(\alpha-1)/\alpha}$$

$$= \frac{\alpha-1}{\alpha} x^{\alpha-2}e^{-x(\alpha-1)/\alpha}(\alpha - x).$$

This derivative equals 0 when $x = \alpha$ and it is clear that this is a local maximum. We take K as the value at this maximum:

$$K = \frac{f(\alpha)}{h(\alpha)} = \frac{\alpha}{\Gamma(\alpha)}\alpha^{\alpha-1}e^{-\alpha(\alpha-1)/\alpha} = \frac{\alpha^\alpha}{\Gamma(\alpha)}e^{-(\alpha-1)}.$$

The improved rejection method now works as follows:

1. Generate a $Un(0, 1)$ random variate, u_1.

2. Use the probability inverse transform method to generate a random variate from the $Ex(1/\alpha)$ distribution, x (say).

3. Generate another $Un(0, 1)$ random variate, u_2.

4. If $uKh(x) \leq f(x)$, accept x as the simulated value from $f(x)$. Otherwise, reject the simulation.

Exercises

1 Explain how to use the rejection method to generate pseudo-random variates from the $Be(2, 2)$ distribution, i.e. the distribution with probability

density function

$$f_X(x) = 6x(1 - x), \qquad 0 < x < 1.$$

Use this method to generate three pseudo-random variates. Find the probability of rejection in this case.

2 Consider the distribution with probability density function

$$f(x) = \sqrt{\frac{2}{\pi}} \exp\left(-\frac{x^2}{2}\right), \qquad x > 0.$$

The point of this exercise is to construct a method of simulating from this distribution using the improved rejection method.

(a) The basic function to be used is $h(x) = \exp[-x]$. Show that $f(x)/h(x)$ takes its maximum value at $x = 1$. Find a constant K such that $g(x) = Kh(x) \geq f(x)$ at all values of $x > 0$.

(b) Write down a method of the improved rejection type that allows simulation from $f(x)$. Find the probability of rejection when using this method.

(c) Use this method to generate three pseudo-random variates from $f(x)$.

Summary

In this chapter, we have discussed general approaches to obtaining the probability distribution of functions of random variables. This has led on to a discussion of the order statistics of a collection of random variables. Rejection methods of simulation have also been discussed.

8
The multivariate normal distribution

Applications of multivariate probability have historically been dominated by the use of the multivariate normal distribution for continuous data. This chapter describes this important distribution in some detail and the following chapter shows some of the ways it is used in applications.

8.1 Characterizing the multivariate normal distribution

Example 8.1

Consider the random vector $\mathbf{X} = [X_1 \, X_2 \ldots X_m]^{\mathrm{T}}$, where $X_1, \ldots, X_m$ are independent random variables and each $X_i \sim N(\mu_i, \, \sigma_i^2)$. Then, because of the independence of the random variables,

$$f_{\mathbf{X}}(\mathbf{x}) = \prod_{i=1}^{m} \frac{1}{\sqrt{2\pi\sigma_i^2}} \exp\left\{ -\frac{1}{2} \left(\frac{x_i - \mu_i}{\sigma_i} \right)^2 \right\}$$

$$= (2\pi)^{-m/2} \frac{1}{\sqrt{\prod_{i=1}^{m} \sigma_i^2}} \exp\left\{ -\frac{1}{2} \sum_{i=1}^{m} \left(\frac{x_i - \mu_i}{\sigma_i} \right)^2 \right\}.$$

The mean vector of $\mathbf{X}$ is $\mathbb{E}(\mathbf{X}) = [\mu_1 \, \mu_2 \ldots \mu_m]^{\mathrm{T}} = \boldsymbol{\mu}$ (say). The covariance matrix of $\mathbf{X}$ is $\mathrm{cov}(\mathbf{X}) = \mathrm{diag}(\sigma_1^2 \, \sigma_2^2 \ldots \sigma_m^2) = \Sigma$ (say).

This means that

$$\sum_{i=1}^{m} \left(\frac{x_i - \mu_i}{\sigma_i} \right)^2 = [\mathbf{x} - \boldsymbol{\mu}]^{\mathrm{T}} \, \Sigma^{-1} \, [\mathbf{x} - \boldsymbol{\mu}]$$

and

$$\prod_{i=1}^{m} \sigma_i^2 = |\Sigma|.$$

So the (joint) probability density function of $\mathbf{X}$ can be written in the form

$$f_{\mathbf{X}}(\mathbf{x}) = (2\pi)^{-m/2} |\Sigma|^{-1/2} \exp\left\{ -\frac{1}{2} (\mathbf{x} - \boldsymbol{\mu})^{\mathrm{T}} \Sigma^{-1} (\mathbf{x} - \boldsymbol{\mu}) \right\}, \qquad \mathbf{x} \in \mathbb{R}^m. \quad (8.1)$$

where $\boldsymbol{\mu} \in \mathbb{R}^m$ and $\Sigma \in M_{mm}$ is a positive definite matrix.

Suppose that an m-dimensional random vector, $\mathbf{X}$, has joint probability density function (8.1). Then $\mathbf{X}$ is said to follow a (non-singular) *multivariate normal (MVN)* distribution, sometimes written $\mathbf{X} \sim N_m(\boldsymbol{\mu}, \Sigma)$.

We shall show later that $\boldsymbol{\mu} = \mathbb{E}(\mathbf{X})$ and $\Sigma = \mathrm{cov}(\mathbf{X})$. This explains the restriction of Σ to positive definite matrices in the above definition, but there is no general requirement for Σ to be diagonal, as it was in Example 8.1. In fact, as we shall show, it is possible to extend the class of MVN distributions to include cases where Σ is positive semi-definite but not positive definite, so that Σ is singular and the joint probability density function does not exist. For the moment, though, we will restrict attention to cases where Σ is positive definite and, hence, non-singular.

When Σ is positive definite, a result from matrix algebra tells us that there exists another positive definite matrix $C \in M_{mm}$ such that $\Sigma = CC^{\mathrm{T}}$ (Result A4.2). Since positive definite matrices are non-singular, this means that Σ^{-1}, C^{-1} and $(C^{\mathrm{T}})^{-1} = (C^{-1})^{\mathrm{T}}$ all exist. We will use the matrix C and its properties in the proofs throughout this chapter.

Proposition 8.1

The function $f_{\mathbf{X}}(\mathbf{x})$ in (8.1) is a valid probability density function.

Proof. Clearly, $f_{\mathbf{X}}(\mathbf{x}) \geq 0$ for all $\mathbf{x} \in \mathbb{R}^m$. We must now show that

$$\int_{-\infty}^{\infty} \cdots \int_{-\infty}^{\infty} f_{\mathbf{X}}(\mathbf{x})\, d\mathbf{x} = 1.$$

We work with the integral

$$\int_{-\infty}^{\infty} \cdots \int_{-\infty}^{\infty} \exp\left\{ -\frac{1}{2}(\mathbf{x} - \boldsymbol{\mu})^{\mathrm{T}} \Sigma^{-1}(\mathbf{x} - \boldsymbol{\mu}) \right\} d\mathbf{x} \qquad (8.2)$$

Since Σ is positive definite, there is a positive definite, and hence non-singular, matrix $C \in M_{mm}$ such that $\Sigma = CC^{\mathrm{T}}$, i.e. $\Sigma^{-1} = (C^{\mathrm{T}})^{-1}C^{-1}$ or $C^{\mathrm{T}}\Sigma^{-1}C = I_m$.

In the integral (8.2), change the variable to $\mathbf{y}$, where $\mathbf{x} - \boldsymbol{\mu} = C\mathbf{y}$. Then

$$\frac{\partial x_i}{\partial y_i} = c_{ij}, \qquad \text{the } (i, j)\text{th element of } C,$$

so the Jacobian of this transformation is $J = |C|$ (noting that $|C| > 0$ since C is positive definite, Result A4.1).

Now,

$$\Sigma = CC^{\mathrm{T}} \Rightarrow |\Sigma| = |C||C^{\mathrm{T}}| = |C|^2 \Rightarrow |\Sigma|^{1/2} = |C| \Rightarrow J = |\Sigma|^{1/2}.$$

Also

$$(\mathbf{x} - \boldsymbol{\mu})^{\mathrm{T}} \Sigma^{-1} (\mathbf{x} - \boldsymbol{\mu}) = (C\mathbf{y})^{\mathrm{T}} \Sigma^{-1} (C\mathbf{y})$$
$$= \mathbf{y}^{\mathrm{T}} (C^{\mathrm{T}} \Sigma^{-1} C) \mathbf{y}$$
$$= \mathbf{y}^{\mathrm{T}} \mathbf{y} \quad (\text{since } C^{\mathrm{T}} \Sigma^{-1} C = I_m)$$
$$= \sum_{i=1}^{m} y_i^2$$

So, the integral (8.2) is

$$\int_{-\infty}^{\infty} \cdots \int_{-\infty}^{\infty} |\Sigma|^{1/2} \exp \left(-\frac{1}{2} \sum_{i=1}^{m} y_i^2 \right) d\mathbf{y} = |\Sigma|^{1/2} \cdot \prod_{i=1}^{m} \left\{ \int_{-\infty}^{\infty} \exp \left(-\frac{y_i^2}{2} \right) dy_i \right\}.$$

But

$$\int_{-\infty}^{\infty} \frac{1}{\sqrt{2\pi}} \exp \left(-\frac{y_i^2}{2} \right) dy = 1 \quad (\text{the } N(0,1) \text{ p.d.f. integrates to 1})$$

So

$$\int_{-\infty}^{\infty} \exp \left(-\frac{y_i^2}{2} \right) dy = \sqrt{2\pi}.$$

This shows that the integral (8.2) is equal to $|\Sigma|^{1/2} (2\pi)^{m/2}$, and the proof is complete. ∎

Proposition 8.2

Suppose $\mathbf{X} \sim N_m (\boldsymbol{\mu}, \Sigma)$, where Σ is a positive definite matrix. Then the contours of constant density

$$(2\pi)^{-m/2} |\Sigma|^{-1/2} \exp \left\{ -\frac{1}{2} (\mathbf{x} - \boldsymbol{\mu})^{\mathrm{T}} \Sigma^{-1} (\mathbf{x} - \boldsymbol{\mu}) \right\} = k \qquad (8.3)$$

are concentric ellipsoids with centre $\boldsymbol{\mu}^{\mathrm{T}}$.

Proof. The canonical equation of an ellipsoid in $\mathbb{R}^m$ is

$$\frac{y_1^2}{a_1^2} + \cdots + \frac{y_m^2}{a_m^2} = 1,$$

where $a_1, \ldots, a_m$ are non-zero real constants. We must rewrite (8.3) in the required form, using a suitable transformation of variables.

As in the proof of Proposition 8.1, write $\mathbf{x} - \boldsymbol{\mu} = C\mathbf{y}$ (where $\Sigma = CC^{\mathrm{T}}$). Then

$$\exp \left\{ -\frac{1}{2} (\mathbf{x} - \boldsymbol{\mu})^{\mathrm{T}} \Sigma^{-1} (\mathbf{x} - \boldsymbol{\mu}) \right\} = \exp \left\{ -\frac{1}{2} \mathbf{y}^{\mathrm{T}} \mathbf{y} \right\} = \exp \left\{ -\frac{1}{2} \sum_{i=1}^{m} y_i^2 \right\}$$

$$\therefore (2\pi)^{-m/2} |\Sigma|^{-1/2} \exp\left\{-\frac{1}{2}(\mathbf{x}-\boldsymbol{\mu})^{\mathrm{T}}\Sigma^{-1}(\mathbf{x}-\boldsymbol{\mu})\right\} = k \Leftrightarrow \sum_{i=1}^{p} y_i^2$$

$$= -2\log_e\left\{k(2\pi)^{m/2}|\Sigma|^{1/2}\right\}.$$

This is an ellipsoid (in fact a spheroid) in **y**-space. But **y**-space is just **x**-space, with the origin shifted to $\boldsymbol{\mu}^{\mathrm{T}}$ and the axes rotated in accordance with C. So, (8.3) represents an ellipsoid in $\mathbb{R}^m$ with origin at $\boldsymbol{\mu}^{\mathrm{T}}$. ∎

Proposition 8.3

Suppose $\mathbf{X} \sim \mathrm{N}_m(\boldsymbol{\mu}, \Sigma)$, where Σ is a positive definite matrix. Then $\mathbf{X}$ has moment-generating function

$$M_{\mathbf{X}}(\mathbf{t}) = \exp\left\{\boldsymbol{\mu}^{\mathrm{T}}\mathbf{t} + \frac{1}{2}\mathbf{t}^{\mathrm{T}}\Sigma\mathbf{t}\right\}, \qquad \mathbf{t} \in \mathbb{R}^m.$$

Proof. By the definition of moment-generating functions,

$$M_{\mathbf{X}}(\mathbf{t}) = \mathbb{E}(\exp\{\mathbf{x}^{\mathrm{T}}\mathbf{t}\})$$

$$= (2\pi)^{-m/2}|\Sigma|^{-1/2}\int_{-\infty}^{\infty}\cdots\int_{-\infty}^{\infty}\exp\{\mathbf{x}^{\mathrm{T}}\mathbf{t}\}.\exp\left\{-\frac{1}{2}(\mathbf{x}-\boldsymbol{\mu})^{\mathrm{T}}\Sigma^{-1}(\mathbf{x}-\boldsymbol{\mu})\right\}d\mathbf{x}.$$

Again make the change of variables $\mathbf{x} - \boldsymbol{\mu} = C\mathbf{y}$ (where $\Sigma = CC^{\mathrm{T}}$), so that $J = |\Sigma|^{1/2}$. Then,

$$\exp\left\{-\frac{1}{2}(\mathbf{x}-\boldsymbol{\mu})^{\mathrm{T}}\Sigma^{-1}(\mathbf{x}-\boldsymbol{\mu})\right\} = \exp\left\{-\frac{1}{2}\sum_{i=1}^{m}y_i^2\right\}$$

and

$$\exp\left\{\mathbf{x}^{\mathrm{T}}\mathbf{t}\right\} = \exp\left\{\boldsymbol{\mu}^{\mathrm{T}}\mathbf{t}\right\}.\exp\left\{\mathbf{y}^{\mathrm{T}}C^{\mathrm{T}}\mathbf{t}\right\}.$$

Hence,

$$M_{\mathbf{X}}(\mathbf{t}) = (2\pi)^{-m/2}\exp\{\boldsymbol{\mu}^{\mathrm{T}}\mathbf{t}\}\int_{-\infty}^{\infty}\cdots\int_{-\infty}^{\infty}\exp\left\{\sum_{i=1}^{m}y_i\sum_{j=1}^{m}c_{ji}t_j - \frac{1}{2}\sum_{i=1}^{m}y_i^2\right\}d\mathbf{y}$$

$$= (2\pi)^{-m/2}\exp\{\boldsymbol{\mu}^{\mathrm{T}}\mathbf{t}\}\prod_{i=1}^{m}\left\{\int_{-\infty}^{\infty}\exp\left\{y_i\sum_{j=1}^{m}c_{ji}t_j - \frac{1}{2}y_i^2\right\}dy_i\right\}.$$

Letting $d_i = \sum_{j=1}^{m}c_{ji}t_j$ (the jth element of $C^{\mathrm{T}}\mathbf{t}$), then

$$y_i d_i - \frac{1}{2}y_i^2 = -\frac{1}{2}\left\{(y_i - d_i)^2 - d_i^2\right\}$$

and

$$\int_{-\infty}^{\infty} \exp\left\{y_i d_i - \frac{1}{2}y_i^2\right\} dy_i = \exp\left\{\frac{d_i^2}{2}\right\} \int_{-\infty}^{\infty} \exp\left\{\frac{(y_i - d_i)^2}{2}\right\} dy_i$$

$$= \exp\left\{\frac{d_i^2}{2}\right\} \sqrt{2\pi}.$$

So

$$M_{\mathbf{X}}(\mathbf{t}) = \exp\{\boldsymbol{\mu}^{\mathsf{T}}\mathbf{t}\} \prod_{i=1}^{m} \exp\left\{\frac{d_i^2}{2}\right\} = \exp\{\boldsymbol{\mu}^{\mathsf{T}}\mathbf{t}\} \exp\left\{\frac{1}{2}\sum_{i=1}^{m} d_i^2\right\}$$

But

$$\sum_{i=1}^{m} d_i^2 = (C^{\mathsf{T}}\mathbf{t})^{\mathsf{T}}(C^{\mathsf{T}}\mathbf{t}) = \mathbf{t}^{\mathsf{T}}\Sigma\mathbf{t},$$

so

$$M_{\mathbf{X}}(\mathbf{t}) = \exp\left\{\boldsymbol{\mu}^{\mathsf{T}}\mathbf{t} + \frac{1}{2}\mathbf{t}^{\mathsf{T}}\Sigma\mathbf{t}\right\}. \qquad \blacksquare$$

The above moment-generating function exists even for singular matrices Σ, when the joint probability density function itself does not exist. Since $\Sigma = \text{cov}(\mathbf{X})$, as we shall shortly prove, we restrict Σ to be positive semi-definite and give the following, broadest possible, definition of the multivariate normal distribution.

The random vector $\mathbf{X}$ is said to follow a *multivariate normal* (MVN) distribution if $\mathbf{X}$ has moment-generating function

$$M_{\mathbf{X}}(\mathbf{t}) = \exp\left\{\boldsymbol{\mu}^{\mathsf{T}}\mathbf{t} + \frac{1}{2}\mathbf{t}^{\mathsf{T}}\Sigma\mathbf{t}\right\}$$

for some $\boldsymbol{\mu} \in \mathbb{R}^m$ and some positive semi-definite matrix $\Sigma \in M_{mm}$. We write $\mathbf{X} \sim N_m(\boldsymbol{\mu}, \Sigma)$.

When Σ is positive semi-definite (but not positive definite) the MVN is a random vector whose moment-generating function exists but whose joint probability density function does not exist. The moments of this random vector also exist even though the probability density function does not.

Example 8.2

Consider the random vector $\mathbf{X} = [X_1 \ X_2 \ X_3]^{\mathsf{T}}$, where X_1, X_2 are independent $N(0, \sigma^2)$ random variables and $X_3 = X_1 + X_2$.

By a reproductive property of the normal distribution (Chapter 6), $X_3 \sim N(0, 2\sigma^2)$. Using the independence of X_1 and X_2, we may write

$$\text{cov}(X_1, X_2) = 0,$$
$$\text{cov}(X_1, X_3) = \text{cov}(X_1, X_1 + X_2) = \text{var}(X_1) + \text{cov}(X_1, X_2) = \sigma^2 + 0 = \sigma^2,$$
$$\text{cov}(X_2, X_3) = \text{cov}(X_2, X_1 + X_2) = \text{var}(X_2) + \text{cov}(X_1, X_2) = \sigma^2 + 0 = \sigma^2.$$

Therefore, $\mathbf{X}$ has expected value and covariance matrix

$$\mu = \mathbb{E}(\mathbf{X}) = \begin{bmatrix} 0 \\ 0 \\ 0 \end{bmatrix}, \qquad \Sigma = \text{cov}(\mathbf{X}) = \begin{bmatrix} \sigma^2 & 0 & \sigma^2 \\ 0 & \sigma^2 & \sigma^2 \\ \sigma^2 & \sigma^2 & 2\sigma^2 \end{bmatrix}.$$

The moment-generating function of $\mathbf{X}$ is obtained from the usual definition as follows:

$$\begin{aligned}
M_{\mathbf{X}}(\mathbf{t}) &= \mathbb{E}\{\exp(X_1 t_1 + X_2 t_2 + X_3 t_3)\} \\
&= \mathbb{E}\{\exp(X_1 t_1 + X_2 t_2 + X_1 t_3 + X_2 t_3)\} \\
&= \mathbb{E}\{\exp(X_1(t_1 + t_3)) \cdot \exp(X_2(t_2 + t_3))\} \\
&= \mathbb{E}\{\exp(X_1(t_1 + t_3))\}\mathbb{E}\{\exp(X_2(t_2 + t_3))\} \\
&\quad \text{(independence of } X_1 \text{ and } X_2) \\
&= M_1(t_1 + t_3)\, M_2(t_2 + t_3) \\
&= \exp\left\{ 0.\,(t_1 + t_3) + \frac{1}{2}\sigma^2 (t_1 + t_3)^2 \right\} \exp\left\{ 0 \cdot (t_2 + t_3) + \frac{1}{2}\sigma^2 (t_2 + t_3)^2 \right\} \\
&= \exp\left\{ [0 \;\; 0 \;\; 0] \begin{bmatrix} t_1 \\ t_2 \\ t_3 \end{bmatrix} + \frac{1}{2}[t_1 \;\; t_2 \;\; t_3] \begin{bmatrix} \sigma^2 & 0 & \sigma^2 \\ 0 & \sigma^2 & \sigma^2 \\ \sigma^2 & \sigma^2 & 2\sigma^2 \end{bmatrix} \begin{bmatrix} t_1 \\ t_2 \\ t_3 \end{bmatrix} \right\}.
\end{aligned}$$

This is the moment-generating function of an $N_3(\mu, \Sigma)$ random vector.

However, it is easy to check that $|\Sigma| = 0$, i.e. Σ is singular. This means that we cannot write down the probability density function of $\mathbf{X}$. This is because the probability is concentrated on the plane $X_3 = X_1 + X_2$.

Proposition 8.4

Suppose $\mathbf{X} \sim N_m(\mu, \Sigma)$. Then $\mathbb{E}(\mathbf{X}) = \mu$ and $\text{cov}(\mathbf{X}) = \Sigma$.

Proof.

$$M_{\mathbf{T}}(\mathbf{t}) = \exp\{\mu^T \mathbf{t} + \tfrac{1}{2}\mathbf{t}^T \Sigma \mathbf{t}\}.$$

This means that X_i has the moment-generating function

$$M_i(t_i) = M_{\mathbf{X}}\{[0\cdots 0\ t_i\ 0\cdots 0]^{\mathrm{T}}\} = \exp\left\{\mu_i t_i + \frac{1}{2}t_i^2\sigma_{ii}\right\}.$$

This is the moment-generating function of an $N(\mu_i, \sigma_{ii})$ distribution. Hence, by the uniqueness property of moment-generating functions, $X_i \sim N(\mu_i, \sigma_{ii})$. It follows that $\mathbb{E}(X_i) = \mu_i$ and $\mathrm{var}(X_i) = \sigma_{ii}$.

In a similar way, X_i and X_j $(i < j)$ have the (joint) moment-generating function

$$M_{ij}(t_i, t_j) = M_{\mathbf{X}}\{[0\cdots 0\ t_i 0\cdots 0\ t_j 0\cdots 0]\}$$

$$= \exp\{\mu_i t_i + \mu_j t_j + \tfrac{1}{2}t_i^2\sigma_{ii} + \tfrac{1}{2}t_j^2\sigma_{jj} + t_i t_j \sigma_{ij}\}$$

and

$$\frac{\partial^2}{\partial t_i \partial t_j}M_{ij}(t_i, t_j) = \exp\left\{\mu_i t_i + \mu_j t_j + \tfrac{1}{2}t_i^2\sigma_{ii} + \tfrac{1}{2}t_j^2\sigma_{jj} + t_i t_j\sigma_{ij}\right\} \times \sigma_{ij}$$

$$+ \exp\left\{\mu_i t_i + \mu_j t_j + \tfrac{1}{2}t_i^2\sigma_{ii} + \tfrac{1}{2}t_j^2\sigma_{jj} + t_i t_j\sigma_{ij}\right\}$$

$$\times [\mu_i + t_i\sigma_{ii} + t_j\sigma_{ij}][\mu_j + t_j\sigma_{jj} + t_i\sigma_{ij}]$$

$$\therefore \quad \mathbb{E}(X_i X_j) = \frac{\partial^2}{\partial t_i \partial t_j}M_{ij}(t_i, t_j)\big|_{t_i=0, t_j=0} = \sigma_{ij} + \mu_i\mu_j$$

$$\therefore \quad \mathrm{cov}(X_i X_j) = \mathbb{E}(X_i X_j) - \mathbb{E}(X_i)\mathbb{E}(X_j) = \sigma_{ij}.$$

So $\mathbb{E}(\mathbf{X}) = \mu$ and $\mathrm{cov}(\mathbf{X}) = \Sigma$ as required. ∎

Proposition 8.5

The m-dimensional random vector $\mathbf{X}$ has an MVN distribution if and only if $\mathbf{a}^{\mathrm{T}}\mathbf{X}$ has a (univariate) normal distribution, for all $\mathbf{a} \in \mathbb{R}^m$ $(\mathbf{a} \neq \mathbf{0})$.

Proof. Suppose $\mathbf{X} \sim N_m(\mu, \Sigma)$, so that $M_{\mathbf{X}}(\mathbf{t}) = \exp\left\{\mu^{\mathrm{T}}\mathbf{t} + \frac{1}{2}\mathbf{t}^{\mathrm{T}}\Sigma\mathbf{t}\right\}$. Choose any $\mathbf{a} \in \mathbb{R}^m$. Then

$$M_{\mathbf{a}^{\mathrm{T}}\mathbf{X}}(t) = \mathbb{E}\{\exp\{(\mathbf{a}^{\mathrm{T}}\mathbf{X})t\}\}$$

$$= \mathbb{E}\{\exp\{(\mathbf{a}^{\mathrm{T}}\mathbf{X})^{\mathrm{T}}t\}\}$$

$$= \mathbb{E}\{\exp\{\mathbf{X}^{\mathrm{T}}\mathbf{a}\,t\}\}$$

$$= M_{\mathbf{X}}(\mathbf{a}\,t)$$

$$= \exp\{\mu^{\mathrm{T}}\mathbf{a}\,t + \tfrac{1}{2}(\mathbf{a}\,t)^{\mathrm{T}}\Sigma(\mathbf{a}\,t)\}$$

$$= \exp\{(\mu^{\mathrm{T}}\mathbf{a})t + \tfrac{1}{2}(\mathbf{a}^{\mathrm{T}}\Sigma\mathbf{a})t^2\}$$

Σ is positive semi-definite, so $\mathbf{a}^T\Sigma\mathbf{a} \geq 0$ and this is the moment-generating function of an $N(\mathbf{a}^T\mu, \mathbf{a}^T\Sigma\mathbf{a})$ distribution, i.e. $\mathbf{a}^T\mathbf{X} \sim N(\mathbf{a}^T\mu, \mathbf{a}^T\Sigma\,\mathbf{a})$ by the uniqueness property of moment-generating functions.

To prove the reverse direction, suppose $\mathbf{a}^T\mathbf{X}$ is normally distributed for all $\mathbf{a} \in \mathbb{R}^m$. Taking $\mathbf{a}_i^T = [0\cdots010\cdots0]$, $X_i = \mathbf{a}_i^T\mathbf{X}$ has a normal distribution, say $X_i \sim N(\mu_i, \sigma_{ii})$. So μ_i and σ_{ii} exist and are finite. This means that $\mathbb{E}(\mathbf{X}) = [\mu_1 \dots \mu_m]^T = \mu$ (say) exists and is finite.

Since $\rho_{ij} \leq 1$ for all i and j ($i \neq j$),

$$\sigma_{ij} = \{\text{cov}(X_i, X_j)\} \leq \sqrt{\text{var}(X_i).\text{var}(X_j)} = \sqrt{\sigma_{ii}\sigma_{jj}} < \infty$$

i.e. $\text{cov}(\mathbf{X}) = [\sigma_{ij}] = \Sigma$ (say) exists and is finite.

The moment-generating function of $\mathbf{X}$ is

$$M_{\mathbf{X}}(\mathbf{t}) = \mathbb{E}(e^{\mathbf{X}^T\mathbf{t}}) = M_{\mathbf{t}^T\mathbf{X}}(1) = \exp\{\mathbf{t}^T\mu + \frac{1}{2}\mathbf{t}^T\Sigma\mathbf{t}\}$$

The last step in the line above follows because $\mathbf{t}^T\mathbf{X}$ has a normal distribution by assumption, and $\mathbb{E}(\mathbf{t}^T\mathbf{X}) = \mathbf{t}^T\mathbb{E}(\mathbf{X}) = \mathbf{t}^T\mu$ while $\text{cov}(\mathbf{t}^T\mathbf{X}) = \mathbf{t}^T\text{cov}(\mathbf{X}).\mathbf{t} = \mathbf{t}^T\Sigma\mathbf{t}$.

Hence, by the uniqueness properties of moment-generating functions, $\mathbf{X} \sim N_m(\mu, \Sigma)$. ∎

In particular, Proposition 8.5 tells us that, when $\mathbf{X} \sim N_m(\mu, \Sigma)$, it follows that each X_i is marginally distributed as an $N(\mu_i, \sigma_{ii})$ random variable. However, even when every X_i is marginally normally distributed, it does not follow that $\mathbf{X}$ follows a MVN distribution. The next example shows this.

Example 8.3

The two-dimensional random vector $\mathbf{X}$ has probability density function

$$f(x_1, x_2) = \frac{1}{2\pi} \exp\left\{-\tfrac{1}{2}(x_1^2 + x_2^2)\right\}\{1 + x_1 x_2 \exp\left(-\tfrac{1}{2}(x_1^2 + x_2^2)\right)\}, \qquad x_1, x_2 \in \mathbb{R}.$$

This is clearly *not* a bivariate normal distribution. However, we shall show that both X_1 and X_2 marginally follow the standard normal distribution:

$$f_1(x_1) = \int_{-\infty}^{\infty} f(x_1, x_2)\, dx_2$$

$$= \frac{1}{2\pi} \exp\left\{-\frac{x_1^2}{2}\right\} \int_{-\infty}^{\infty} \exp\left\{-\frac{x_2^2}{2}\right\} (1 + x_1 x_2 \exp\left[-\tfrac{1}{2}(x_1^2 + x_2^2)\right])\, dx_2$$

$$= \frac{1}{2\pi} \exp\left\{-\frac{x_1^2}{2}\right\} \left\{ \int_{-\infty}^{\infty} \exp\left\{-\frac{x_2^2}{2}\right\} dx_2 \right.$$

$$\left. + x_1 \int_{-\infty}^{\infty} x_2 \exp\left\{-\tfrac{1}{2}x_2^2 - \tfrac{1}{2}(x_1^2 + x_2^2)\right\} dx_2 \right\}$$

$$= \frac{1}{2\pi} \exp\left\{-\frac{x_1^2}{2}\right\} \left\{ \sqrt{2\pi} + x_1 \exp\left\{-\frac{x_1^2}{2}\right\} \underbrace{\int_{-\infty}^{\infty} x_2 \, e^{-x_2^2} \, dx_2}_{\substack{= 0, \text{ since integrand} \\ \text{is an odd function}}} \right\}$$

$$= \frac{1}{\sqrt{2\pi}} \exp\left\{-\frac{x_1^2}{2}\right\}$$

i.e. $X_1 \sim N(0, 1)$. By symmetry, $X_2 \sim N(0, 1)$. So both the marginal distributions are normal but the joint distribution is not MVN. In light of Proposition 8.5, there must be real, non-zero constants a_1 and a_2 such that $a_1 X_1 + a_2 X_2$ does not (marginally) follow a normal distribution.

Proposition 8.6

Suppose that the m-dimensional random vector $\mathbf{X}$ follows the $N_m(\mu, \Sigma)$ distribution. If $A \in M_{qm}$ is a matrix of constants, and $\mathbf{b} \in \mathbb{R}^q$ is a vector of constants, then

$$A\mathbf{X} + \mathbf{b} \sim N_q(A\mu + \mathbf{b}, \, A\Sigma A^T).$$

If $\mathbf{X}$ follows a non-singular MVN distribution, then the distribution of $A\mathbf{X} + \mathbf{b}$ is also non-singular if and only if A has rank q. (Among other things, this requires that $q \leq m$.)

Proof.

$$\begin{aligned} M_{A\mathbf{X}+\mathbf{b}}(\mathbf{t}) &= e^{\mathbf{b}^T \mathbf{t}} M_{\mathbf{X}}(A^T \mathbf{t}) \\ &= e^{\mathbf{b}^T \mathbf{t}} \exp\left\{\mu^T(A^T \mathbf{t}) + \tfrac{1}{2}(A^T \mathbf{t})^T \Sigma (A^T \mathbf{t})\right\} \\ &= \exp\left\{[A\mu + \mathbf{b}]^T \mathbf{t} + \tfrac{1}{2}\mathbf{t}^T[A\Sigma A^T]\mathbf{t}\right\}. \end{aligned}$$

This is the moment-generating function of the $N_q(A\mu + \mathbf{b}, \, A\Sigma A^T)$ distribution. Hence, by the uniqueness property of moment-generating functions, $A\mathbf{X} + \mathbf{b} \sim N_q(A\mu + \mathbf{b}, A\Sigma A^T)$.

So long as Σ is non-singular (and hence positive definite), $A\Sigma A^T$ is positive definite if and only if A has rank q. ∎

Example 8.4

Suppose that $\mathbf{X}$ follows a *spherical normal* distribution, i.e. $\mathbf{X} \sim N_m(\mu, \sigma^2 I_m)$. Let Q be any orthogonal, $m \times m$ matrix. It follows from Proposition 8.6 that $Q\mathbf{X} \sim N_m(Q\mu, \sigma^2 I_m)$. In particular, if $\mu = \mathbf{0}$, then $\mathbf{X}$ and $Q\mathbf{X}$ follow the same distribution, namely $N_m(\mathbf{0}, \sigma^2 I_m)$.

Corollary 8.7

If $\mathbf{X} \sim N_m(\boldsymbol{\mu}, \Sigma)$, then any sub-vector of $\mathbf{X}$ has a (joint) marginal distribution that is also MVN. If $\mathbf{X}$ has a non-singular distribution, then so too has any sub-vector of $\mathbf{X}$.

Proof. We can permute the elements of $\mathbf{X}$ as we please and still obtain an MVN distribution, so we shall assume (without loss of generality) that the sub-vector whose distribution we wish to find is $\mathbf{X}^{(1)}$, consisting of the first r $(1 \le r \le m-1)$ elements of $\mathbf{X}$. ∎

We partition $\mathbf{X}$, $\boldsymbol{\mu}$ and Σ conformably as follows:

$$\mathbf{X} = \begin{bmatrix} \mathbf{X}^{(1)} \\ \mathbf{X}^{(2)} \end{bmatrix}, \qquad \boldsymbol{\mu} = \begin{bmatrix} \boldsymbol{\mu}^{(1)} \\ \boldsymbol{\mu}^{(2)} \end{bmatrix}, \qquad \Sigma = \begin{bmatrix} \Sigma_{11} & \Sigma_{12} \\ \Sigma_{12}^T & \Sigma_{22} \end{bmatrix}.$$

Let $B = [I_r \mid 0_{r,m-r}]$. Then B is an $r \times m$ matrix of rank r. Also,

$$B\mathbf{X} = \mathbf{X}^{(1)}, \qquad B\boldsymbol{\mu} = \boldsymbol{\mu}^{(1)}, \qquad B\Sigma B^T = \Sigma_{11}.$$

By Proposition 8.6, then, $\mathbf{X}^{(1)} \sim N_r(\boldsymbol{\mu}^{(1)}, \Sigma_{11})$. If Σ is non-singular (i.e. positive definite), so too is $B \Sigma B^T = \Sigma_{11}$.

Example 8.5

Suppose that

$$\mathbf{X} \sim N_3 \left(\begin{bmatrix} 1 \\ 2 \\ 3 \end{bmatrix}, \begin{bmatrix} 1 & 0.2 & 0.3 \\ 0.2 & 1 & 0.4 \\ 0.3 & 0.4 & 1 \end{bmatrix} \right).$$

It follows from Proposition 8.5 that $X_1 \sim N(1, 1)$, $X_2 \sim N(2, 1)$, $X_3 \sim N(3, 1)$.
From Corollary 8.7, we can obtain the distribution of (X_1, X_2) as follows:

$$\mathbf{X} = \begin{bmatrix} X_1 \\ X_2 \\ X_3 \end{bmatrix}, \qquad \boldsymbol{\mu} = \begin{bmatrix} 1 \\ 2 \\ 3 \end{bmatrix}, \qquad \Sigma = \begin{bmatrix} 1 & 0.2 & 0.3 \\ 0.2 & 1 & 0.4 \\ 0.3 & 0.4 & 1 \end{bmatrix}$$

and so

$$\begin{bmatrix} X_1 \\ X_2 \end{bmatrix} \sim N_2 \left(\begin{bmatrix} 1 \\ 2 \end{bmatrix}, \begin{bmatrix} 1 & 0.2 \\ 0.2 & 1 \end{bmatrix} \right).$$

Similarly (permuting the components of $\mathbf{X}$ as necessary),

$$\begin{bmatrix} X_1 \\ X_3 \end{bmatrix} \sim N_2 \left(\begin{bmatrix} 1 \\ 3 \end{bmatrix}, \begin{bmatrix} 1 & 0.3 \\ 0.3 & 1 \end{bmatrix} \right)$$

$$\begin{bmatrix} X_2 \\ X_3 \end{bmatrix} \sim N_2\left(\begin{bmatrix} 2 \\ 3 \end{bmatrix}, \begin{bmatrix} 1 & 0.4 \\ 0.4 & 1 \end{bmatrix}\right).$$

Exercises

1 The random vector $\mathbf{X} = [X_1\ X_2\ X_3]^T$ follows an MVN distribution with

$$\boldsymbol{\mu} = \mathbb{E}(\mathbf{X}) = \begin{bmatrix} -1 \\ 0 \\ 1 \end{bmatrix}, \qquad \Sigma = \mathrm{cov}(\mathbf{X}) = \begin{bmatrix} 9 & 6 & 3 \\ 6 & 9 & 6 \\ 3 & 6 & 9 \end{bmatrix}.$$

Write down the marginal distributions of $X_i(i = 1, 2, 3)$ and $(X_i, X_j), i < j$.

2 Suppose that $X_1, X_2, \ldots, X_n$ are independent random variables all following the standard normal distribution. Define the MVN random vector $\mathbf{Y}$ by $\mathbf{Y} = Q\mathbf{X}$, where the $n \times n$ matrix Q is Helmert's matrix

$$Q = \begin{bmatrix} \frac{1}{\sqrt{2}} & \frac{-1}{\sqrt{2}} & 0 & 0 & \cdots & 0 & 0 \\ \frac{1}{\sqrt{6}} & \frac{1}{\sqrt{6}} & \frac{-2}{\sqrt{6}} & 0 & \cdots & 0 & 0 \\ \frac{1}{\sqrt{12}} & \frac{1}{\sqrt{12}} & \frac{1}{\sqrt{12}} & \frac{-3}{\sqrt{12}} & \cdots & 0 & 0 \\ \vdots & \vdots & \vdots & \vdots & \ddots & \vdots & \vdots \\ \frac{1}{\sqrt{n(n-1)}} & \frac{1}{\sqrt{n(n-1)}} & \frac{1}{\sqrt{n(n-1)}} & \frac{1}{\sqrt{n(n-1)}} & \cdots & \frac{1}{\sqrt{n(n-1)}} & \frac{-(n-1)}{\sqrt{n(n-1)}} \\ \frac{1}{\sqrt{n}} & \frac{1}{\sqrt{n}} & \frac{1}{\sqrt{n}} & \frac{1}{\sqrt{n}} & \cdots & \frac{1}{\sqrt{n}} & \frac{1}{\sqrt{n}} \end{bmatrix}$$

(a) Show that $\mathbf{Y}$ follows the same distribution as $\mathbf{X}$, so that $Y_1, \ldots, Y_n$ are also independent standard normal random variables.

(b) Define the sample mean and sample variance of the X_i by

$$\overline{X} = \frac{1}{n}\sum_{i=1}^{n} X_i, \qquad S^2 = \frac{1}{n-1}\sum_{i=1}^{n}(X_i - \overline{X})^2.$$

Show that

$$n\overline{X}^2 = Y_n^2, \qquad S^2 = \frac{1}{n-1}\sum_{i=1}^{n-1} Y_i^2.$$

Deduce that the sample mean and variance are independent random variables.

3 Suppose that $\mathbf{X} = [X_1\ X_2\ X_3]^T$ follows a (non-singular) MVN distribution such that each X_i has expected value μ and variance $\sigma^2(>0)$ and that $\rho_{ij} = \rho$ for all pairs $i \neq j$. Let $\overline{X}$ denote the sample mean of X_1, X_2 and X_3.

(a) Write down the distribution of $\mathbf{Y} = [X_1 \ X_2 \ X_3 \ \overline{X}]^{\mathrm{T}}$.

(b) Show that $\mathrm{cov}(\mathbf{Y})$ is singular and explain why this is the case.

(c) Find the marginal distribution of $\overline{X}$.

8.2 Independence and association

Proposition 8.8

Suppose that the m-dimensional random vector $\mathbf{X}$ follows the $N_m(\mu, \Sigma)$ distribution. Write

$$\mathbf{X} = \begin{bmatrix} \mathbf{X}^{(1)} \\ \mathbf{X}^{(2)} \end{bmatrix},$$

where $\mathbf{X}^{(1)}$ consists of the first r $(1 \le r \le m-1)$ elements of $\mathbf{X}$. Partition μ and Σ conformably as

$$\mu = \begin{bmatrix} \mu^{(1)} \\ \mu^{(2)} \end{bmatrix}, \qquad \Sigma = \begin{bmatrix} \Sigma_{11} & \Sigma_{12} \\ \Sigma_{12}^{\mathrm{T}} & \Sigma_{22} \end{bmatrix}.$$

Then $\mathbf{X}^{(1)}$ and $\mathbf{X}^{(2)}$ are independent random vectors if and only if $\Sigma_{12} = 0$.

Proof. If $\mathbf{X}^{(1)}$ and $\mathbf{X}^{(2)}$ are independent, then

$$\mathrm{cov}(\mathbf{X}^{(1)}, \mathbf{X}^{(2)}) = 0 \Rightarrow \Sigma_{12} = 0.$$

To prove the reverse direction, suppose that $\Sigma_{12} = 0$. Then

$$\mathbf{t}^{\mathrm{T}} \Sigma \mathbf{t} = \mathbf{t}^{(1)\mathrm{T}} \Sigma_{11} \mathbf{t}^{(1)} + \mathbf{t}^{(2)\mathrm{T}} \Sigma_{22} \mathbf{t}^{(2)} \quad \forall \mathbf{t} = \begin{bmatrix} \mathbf{t}^{(1)} \\ \mathbf{t}^{(2)} \end{bmatrix} \in \mathbb{R}^m$$

Therefore,

$$\begin{aligned}
M_{\mathbf{X}}(\mathbf{t}) &= \exp\left\{\mu^{\mathrm{T}} \mathbf{t} + \tfrac{1}{2} \mathbf{t}^{\mathrm{T}} \Sigma \mathbf{t}\right\} \\
&= \exp\left\{\mu^{(1)\mathrm{T}} \mathbf{t}^{(1)} + \mu^{(2)\mathrm{T}} \mathbf{t}^{(2)} + \tfrac{1}{2} \mathbf{t}^{(1)\mathrm{T}} \Sigma_{11} \mathbf{t}^{(1)} + \tfrac{1}{2} \mathbf{t}^{(2)\mathrm{T}} \Sigma_{22} \mathbf{t}^{(2)}\right\} \\
&= \exp\left\{\mu^{(1)\mathrm{T}} \mathbf{t}^{(1)} + \tfrac{1}{2} \mathbf{t}^{(1)\mathrm{T}} \Sigma_{11} \mathbf{t}^{(1)}\right\} \cdot \exp\left\{\mu^{(2)\mathrm{T}} \mathbf{t}^{(2)} + \tfrac{1}{2} \mathbf{t}^{(2)\mathrm{T}} \Sigma_{22} \mathbf{t}^{(2)}\right\} \\
&= M_{\mathbf{X}^{(1)}}(\mathbf{t}^{(1)}) \cdot M_{\mathbf{X}^{(2)}}(\mathbf{t}^{(2)}).
\end{aligned}$$

So, by the factorization theorem for moment-generating functions, $\mathbf{X}^{(1)}$ and $\mathbf{X}^{(2)}$ are independent. ∎

Corollary 8.9

Suppose

$$\mathbf{X} = \begin{bmatrix} \mathbf{X}^{(1)} \\ \mathbf{X}^{(2)} \\ \vdots \\ \mathbf{X}^{(k)} \end{bmatrix} \sim N_m(\boldsymbol{\mu}, \boldsymbol{\Sigma}),$$

and that

$$\mathrm{cov}(\mathbf{X}^{(i)}, \mathbf{X}^{(j)}) = 0 \qquad \text{for all } i, j \ (i \neq j).$$

Then $\mathbf{X}^{(1)}, \ldots, \mathbf{X}^{(k)}$ are independent (not just pairwise independent).

Corollary 8.10

If $\mathbf{X} \sim N_m(\boldsymbol{\mu}, \boldsymbol{\Sigma})$, then $X_1, \ldots, X_m$ are independent if and only if $\boldsymbol{\Sigma} = \mathrm{diag}(\sigma_{11} \ldots \sigma_{mm})$.

For the MVN distribution, then, random variables that are uncorrelated must also be independent. This is not true in general for other distributions, as we have already seen.

Proposition 8.11

Suppose that the m-dimensional random vector $\mathbf{X}$ follows the $N_m(\boldsymbol{\mu}, \boldsymbol{\Sigma})$ distribution. Write

$$\mathbf{X} = \begin{bmatrix} \mathbf{X}^{(1)} \\ \mathbf{X}^{(2)} \end{bmatrix},$$

where $\mathbf{X}^{(1)}$ consists of the first r $(1 \leq r \leq m - 1)$ elements of $\mathbf{X}$. Partition $\boldsymbol{\mu}$ and $\boldsymbol{\Sigma}$ conformably as

$$\boldsymbol{\mu} = \begin{bmatrix} \boldsymbol{\mu}^{(1)} \\ \boldsymbol{\mu}^{(2)} \end{bmatrix}, \qquad \boldsymbol{\Sigma} = \begin{bmatrix} \Sigma_{11} & \Sigma_{12} \\ \Sigma_{12}^{\mathrm{T}} & \Sigma_{22} \end{bmatrix}.$$

Provided that Σ_{11} is non-singular, then the conditional distribution of $\mathbf{X}^{(2)}$ given $\mathbf{X}^{(1)} = \mathbf{x}^{(1)}$ is

$$N_{m-r}(\boldsymbol{\mu}^{(2)} + \Sigma_{12}^{\mathrm{T}} \Sigma_{11}^{-1}(\mathbf{x}^{(1)} - \boldsymbol{\mu}^{(1)}), \Sigma_{22} - \Sigma_{12}^{\mathrm{T}} \Sigma_{11}^{-1} \Sigma_{12}).$$

Proof. $f_{\mathbf{X}}(\mathbf{x})$ is given by equation (8.1). By Corollary 8.7, $\mathbf{X}^{(1)} \sim N_r(\boldsymbol{\mu}^{(1)}, \Sigma_{11})$ and so

$$f_{\mathbf{X}^{(1)}}(\mathbf{x}^{(1)}) = (2\pi)^{-r/2} |\Sigma|^{-1/2} \exp \left\{ -\frac{1}{2} (\mathbf{x}^{(1)} - \boldsymbol{\mu}^{(1)})^{\mathrm{T}} \Sigma_{11}^{-1} (\mathbf{x}^{(1)} - \boldsymbol{\mu}^{(1)}) \right\}, \quad \mathbf{x}^{(1)} \in \mathbb{R}^r.$$

Therefore,

$$f_{(2)|(1)}\left(\mathbf{x}^{(2)}|\mathbf{x}^{(1)}\right) = \frac{f_{\mathbf{X}}(\mathbf{x})}{f_{\mathbf{X}^{(1)}}\left(\mathbf{x}^{(1)}\right)}$$

$$= (2\pi)^{-(n-r)/2} \frac{|\Sigma|^{-1/2}}{|\Sigma_{11}|^{-1/2}}$$

$$\times \exp\left\{-\frac{1}{2}\left[(\mathbf{x} - \boldsymbol{\mu})^{\mathrm{T}}\Sigma^{-1}(\mathbf{x} - \boldsymbol{\mu})\right.\right.$$

$$\left.\left. -(\mathbf{x}^{(1)} - \boldsymbol{\mu}^{(1)})^{\mathrm{T}}\Sigma_{11}^{-1}\left(\mathbf{x}^{(1)} - \boldsymbol{\mu}^{(1)}\right)\right]\right\}.$$

The required result now follows using Result A6.1.

This proposition shows that conditional distributions in the MVN case have two important features. First, the regression of $\mathbf{X}^{(2)}$ on $\mathbf{X}^{(1)}$ (conditional expected value) is a linear function of the elements of $\mathbf{X}^{(1)}$. Second, the covariance matrix of $\mathbf{X}^{(2)}$ conditional on $\mathbf{X}^{(1)}$ does not depend on the particular value of $\mathbf{x}^{(1)}$, i.e. it is the same whatever the choice of $\mathbf{x}^{(1)}$. ∎

Example 8.5 – continued

We will now find the conditional distribution of X_2 given that $X_1 = x_1$ and $X_3 = x_3$.

In order to use Proposition 8.11 directly, we can permute the elements of $\mathbf{X}$ as follows.

$$\begin{bmatrix} X_1 \\ X_2 \\ X_3 \end{bmatrix} \sim N_3 \left(\begin{bmatrix} 1 \\ 2 \\ 3 \end{bmatrix}, \begin{bmatrix} 1 & 0.2 & 0.3 \\ 0.2 & 1 & 0.4 \\ 0.3 & 0.4 & 1 \end{bmatrix} \right)$$

$$\Leftrightarrow \begin{bmatrix} X_1 \\ X_3 \\ X_2 \end{bmatrix} \sim N_3 \left(\begin{bmatrix} 1 \\ 3 \\ 2 \end{bmatrix}, \begin{bmatrix} 1 & 0.3 & 0.2 \\ 0.3 & 1 & 0.4 \\ 0.2 & 0.4 & 1 \end{bmatrix} \right)$$

i.e.

$$\mathbf{X}^{(2)} = X_2, \qquad \mu^{(2)} = 2, \qquad \Sigma_{22} = 1,$$

$$\mathbf{X}^{(1)} = \begin{bmatrix} X_1 \\ X_3 \end{bmatrix} \qquad \boldsymbol{\mu}^{(1)} = \begin{bmatrix} 1 \\ 3 \end{bmatrix} \qquad \Sigma_{11} = \begin{bmatrix} 1 & 0.3 \\ 0.3 & 1 \end{bmatrix} \qquad \Sigma_{12}^{\mathrm{T}} = [0.2 \quad 0.4]$$

$$\therefore \quad \mu^{(2)} + \Sigma_{12}^T \Sigma_{11}^{-1}(\mathbf{x}^{(1)} - \mu^{(1)}) = 2 + \begin{bmatrix} 0.2 & 0.4 \end{bmatrix} \frac{1}{0.91} \begin{bmatrix} 1 & -0.3 \\ -0.3 & 1 \end{bmatrix} \begin{bmatrix} x_1 - 1 \\ x_3 - 3 \end{bmatrix}$$

$$= 2 + \frac{1}{9.1} \begin{bmatrix} 0.8 & 3.4 \end{bmatrix} \begin{bmatrix} x_1 - 1 \\ x_3 - 3 \end{bmatrix}$$

$$= 2 + \frac{1}{9.1} \{0.8 x_1 - 0.8 + 3.4 x_3 - 10.2\}$$

$$= 0.088 x_1 + 0.374 x_3 + 0.791$$

and

$$\Sigma_{11} - \Sigma_{12}\Sigma_{22}^{-1}\Sigma_{12}^T = 1 - \frac{1}{9.1}\begin{bmatrix} 0.8 & 3.4 \end{bmatrix}\begin{bmatrix} 0.2 \\ 0.4 \end{bmatrix} = 1 - \left(\frac{1}{91} \times 15.2\right) = 0.833.$$

That is, given (x_1, x_3), the conditional distribution of X_2 is $N(0.088 x_1 + 0.374 x_3 + 0.791, 0.833)$.

Example 8.6

Consider the special case where $m = 3$ and $r = 1$. Here, $\mathbf{X}$ follows the $N_3(\mu, \Sigma)$ distribution. In order to find the parameters of the conditional distribution of (X_2, X_3) given X_1, we can write

$$\mathbf{X} = \begin{bmatrix} X_1 \\ X_2 \\ X_3 \end{bmatrix} \quad \mu = \begin{bmatrix} \mu_1 \\ \mu_2 \\ \mu_3 \end{bmatrix} \quad \Sigma = \begin{bmatrix} \sigma_1^2 & \rho_{12}\sigma_1\sigma_2 & \rho_{13}\sigma_1\sigma_3 \\ \rho_{12}\sigma_1\sigma_2 & \sigma_2^2 & \rho_{23}\sigma_2\sigma_3 \\ \rho_{13}\sigma_1\sigma_3 & \rho_{23}\sigma_2\sigma_3 & \sigma_3^2 \end{bmatrix}.$$

So

$$\mu^{(2)} + \Sigma_{12}^T \Sigma_{11}^{-1}(\mathbf{x}^{(1)} - \mu^{(1)}) = \begin{bmatrix} \mu_2 \\ \mu_3 \end{bmatrix} + \begin{bmatrix} \rho_{12}\sigma_1\sigma_2 \\ \rho_{13}\sigma_1\sigma_3 \end{bmatrix}\frac{1}{\sigma_1^2}(x_1 - \mu_1)$$

$$= \begin{bmatrix} \mu_2 + \frac{\rho_{12}\sigma_2}{\sigma_1}x_1 - \frac{\rho_{12}\sigma_2}{\sigma_1}\mu_1 \\ \mu_3 + \frac{\rho_{13}\sigma_3}{\sigma_1}x_1 - \frac{\rho_{13}\sigma_3}{\sigma_1}\mu_1 \end{bmatrix}$$

Also

$$\Sigma_{22} - \Sigma_{12}^T\Sigma_{11}^{-1}\Sigma_{12} = \begin{bmatrix} \sigma_2^2 & \rho_{23}\sigma_2\sigma_3 \\ \rho_{23}\sigma_2\sigma_3 & \sigma_3^2 \end{bmatrix} - \begin{bmatrix} \rho_{12}\sigma_1\sigma_2 \\ \rho_{13}\sigma_1\sigma_3 \end{bmatrix}\frac{1}{\sigma_1^2}\begin{bmatrix} \rho_{12}\sigma_1\sigma_2 & \rho_{13}\sigma_1\sigma_3 \end{bmatrix}$$

$$= \begin{bmatrix} \sigma_2^2\left(1 - \rho_{12}^2\right) & \sigma_2\sigma_3\left(\rho_{23} - \rho_{12}\rho_{13}\right) \\ \sigma_2\sigma_3\left(\rho_{23} - \rho_{12}\rho_{13}\right) & \sigma_3^2\left(1 - \rho_{13}^2\right) \end{bmatrix}.$$

Each of the conditional expectations is a linear function of x_1.
   The conditional correlation of X_2 and X_3 given $X_1 = x_1$ is

$$\rho_{23|1} = \frac{\sigma_2\sigma_3\left(\rho_{23} - \rho_{12}\rho_{13}\right)}{\sqrt{\sigma_2^2\left(1 - \rho_{12}^2\right)\sigma_3^2\left(1 - \rho_{13}^2\right)}} = \frac{\rho_{23} - \rho_{12}\rho_{13}}{\sqrt{\left(1 - \rho_{12}^2\right)\left(1 - \rho_{13}^2\right)}}.$$

We recognize this expression as the partial correlation coefficient of X_2 and X_3 allowing for the effects of X_1 (Chapter 5). In the case of an MVN distribution, then, the partial correlation coefficient is an unambiguous and useful measure; it is simply the conditional correlation (which does not change with changing values of the variable being conditioned upon).

In the MVN context, it also makes sense to define another measure of correlation – the multiple correlation. The *multiple correlation* between X_m and all the other random variables, $X_1, \ldots, X_{m-1}$ is defined as follows:

$$R^2_{m(1,2,\ldots,m-1)} = 1 - \frac{\text{var}(X_m | X_1, \ldots, X_{m-1})}{\text{var}(X_m)}.$$

In other words, the multiple correlation is the proportional reduction in the variance of X_m due to its regression on $X_1, \ldots, X_{m-1}$. Proposition 4.10 gave a similar interpretation of the correlation between two random variables in a bivariate distribution.

Example 8.5 – continued

Here, $\text{var}(X_2) = 1$ but $\text{var}(X_2 | X_1, X_3) = 0.833$. So, the multiple correlation of X_2 on X_1 and X_3 is

$$R^2_{2(1,3)} = 1 - \frac{\text{var}(X_2 | X_1, X_3)}{\text{var}(X_2)} = 1 - \frac{0.833}{1} = 0.167.$$

Therefore, knowledge of X_1 and X_3 leads to only a small reduction in uncertainty about X_2.

A more detailed discussion of partial and multiple correlation can be found in Stuart and Ord (1996, Chapter 27).

Proposition 8.12

(a) Let $\mathbf{X}_1, \ldots, \mathbf{X}_n$ be independent, m-dimensional random vectors, with each $\mathbf{X}_i \sim N_m(\boldsymbol{\mu}_i, \Sigma_i)$. Then, for any real constants $a_1, \ldots, a_n$,

$$a_1 \mathbf{X}_1 + \cdots a_n \mathbf{X}_n \sim N_p \left(\sum_{i=1}^n a_i \boldsymbol{\mu}_i, \sum_{i=1}^n (a_i^2 \Sigma_i) \right).$$

(b) Suppose now that $\mathbf{X}_1, \ldots, \mathbf{X}_n$ are independent random vectors, with each $\mathbf{X}_i$ following the $N_m(\boldsymbol{\mu}, \Sigma)$ distribution. Then the sample mean vector, $\overline{\mathbf{X}} \equiv (1/n) \sum_{i=1}^n \mathbf{X}_i$, has distribution

$$\overline{\mathbf{X}} \equiv \frac{1}{n} \sum_{i=1}^n \mathbf{X}_i \sim N_m \left(\boldsymbol{\mu}, \frac{1}{n} \Sigma \right).$$

The proof of this proposition is left as an exercise.

Proposition 8.13

Suppose that $\mathbf{X} \sim N_m(\boldsymbol{\mu}, \Sigma)$, where Σ is non-singular. Then

$$(\mathbf{X} - \boldsymbol{\mu})^T \Sigma^{-1} (\mathbf{X} - \boldsymbol{\mu}) \sim \chi_m^2.$$

Proof. Since Σ is specified to be non-singular, then it must be positive definite. As before, then, there is another positive definite matrix $C \in M_{mm}$ such that $\Sigma = CC^T$.

Define the random vector $\mathbf{Z}$ by $\mathbf{Z} = C^{-1}(\mathbf{X} - \boldsymbol{\mu})$. Then

$$\mathbf{Z} \sim N_m(\mathbf{0}, I_m).$$

It follows that $Z_1, \ldots, Z_m$ are independent random variables (since I_m is a diagonal matrix). Also

$$Z_i \sim N(0, 1).$$

But

$$(\mathbf{X} - \boldsymbol{\mu})^T \Sigma^{-1} (\mathbf{X} - \boldsymbol{\mu}) = \mathbf{Z}^T \mathbf{Z} = \sum_{i=1}^{m} Z_i^2 \sim \chi_m^2$$

since the sum of squares of independent $N(0, 1)$ random variables is a χ^2 random variable (Chapter 6). ∎

Exercises

1 Consider again the random vector introduced in Exercise 1 on Section 8.1.

   (a) Find the conditional distribution of X_3 given X_1 and X_2. Find and interpret the multiple correlation of X_3 on X_1 and X_2.

   (b) Find the (joint) conditional distribution of X_2 and X_3 given $X_1 = x_1$. Use this to write down the partial correlation $\rho_{23 \cdot 1}$. Check that you would get the same answer from the usual formula.

2 Suppose that $\mathbf{X} = [X_1 \ X_2 \ X_3 \ X_4]^T$ is a (non-singular) MVN random vector. Find the conditional distribution of (X_1, X_2) given that $X_3 = x_3$ and $X_4 = x_4$. Hence write down $\rho_{12 \cdot 34}$, the conditional (or partial) correlation of X_1 and X_2 given X_3 and X_4. Show that

$$\rho_{12 \cdot 34} = \frac{\rho_{12 \cdot 4} - \rho_{13 \cdot 4} \rho_{23 \cdot 4}}{\sqrt{\left(1 - \rho_{13 \cdot 4}^2\right)\left(1 - \rho_{23 \cdot 4}^2\right)}}.$$

3 Suppose that $\mathbf{X} = [X_1\ X_2\ X_3]^T$ is a (non-singular) MVN random vector and that X_1 and X_2 are independent random variables.

(a) Write down an expression for the regression of X_3 on X_1 and X_2. Find the conditional variance of X_3 given X_1 and X_2.

(b) Now consider the marginal distribution of X_1 and X_3. Write down the regression of X_3 on X_1 and the conditional variance of X_3 given X_1. Compare these results with the results in (a).

4 As in Exercise 3, suppose that $\mathbf{X} = [X_1\ X_2\ X_3]^T$ is a (non-singular) MVN random vector, but this time assume that X_2 and X_3 are independent random variables. Write down an expression for the regression of X_3 on X_1 and X_2 and find the conditional variance of X_3 given X_1 and X_2. Compare these results with those you obtained in Exercise 3.

5 The following is a general method, based on the inverse distribution function, for simulating values of a random vector $\mathbf{X} = [X_1\ X_2 \ldots X_m]^T$. Begin by generating independent $Un(0, 1)$ random variates, denoted $U_1, \ldots, U_m$. Then set

$$u_1 = F_1(x_1)$$
$$u_2 = F_{2|1}(x_2|x_1),$$
$$\vdots$$
$$u_m = F_{m|1,2,\ldots,m-1}(x_m|x_1,\ldots,x_{m-1}).$$

Explain how to implement this method of simulation for the MVN case.

Summary

This chapter has reviewed many important properties of the multivariate normal distribution, which is still the most used multivariate probability model. In the next chapter, we will deal with sampling distributions related to the multivariate normal distribution.

9
Sampling

One of the most important areas of application for multivariate probability is statistical inference. Since statistics is basically concerned with drawing conclusions about populations based on information collected from samples, a key link between the topics of probability and statistics is provided by the theory of sampling. A 'sampling distribution' is a probability distribution that is required for making inferences based on a sample of data from a (univariate or multivariate) population. It is reasonably well known that the t, χ^2 and F distributions arise as sampling distributions in the case of univariate normal data; we have already seen that the t and F distributions are intimately related (since the square of a t_k random variable follows an $F_{1,k}$ distribution). In this chapter, we also introduce the analogues of these distributions which arise as sampling distributions in the case of multivariate normal data, namely the T^2 and Wishart distributions. The same sampling distributions also arise in the theory of linear regression models.

9.1 Sampling with and without replacement

Suppose we wish to investigate some feature of a *population*, which is defined to be a group of units about which we wish to form some conclusion. In the simplest case, we are only interested in one characteristic of each individual unit in the population. This might be a binary characteristic, such as whether the individual has or has not ever bought a particular product, in which case we would probably like to find out about a population proportion, e.g. the proportion of individuals in the population who have bought this product at some time. On the other hand, we might be interested in a measured characteristic, such as the individual's length or systolic blood pressure, possibly with a view to finding the mean length or median systolic blood pressure in the population as a whole.

In general, we will assume that the characteristic of interest can always be described numerically. This is naturally true for measurement data. Binary data could be converted to zeros and ones in the way we have discussed when introducing the binomial distribution, with a 1 denoting a 'success' and a 0 a 'failure'. In a population of size N, let the population values of the characteristic be denoted $u_1, u_2, \ldots, u_N$. This means that the *population mean* is

$$\mu = \frac{1}{N} \sum_{j=1}^{N} u_j.$$

The *population variance* is

$$\sigma^2 = \frac{1}{N} \sum_{j=1}^{N} u_j^2 - \left\{ \frac{1}{N} \sum_{j=1}^{N} u_j \right\}^2.$$

Since populations are often large and costly to investigate, it is usual to conduct such investigations on a *sample*, a relatively small sub-group of units drawn from the population.

In order to avoid sampling bias, where consciously or subconsciously we sample units from the population that are unrepresentative of the population as a whole (e.g. prettier, more interesting), it is usual for statisticians to recommend that some random selection process be used. The most straightforward method of this kind is *simple random sampling*, where every group of n different individuals in the population has the same probability of becoming the sample.

The observed characteristic of the ith individual in the sample is a random variable that we will denote by X_i. In order to simplify the terminology, we will now simply refer to $X_1, X_2, \ldots, X_n$ as a sample from a population, though this is actually a sample from the probability distribution that describes the variation in a particular characteristic in the whole population.

To begin with, we shall assume that we draw our sample *with replacement*. This means that, once a unit from the population has been chosen for the sample and observed, it is returned to the pool and may be chosen and observed again. In this case, each X_i is equally likely to be any one of the population values, u_1, $u_2, \ldots, u_N$. So $\mathbb{E}(X_i) = \mu$ and $\mathrm{var}(X_i) = \sigma^2$. The choice available for X_i is not affected in any way by the values of the other random variables, X_j ($j \neq i$) and so $X_1, X_2, \ldots, X_n$ are independent and identically distributed random variables. This means we are in the position discussed fully in Chapter 6. In particular, if S denotes the sum of the sample values and $\overline{X}$ their average, then

$$\mathbb{E}(S) = n\mu, \qquad \mathrm{var}(S) = n\sigma^2$$

and

$$\mathbb{E}(\overline{X}) = \mu, \qquad \mathrm{var}(\overline{X}) = \frac{\sigma^2}{n}.$$

We shall return to this situation later, but in the meantime we shall consider the more usual case of *sampling without replacement*. In this case, all n sample values must come from different members of the population. Since there are

$\binom{N}{n}$ different collections of n individuals in a population of size N, then each possible sample has probability $1 \big/ \binom{N}{n}$ of being selected.

In this case, $X_1, X_2, \ldots, X_n$ are not independent. Knowledge of X_i changes the probabilities associated with X_j. For example, knowing that X_i is u_1 means that X_j $(j \neq i)$ cannot be u_1. So, $X_1, X_2, \ldots, X_n$ have a full, n-dimensional probability distribution.

It is still true that, marginally, $\mathbb{E}(X_i) = \mu$ and $\mathrm{var}(X_i) = \sigma^2$. This is because (with simple random sampling) X_i is equally likely to be any of the values $u_1, u_2, \ldots, u_N$. But we also need to consider the covariance between X_i and X_j. Suppose that X_i is known to be u_p (for some p in the range $1, 2, \ldots, N$); conditional on this value of X_i, the random variable X_j is equally likely to take any of the other values in the population. So

$$\mathbb{E}(X_j | X_i = u_p) = \frac{1}{N-1} \sum_{r \neq p} u_r = \frac{1}{N-1} \left\{ \sum_{r=1}^{N} u_r - u_p \right\}$$

$$= \frac{1}{N-1} \{ N\mu - u_p \},$$

$$\mathbb{E}(X_j^2 | X_i = u_p) = \frac{1}{N-1} \sum_{r \neq p} u_r^2 = \frac{1}{N-1} \left\{ \sum_{r=1}^{N} u_r^2 - u_p^2 \right\}$$

$$= \frac{1}{N-1} \left\{ N(\sigma^2 + \mu^2) - u_p^2 \right\},$$

$$\mathrm{var}(X_j | X_i = u_p) = \frac{1}{N-1} \left\{ N(\sigma^2 + \mu^2) - u_p^2 \right\} - \frac{1}{(N-1)^2} \{ N\mu - u_p \}^2$$

$$= \frac{N}{(N-1)^2} \left\{ (N-1)\sigma^2 - (u_p - \mu)^2 \right\}.$$

Using Proposition 4.6, which extended the formula for iterated expectation,

$$\mathbb{E}(X_i X_j) = \mathbb{E} \left\{ X_i \mathbb{E} (X_j | X_i) \right\}$$

$$= \mathbb{E} \left\{ X_i \left[\frac{1}{N-1} (N\mu - X_i) \right] \right\}$$

$$= \frac{N\mu}{N-1} \mathbb{E} (X_i) - \frac{1}{N-1} \mathbb{E}(X_i^2)$$

$$= \frac{N\mu^2}{N-1} - \frac{1}{N-1} (\sigma^2 + \mu^2)$$

$$= \mu^2 - \frac{\sigma^2}{N-1}.$$

Therefore,

$$\text{cov}(X_i, X_j) = \mathbb{E}(X_i X_j) - \mathbb{E}(X_i)\mathbb{E}(X_j) = \mu^2 - \frac{\sigma^2}{N-1} - \mu^2 = -\frac{\sigma^2}{N-1}.$$

The negative sign here is to be expected. If X_i is one of the larger values in the population, then X_j must be chosen from among somewhat smaller values than if X_i is a relatively small value in the population.

Letting $S = X_1 + X_2 + \cdots + X_n$, then, using Proposition 5.4,

$$\mathbb{E}(S) = \mathbb{E}(X_1) + \cdots + \mathbb{E}(X_n) = n\mu,$$

$$\text{var}(S) = \sum_{i=1}^{n} \text{var}(X_i) + 2 \sum_{i=1}^{n-1} \sum_{j=i+1}^{n} \text{cov}(X_i, X_j)$$

$$= n\sigma^2 - 2\frac{n(n-1)}{2}\frac{\sigma^2}{N-1}$$

$$= n\sigma^2 \left(\frac{N-n}{N-1}\right).$$

Also,

$$\mathbb{E}(\overline{X}) = \frac{1}{n}\mathbb{E}(S) = \mu,$$

$$\text{var}(\overline{X}) = \frac{1}{n^2}\text{var}(S) = \frac{\sigma^2}{n}\left(\frac{N-n}{N-1}\right).$$

These results for the expected values are identical to those obtained for sampling with replacement. The results for the variances are modified by the inclusion of an extra term, $(N-n)/(N-1)$. This term is known as *the finite population correction*.

For a large population N, or a small sampling fraction n/N, the term $(N-n)/(N-1)$ is close to 1 and so its inclusion makes no effective difference to the formulae for the variances. This makes intuitive sense; when sampling from a large population or drawing a sample that is small relative to the population size, it is very unlikely that the same unit will be selected more than once even if sampling with replacement is used. So the two approaches to sampling are difficult to distinguish in these cases.

This is an important point because probability calculations are usually a lot easier in the case of sampling with replacement than when sampling without replacement. For example, when sampling from a population of binary values, sampling with replacement leads to a binomial distribution while sampling without replacement leads to a hypergometric distribution. The binomial is much the easier distribution to deal with. The above results suggest that, in the limit, the hypergeometric is well approximated by the binomial. Since the binomial distribution is well approximated by the normal for large sample sizes,

n (using the central limit theorem) this suggests that there is a good normal approximation to the hypergeometric as well.

As has already been emphasized, the purpose of sampling is usually to find out about some feature of the population such as the population mean or population variance. Consider a simple random sample of data, $X_1, \ldots, X_n$, from an $N(\mu, \sigma^2)$ population. A continuous distribution, such as the normal, cannot exactly fit the distribution of values in a finite population; only an infinite population has sufficient continuity of values. Nevertheless, continuous probability models do provide adequate approximations to large (but finite) population distributions. Using the phrase 'simple random sample' in this context is equivalent to saying that $X_1, \ldots, X_n$ are independent and identically distributed $N(\mu, \sigma^2)$ random variables.

Important summary statistics from the sample are the *sample mean*,

$$\overline{X} = \frac{1}{n} \sum_{i=1}^{n} X_i,$$

and the *sample variance*,

$$S^2 = \frac{1}{n-1} \sum_{i=1}^{n} \left(X_i - \overline{X} \right)^2.$$

An *estimator* is a function of the sample values that gives an indication of the value of a population parameter. The sample mean is an estimator of the population mean, μ, while the sample variance is an estimator of the population variance, σ^2. It is left as an exercise to show that these are both *unbiased* estimators, in the sense that their expected value is the population parameter they are intended to estimate.

Statistical inference is based on the result (proved in Chapter 8) that

$$\overline{X} \sim N\left(\mu, \frac{1}{n}\sigma^2\right) \text{ independently of } \frac{(n-1)S^2}{\sigma^2} \sim \chi^2_{n-1}.$$

The distribution of $\overline{X}$ depends not only on μ but on σ^2 (which is also unknown). This has the implication that $\overline{X}$ is not an appropriate statistic from which to draw inferences about μ. A more appropriate statistic is constructed as follows, using the Student's t distribution that was introduced in Chapter 7:

$$\frac{\overline{X} - \mu}{\sqrt{\sigma^2/n}} \sim N(0,1) \text{ independently of } \frac{(n-1)S^2}{\sigma^2} \sim \chi^2_{n-1}$$

therefore

$$\frac{\overline{X} - \mu}{\sqrt{\sigma^2/n}} \div \sqrt{\frac{(n-1)S^2}{\sigma^2(n-1)}} \sim t_{n-1},$$

that is,

$$\frac{\overline{X} - \mu}{\sqrt{S^2/n}} \sim t_{n-1}.$$

An equivalent result uses the square of this statistic,

$$\frac{n(\overline{X} - \mu)^2}{S^2} \sim F_{1,n-1}.$$

Exercises

1 M red beads and $N - M$ blue beads (where $M \geq 1$ and $N - M \geq 1$) are placed in a bag and mixed together thoroughly. n of the beads (where $n \leq M$ and $n \leq N - M$) are removed from the bag, without replacement; each bead to be removed is selected at random from all the beads that remain in the bag. The random variable X_i $(i = 1, 2, \ldots, n)$ takes the value 1 if the ith bead removed from the bag is red and the value 0 if it is blue.

(a) For $i = 1, 2, \ldots, n$ and $j = 1, 2, \ldots, n$, where $i \neq j$, explain why $P(X_i = 1) = M/N$ and $P(X_i = 1 \text{ and } X_j = 1) = M(M - 1)/N(N - 1)$. Use these results to find the expected value and variance of X_i and X_j, and the covariance between X_i and X_j $(i \neq j)$.

(b) The random variable S is the total number of red beads removed from the bag. What type of distribution does S follow? Use the results derived in (a) to find the expected value and variance of S.

2 Suppose that a simple random sample $X_1, \ldots, X_n$ is drawn from an $N(\mu_1, \sigma^2)$ population and that, independently, a simple random sample $Y_1, \ldots, Y_m$ is drawn from an $N(\mu_2, \sigma^2)$ population. Let $\overline{X}$ and S_1^2 denote the mean and variance of the first sample and $\overline{Y}$ and S_2^2 denote the mean and variance of the second sample.

(a) The pooled estimate of σ^2 is

$$S_p^2 = \frac{(n-1)S_1^2 + (m-1)S_2^2}{n+m-2}.$$

Show that $(n + m - 2) S_p^2/\sigma^2 \sim \chi^2_{n+m-2}$.

(b) Show that $\overline{X} - \overline{Y}$ follows a normal distribution.

(c) Explain why $\overline{X} - \overline{Y}$ and S_p^2 are independent random variables. Hence find a random variable, following a Student's t distribution, that would be useful for estimating the difference in population means, $\mu_1 - \mu_2$.

3 Suppose that $X_1, X_2, \ldots, X_n$ is a random sample from an $\text{Ex}(\theta)$ distribution. In Chapter 6, we showed that $\overline{X} \sim \text{Ga}(n, n\theta)$. This means that $\mathbb{E}(\overline{X}) = 1/\theta$, so it seems plausible that $1/\overline{X}$ might be a sensible estimator of θ. Show that

$$\mathbb{E}\left(\frac{1}{\overline{X}}\right) = \frac{n\theta}{n-1}, \qquad \text{var}\left(\frac{1}{\overline{X}}\right) = \frac{n^2\theta^2}{(n-1)^2(n-2)}.$$

9.2 Multivariate sampling distributions

Now, in a similar way, consider a random sample of data vectors, $\mathbf{X}_1, \ldots, \mathbf{X}_n$, from a $\text{N}_p(\boldsymbol{\mu}, \Sigma)$ population. In other words, $\mathbf{X}_1, \ldots, \mathbf{X}_n$ are independent and identically distributed $\text{N}_p(\boldsymbol{\mu}, \Sigma)$ random vectors. We usually write the ith random vector as

$$\mathbf{X}_i = [X_{i1} \ X_{i2} \ldots X_{ip}]^{\text{T}}.$$

The data from the sample are grouped in the $n \times p$ *data matrix*

$$X = \begin{bmatrix} \mathbf{X}_1^{\text{T}} \\ \mathbf{X}_2^{\text{T}} \\ \vdots \\ \mathbf{X}_n^{\text{T}} \end{bmatrix} = \begin{bmatrix} X_{11} & X_{12} & \ldots & X_{1p} \\ X_{21} & X_{22} & \ldots & X_{2p} \\ \vdots & \vdots & & \vdots \\ X_{n1} & X_{n2} & \ldots & X_{np} \end{bmatrix}.$$

Each row of X corresponds to a data vector (usually a specific subject in the sample), whereas each column of X corresponds to a random variable (usually a specific measurement).

Appropriate summary statistics for the jth random variable in $\mathbf{X}$ $(j = 1, 2, \ldots, p)$ are

$$\overline{X}_j = \frac{1}{n}\sum_{i=1}^{n} X_{ij}, \qquad S_{jj} = \sum_{i=1}^{n}\left(X_{ij} - \overline{X}_j\right)^2$$

Restricting attention to the jth random variable in X, then unbiased estimators of μ_j and σ_j^2 are

$$\hat{\mu}_j = \overline{X}_j, \qquad \hat{\sigma}_j^2 = \frac{1}{n-1}S_{jj}.$$

In order to describe the joint sample variation in the jth and kth random variables, we can use the corrected sum of cross products

$$S_{jk} = \sum_{i=1}^{n}\left(X_{ij} - \overline{X}_j\right)\left(X_{ik} - \overline{X}_k\right).$$

An unbiased estimator of the covariance, σ_{jk}, is

$$\hat{\sigma}_{jk} = \frac{1}{n-1} S_{jk}.$$

The p-dimensional vector $\overline{\mathbf{X}} = [\overline{X}_1 \, \overline{X}_2 \dots \overline{X}_p]^{\mathrm{T}}$ is known as the *sample mean vector*.

The $p \times p$ matrix S with (j, k)th element S_{jk} is known as the *sample matrix of corrected sums of squares and cross products* (or *sample CSSP matrix*).

Notice that the estimators previously discussed may be combined in the form

$$\hat{\mu} = \overline{\mathbf{X}}, \qquad \hat{\Sigma} = \frac{1}{n-1} S.$$

The properties of the $p \times p$ matrix S are of considerable importance. Notice that S can be written in the following form:

$$S = \sum_{i=1}^{n} (\mathbf{X}_i - \overline{\mathbf{X}})(\mathbf{X}_i - \overline{\mathbf{X}})^{\mathrm{T}}$$

$$= \left[\mathbf{X}_1 - \overline{\mathbf{X}} \; \mathbf{X}_2 - \overline{\mathbf{X}} \dots \mathbf{X}_n - \overline{\mathbf{X}} \right] \begin{bmatrix} (\mathbf{X}_1 - \overline{\mathbf{X}})^{\mathrm{T}} \\ (\mathbf{X}_2 - \overline{\mathbf{X}})^{\mathrm{T}} \\ \vdots \\ (\mathbf{X}_n - \overline{\mathbf{X}})^{\mathrm{T}} \end{bmatrix}$$

$$= U \, U^{\mathrm{T}}$$

(say). This means that S has rank

$$r(S) = r(UU^{\mathrm{T}}) = r(U).$$

We know that the sum of the columns of U is $\mathbf{0}$. So the maximum possible column rank of U is $n - 1$. Since U has p rows, it follows that U, and hence S, can have rank p only when $n > p$. So S can only be non-singular when $n > p$, i.e. when the sample size is greater than the number of random variables under consideration.

Also,

$$S = \sum_{i=1}^{n} (\mathbf{X}_i - \overline{\mathbf{X}})(\mathbf{X}_i - \overline{\mathbf{X}})^{\mathrm{T}} = \sum_{i=1}^{n} \mathbf{X}_i \mathbf{X}_i^{\mathrm{T}} - n \overline{\mathbf{X}} \overline{\mathbf{X}}^{\mathrm{T}}.$$

Now

$$\sum_{i=1}^{n} \mathbf{X}_i \, \mathbf{X}_i^{\mathrm{T}} = [\mathbf{X}_1 \, \mathbf{X}_2 \; \dots \; \mathbf{X}_n] \begin{bmatrix} \mathbf{X}_1^{\mathrm{T}} \\ \mathbf{X}_2^{\mathrm{T}} \\ \vdots \\ \mathbf{X}_n^{\mathrm{T}} \end{bmatrix} = X^{\mathrm{T}} X$$

and

$$\bar{\mathbf{X}} = \frac{1}{n}(\mathbf{X}_1 + \mathbf{X}_2 + \ldots + \mathbf{X}_n) = \frac{1}{n}[\mathbf{X}_1\,\mathbf{X}_2\ldots\mathbf{X}_n]\begin{bmatrix}1\\1\\\vdots\\1\end{bmatrix} = \frac{1}{n}X^{\mathrm{T}}\mathbf{1}_n$$

$$\therefore \qquad S = X^{\mathrm{T}}X - n\left(\frac{1}{n}X^{\mathrm{T}}\mathbf{1}_n\right)\left(\frac{1}{n}X^{\mathrm{T}}\mathbf{1}_n\right)^{\mathrm{T}} = X^{\mathrm{T}}\left(I_n - \frac{1}{n}\mathbf{1}_n\mathbf{1}_n^{\mathrm{T}}\right)X.$$

These formulae will be of use theoretically, in what follows.

We will now derive the probability distributions of $\bar{\mathbf{X}}$ and S. In Chapter 8, we showed that

$$\bar{\mathbf{X}} \sim \mathrm{N}_p\left(\mu, \frac{1}{n}\Sigma\right).$$

Now S is a random matrix, so we need to define a new type of multivariate probability distribution to deal with it.

Suppose that $\mathbf{U}_1, \ldots, \mathbf{U}_n$ are independent and identically distributed $\mathrm{N}_p(\mathbf{0}, \Sigma)$ random vectors. Then the random matrix

$$A = \sum_{i=1}^{n} \mathbf{U}_i\,\mathbf{U}_i^{\mathrm{T}}$$

is said to have the p-variate *Wishart* distribution on n degrees of freedom with parameter Σ. This is written

$$A \sim \mathrm{W}_p(n, \Sigma).$$

Notice that A is a symmetric, positive semi-definite matrix. The Wishart distribution is analogous to the χ^2 distribution; compare one definition of the χ^2 distribution:

if $U_1, \ldots, U_n$ are independent $\mathrm{N}(0, \sigma^2)$ *random variables, then* $\sum_{i=1}^{n} U_i^2 \sim \sigma^2\chi_n^2.$

Proposition 9.1

Suppose that $A \sim \mathrm{W}_p(n, \Sigma)$ independently of $B \sim \mathrm{W}_p(m, \Sigma)$. Then:

(a) $\mathbb{E}(A) = n\,\Sigma$ and (when A^{-1} exists)

$$\mathbb{E}(A^{-1}) = \frac{1}{n - p - 1}\Sigma^{-1};$$

(b) for any vector of constants $\alpha \in \mathbb{R}^P, \alpha^{\mathrm{T}}A\alpha \sim (\alpha^{\mathrm{T}}\Sigma\alpha)\,\chi_n^2;$

(c) $A + B \sim \mathrm{W}(n + m, \Sigma).$

Proof. (a) By definition, there are random vectors $\mathbf{U}_1, \ldots, \mathbf{U}_n$ which are independent and identically distributed, following the $N_p(\mathbf{0}, \Sigma)$ distribution, such that

$$A = \sum_{i=1}^{n} \mathbf{U}_i \mathbf{U}_i^{\mathrm{T}}.$$

So

$$\mathbb{E}(A) = \sum_{i=1}^{n} \mathbb{E}(\mathbf{U}_i \mathbf{U}_i^{\mathrm{T}}).$$

Now, for any i,

$$\mathrm{cov}(\mathbf{U}_i) = \mathbb{E}(\mathbf{U}_i \mathbf{U}_i^{\mathrm{T}}) - \mathbb{E}(\mathbf{U}_i)\{\mathbb{E}(\mathbf{U}_i)\}^{\mathrm{T}}$$
$$\Longleftrightarrow \Sigma = \mathbb{E}(\mathbf{U}_i \mathbf{U}_i^{\mathrm{T}}) - \mathbf{0}\mathbf{0}^{\mathrm{T}}$$
$$\Longleftrightarrow \mathbb{E}(\mathbf{U}_i \mathbf{U}_i^{\mathrm{T}}) = \Sigma$$

So

$$\mathbb{E}(A) = n\Sigma.$$

We omit the proof that $\mathbb{E}(A^{-1}) = \Sigma^{-1}/(n - p - 1)$.
(b) With the same terminology as above,

$$\mathbf{a}^{\mathrm{T}} A \mathbf{a} = \sum_{i=1}^{n} \mathbf{a}^{\mathrm{T}} \mathbf{U}_i \mathbf{U}_i^{\mathrm{T}} \mathbf{a}$$

$$= \sum_{i=1}^{n} (\mathbf{a}^{\mathrm{T}} \mathbf{U}_i)(\mathbf{a}^{\mathrm{T}} \mathbf{U}_i)^{\mathrm{T}}$$

$$= \sum_{i=1}^{n} (\mathbf{a}^{\mathrm{T}} \mathbf{U}_i)^2.$$

But $\mathbf{a}^{\mathrm{T}} \mathbf{U}_1, \ldots, \mathbf{a}^{\mathrm{T}} \mathbf{U}_n$ are independent and identically distributed $N_p(0, \mathbf{a}^{\mathrm{T}} \Sigma \mathbf{a})$ random variables. So

$$\mathbf{a}^{\mathrm{T}} A \mathbf{a} \sim (\mathbf{a}^{\mathrm{T}} \Sigma \mathbf{a}) \, \chi_n^2.$$

(c) Extending the notation used in parts (a) and (b), we can write

$$B = \sum_{i=n+1}^{n+m} \mathbf{U}_i \mathbf{U}_i^{\mathrm{T}},$$

where $\mathbf{U}_{n+1}, \ldots, \mathbf{U}_{n+m}$ are independent and identically distributed random vectors, following the $N_p(\mathbf{0}, \Sigma)$ distribution. Since A and B are independent, it must follow that $\mathbf{U}_1, \ldots, \mathbf{U}_{n+m}$ are all independent $N_p(\mathbf{0}, \Sigma)$ random vectors. But

$$A + B = \sum_{i=1}^{n+m} \mathbf{U}_i \mathbf{U}_i^{\mathrm{T}}$$

So, by definition, $A + B \sim W_p(n+m, \Sigma)$. ∎

Proposition 9.2

Suppose that $\mathbf{X}_1, \ldots, \mathbf{X}_n$ are independent and identically distributed random vectors, each following the $N_p(\mu, \Sigma)$ distribution. Let $\overline{\mathbf{X}}$ and S denote (respectively) the sample mean vector and CSSP matrix. Then

$$\overline{\mathbf{X}} \sim N_p\left(\mu, \tfrac{1}{n}\Sigma\right) \text{ independently of } S \sim W_p(n-1, \Sigma).$$

Proof. The distribution of $\overline{\mathbf{X}}$ follows from the 'reproductive' properties of the MVN – see Proposition 8.12. Define

$$\mathbf{U}_i = \mathbf{X}_i - \mu, \qquad i = 1, \ldots, n.$$

Then $\mathbf{U}_1, \ldots, \mathbf{U}_n$ are independent $N_p(\mathbf{0}, \Sigma)$ random vectors. So

$$U = \begin{bmatrix} \mathbf{U}_1^{\mathrm{T}} \\ \mathbf{U}_2^{\mathrm{T}} \\ \vdots \\ \mathbf{U}_n^{\mathrm{T}} \end{bmatrix} = \begin{bmatrix} \mathbf{X}_1^{\mathrm{T}} - \mu^{\mathrm{T}} \\ \mathbf{X}_2^{\mathrm{T}} - \mu^{\mathrm{T}} \\ \vdots \\ \mathbf{X}_n^{\mathrm{T}} - \mu^{\mathrm{T}} \end{bmatrix} = X - \mathbf{1}_n \mu^{\mathrm{T}}$$

$$\therefore \qquad X = U + \mathbf{1}_n \mu^{\mathrm{T}}.$$

But

$$\begin{aligned} S &= X^{\mathrm{T}}\left(I_n - \tfrac{1}{n}\mathbf{1}_n\mathbf{1}_n^{\mathrm{T}}\right)X \\ &= U^{\mathrm{T}}\left(I_n - \frac{1}{n}\mathbf{1}_n\mathbf{1}_n^{\mathrm{T}}\right)U + \mu\mathbf{1}_n^{\mathrm{T}}\left(I_n - \frac{1}{n}\mathbf{1}_n\mathbf{1}_n^{\mathrm{T}}\right)\mathbf{1}_n\mu^{\mathrm{T}} \\ &= U^{\mathrm{T}}\left(I_n - \frac{1}{n}\mathbf{1}_n\mathbf{1}_n^{\mathrm{T}}\right)U + \mu\left\{\mathbf{1}_n^{\mathrm{T}}\mathbf{1}_n - \frac{1}{n}\left(\mathbf{1}_n^{\mathrm{T}}\mathbf{1}_n\right)\left(\mathbf{1}_n^{\mathrm{T}}\mathbf{1}_n\right)\right\}\mu^{\mathrm{T}} \\ &= U^{\mathrm{T}}\left(I_n - \frac{1}{n}\mathbf{1}_n\mathbf{1}_n^{\mathrm{T}}\right)U + \mu\left(n - \frac{1}{n}n^2\right)\mu^{\mathrm{T}} \\ &= U^{\mathrm{T}}\left(I_n - \frac{1}{n}\mathbf{1}_n\mathbf{1}_n^{\mathrm{T}}\right)U. \end{aligned} \tag{9.1}$$

Now $J = I_n - (1/n)\mathbf{1}_n\mathbf{1}_n^T$ is a symmetric idempotent matrix. So (Result A5.2(ii)),

$$r(J) = \text{tr}(J) = \text{tr}(I_n) - \frac{1}{n}\text{tr}(\mathbf{1}_n\mathbf{1}_n^T) = n - \frac{1}{n}n = n - 1.$$

This means (Result A5.2(i)) that J has one eigenvalue equal to 0 and $n-1$ eigenvalues equal to 1.

   Since

$$J \cdot \mathbf{1}_n = \mathbf{1}_n - \frac{1}{n}\mathbf{1}_n\mathbf{1}_n^T\mathbf{1}_n = \mathbf{1}_n - \frac{1}{n}\cdot\mathbf{1}_n\cdot n = \mathbf{0} = 0\cdot\mathbf{1}_n,$$

it follows that an eigenvector corresponding to the zero eigenvalue is $\mathbf{1}_n$. Normalizing this, we have the eigenvector $(1/\sqrt{n})\mathbf{1}_n$.

   The spectral decomposition theorem (Result A3.4) tells us that we can find eigenvectors $\mathbf{W}_1, \ldots, \mathbf{W}_{n-1} \in \mathbb{R}^n$, corresponding to the repeated eigenvalue at 1, such that

$$\mathbf{W}_i^T\mathbf{W}_i = 1 \qquad (i = 1, 2, \ldots, n-1),$$
$$\mathbf{W}_i^T\mathbf{W}_j = 0 \qquad (i \neq j),$$
$$\mathbf{W}_i^T\left(\frac{1}{\sqrt{n}}\mathbf{1}_n\right) = 0 \qquad (i = 1, 2, \ldots, n-1)$$

and

$$J = 0\cdot\left(\frac{1}{\sqrt{n}}\mathbf{1}_n\right) + \sum_{i=1}^{n-1}1\cdot\mathbf{W}_i\mathbf{W}_i^T = \sum_{i=1}^{n-1}\mathbf{W}_i\mathbf{W}_i^T.$$

Now, from equation (9.1),

$$S = U^T J U = \sum_{i=1}^{n-1}U^T\mathbf{W}_i\mathbf{W}_i^T U = \sum_{i=1}^{n-1}\mathbf{Z}_i\mathbf{Z}_i^T$$

where

$$\mathbf{Z}_i = U^T\mathbf{W}_i$$

$$= [\mathbf{U}_1 \ \mathbf{U}_2 \ \ldots \ \mathbf{U}_n]\begin{bmatrix} w_{i1} \\ w_{i2} \\ \vdots \\ w_{in} \end{bmatrix}$$

$$= \sum_{k=1}^{n} w_{ik}\mathbf{U}_k$$

$$\therefore \qquad \mathbf{Z}_i \sim N_p\left(\sum_{k=1}^{n} w_{ik}\mathbf{0}, \sum_{k=1}^{n} w_{ik}^2\Sigma\right), \qquad \text{(since each } \mathbf{U}_k \sim N_p(\mathbf{0}, \Sigma))$$

that is,

$$Z_i \sim N_p(\mathbf{0}, \Sigma) \qquad \left(\text{since } \sum_{k=1}^{n} w_{ik}^2 = \mathbf{W}_i^T \mathbf{W}_i = 1\right).$$

When $i \neq j$,

$$\text{cov}(\mathbf{Z}_i, \mathbf{Z}_j) = \text{cov}\left(\sum_{k=1}^{n} w_{ik} \mathbf{U}_k, \sum_{l=1}^{n} w_{jl} \mathbf{U}_l\right)$$

$$= \sum_{k=1}^{n} \sum_{l=1}^{n} w_{ik} w_{jl} \text{cov}(\mathbf{U}_k, \mathbf{U}_l)$$

$$= \sum_{k=1}^{n} w_{ik} w_{jl} \Sigma \qquad (\text{since } \mathbf{U}_k, \mathbf{U}_l \text{ are independent, } k \neq l)$$

$$= 0_{p \times p} \qquad \left(\text{since } \sum_{k=1}^{n} w_{ik} w_{jl} = \mathbf{W}_i^T \mathbf{W}_j = 0\right).$$

Since $\mathbf{Z}_1, \ldots, \mathbf{Z}_{n-1}, \overline{\mathbf{U}}$ are all multivariate normal random vectors, it follows that $\mathbf{Z}_1, \ldots, \mathbf{Z}_{n-1}$ are independent of $\overline{\mathbf{U}}$.

So, by Exercise 11 on Section 7.2, $S = \sum_{i=1}^{n-1} \mathbf{Z}_i \mathbf{Z}_i^T$ is independent of $\overline{\mathbf{X}} = \overline{\mathbf{U}} + \mu$. ∎

Notice again that Proposition 9.2 is analogous to the following univariate result:

$$\overline{X} \sim N_p\left(\mu, \frac{1}{n}\sigma^2\right) \text{ independently of } S^2 \sim \sigma^2 \chi_{n-1}^2.$$

Suppose that $A \sim W_p(n, \Sigma)$ independently of $B \sim W_p(m, \Sigma)$. Then

$$\Lambda \equiv \frac{|A|}{|A + B|}$$

is said to have the *Wilks* $\Lambda(p, n, m)$ distribution. We state the following proposition without proof.

Proposition 9.3

(a) If $\Lambda \sim \Lambda(p, n, 2)$, then

$$\frac{n - p + 1}{p} \times \frac{1 - \sqrt{\Lambda}}{\sqrt{\Lambda}} \sim F_{2p, 2(n-p-1)}.$$

(b) If $\Lambda \sim \Lambda\,(2, n, m)$, then

$$\frac{n-1}{m} \times \frac{1-\sqrt{\Lambda}}{\sqrt{\Lambda}} \sim F_{2m,2(n-1)}.$$

The third and last sampling distribution to be introduced is the multivariate equivalent of the t distribution. Suppose that $\mathbf{U} \sim N_p(\mathbf{0}, \Sigma)$ *independently* of $A \sim W_p\,(n, \Sigma)$, where A is non-singular. Then the random variable

$$T^2 = n\mathbf{U}^T A^{-1}\mathbf{U}$$

is said to have Hotelling's $T^2(p, n)$ distribution. Again, we state the following proposition without proof.

Proposition 9.4

Suppose $T^2 \sim T(p, n)$. Then

$$\frac{n-p+1}{np}\, T^2 \sim F_{p,n-p+1}.$$

Proposition 9.5

Suppose that $\mathbf{X}_1, \ldots, \mathbf{X}_n$ are identical $N_p(\mu, \Sigma)$ random vectors, and let $\overline{\mathbf{X}}$ and S denote (respectively) the sample mean vector and CSSP matrix. Suppose that S is non-singular. Then

$$n(n-1) \left(\overline{\mathbf{X}} - \mu\right)^T S^{-1} \left(\overline{\mathbf{X}} - \mu\right) \sim T^2(p, n-1),$$

that is,

$$\frac{n(n-p)}{p}(\overline{\mathbf{X}} - \mu)^T S^{-1}(\overline{\mathbf{X}} - \mu) \sim F_{p,n-p}.$$

Proof. From Proposition 9.2, $\sqrt{n}(\overline{\mathbf{X}} - \mu) \sim N_p(\mathbf{0}, \Sigma)$ independently of $S \sim W_p(n-1, \Sigma)$. By definition, then,

$$(n-1) \left\{\sqrt{n}\left(\overline{\mathbf{X}} - \mu\right)\right\}^T S^{-1} \left\{\sqrt{n}\left(\overline{\mathbf{X}} - \mu\right)\right\} \sim T^2(p, n-1).$$

that is,

$$n(n-1) \left(\overline{\mathbf{X}} - \mu\right)^T S^{-1} \left(\overline{\mathbf{X}} - \mu\right) \sim T^2(p, n-1).$$

Hence, by Proposition 9.4,

$$\frac{n-1-p+1}{(n-1)p}n(n-1)\left(\overline{\mathbf{X}}-\boldsymbol{\mu}\right)^{\mathrm{T}}S^{-1}\left(\overline{\mathbf{X}}-\boldsymbol{\mu}\right) \sim F_{p,n-1-p+1}$$

that is,

$$\frac{n(n-p)}{p}\left(\overline{\mathbf{X}}-\boldsymbol{\mu}\right)^{\mathrm{T}}S^{-1}\left(\overline{\mathbf{X}}-\boldsymbol{\mu}\right) \sim F_{p,n-p}. \qquad \blacksquare$$

Exercises

1 Suppose that $\mathbf{X}_1, \ldots, \mathbf{X}_n$ are independent $N_p(\boldsymbol{\mu}, \Sigma)$ random vectors, and let $\overline{\mathbf{X}}$ and S denote (respectively) the sample mean vector and CSSP matrix. Suppose that S is non-singular. Then the *Mahalanobis distance* of $\mathbf{X}_i$ from the sample mean vector is

$$D_i = (\mathbf{X}_i - \overline{\mathbf{X}})^{\mathrm{T}}S^{-1}\left(\mathbf{X}_i - \overline{\mathbf{X}}\right).$$

D_i is a scaled measure of how extreme the vector $\mathbf{X}_i$ is relative to the probability distribution. Show that

$$\frac{n(n-p)}{(n+1)(n-1)p}D_i \sim F_{p,n-p}.$$

2 In the context described in Exercise 1, suppose that $\mathbf{X}_i$ is ignored and the sample mean vector, $\overline{\mathbf{X}}_{-i}$, and sample CSSP matrix, S_{-i}, are calculated from the remaining $n - 1$ data vectors. The *jackknifed Mahalanobis distance* of $\mathbf{X}_i$ is

$$JD_i = \left(\mathbf{X}_i - \overline{\mathbf{X}}_{-i}\right)^{\mathrm{T}} S_{-i}^{-1}(\mathbf{X}_i - \overline{\mathbf{X}}_{-i}).$$

JD_i is a scaled measure of how extreme the vector $\mathbf{X}_i$ is relative to the probability distribution. Show that

$$\frac{(n-1)(n-1-p)}{n(n-2)p}JD_i \sim F_{p,n-1-p}.$$

3 Suppose that a simple random sample $\mathbf{X}_1, \ldots, \mathbf{X}_n$ is drawn from an $N_p(\boldsymbol{\mu}_1, \Sigma)$ population and that, independently, a simple random sample $\mathbf{Y}_1, \ldots, \mathbf{Y}_m$ is drawn from an $N_p(\boldsymbol{\mu}_2, \Sigma)$ population. Let $\mathbf{X}$ and S_1 denote the mean vector and CSSP matrix of the first sample and $\mathbf{Y}$ and S_2 denote the mean vector and CSSP matrix of the second sample.

(a) The pooled estimate of Σ is

$$S_{\mathrm{p}} = \frac{n-1}{n+m-2}S_1 + \frac{m-1}{n+m-2}S_2$$

Show that $(n+m-2)S_{\mathrm{p}}$ follows a Wishart distribution.

(b) Show that $\overline{\mathbf{X}} - \overline{\mathbf{Y}}$ follows an MVN distribution.

(c) Explain why $\overline{\mathbf{X}} - \overline{\mathbf{Y}}$ and S_p are independent. Hence find a random vector, following a Hotelling's T^2 distribution, that would be useful for estimating the difference in population mean vectors, $\mu_1 - \mu_2$.

9.3 The normal linear model

In this section, we consider a random vector of the form

$$[Y|X_1 \cdots X_m]^{\mathrm{T}} = [Y|\mathbf{X}]^{\mathrm{T}}.$$

The first of the random variables in this vector, denoted Y, is considered in a rather different light from the others, denoted $\mathbf{X} = [X_1 \ldots X_m]$. Particular interest lies in the conditional distribution of Y given $\mathbf{X}$. One case of great importance arises when the regression of Y on $\mathbf{X}$ is linear, i.e. has the form:

$$\mathbb{E}\{Y|\mathbf{X}\} = \beta_0 + \beta_1 X_1 + \cdots + \beta_m X_m.$$

Here, $\beta_0, \beta_1, \ldots, \beta_m$ are constants, known as the *parameters* of the regression model.

The linear regression model may be rewritten in the form

$$Y = \mathbb{E}\{Y|\mathbf{X}\} + (Y - \mathbb{E}\{Y|\mathbf{X}\}) = \beta_0 + \beta_1 X_1 + \cdots + \beta_m X_m + \varepsilon,$$

where the random variable ε is defined to be the difference between Y and its conditional expected value.

In statistical inference, the values of $\beta_0, \beta_1, \ldots, \beta_m$ are usually unknown. They have to be estimated from data. The experiment that is set up for this purpose is intended to obtain n readings of data. On the ith replicate of this experiment, the values of $(x_1, \ldots, x_m)$ are chosen or found to be exactly $(x_{i1}, \ldots, x_{im})$ and the value Y_i of the response variable is observed. It should be noted that, within this experimental set-up, the values $(x_{i1}, \ldots, x_{im})$ are usually assumed to be constants and not random quantities; the only random variables are $Y_1, \ldots, Y_n$. All inference is conditional on the x values.

A *linear regression model* for the random vector $\mathbf{Y} = [Y_1 \ldots Y_n]$ is obtained by taking together all the regression equations for the individual random vectors:

$$\begin{bmatrix} Y_1 \\ Y_2 \\ \vdots \\ Y_n \end{bmatrix} = \begin{bmatrix} \beta_0 + \beta_1 x_{11} + \cdots + \beta_m x_{1m} + \varepsilon_1 \\ \beta_0 + \beta_1 x_{21} + \cdots + \beta_m x_{2m} + \varepsilon_2 \\ \vdots \\ \beta_0 + \beta_1 x_{n1} + \cdots + \beta_m x_{nm} + \varepsilon_n \end{bmatrix}$$

that is,

$$
\begin{bmatrix} Y_1 \\ Y_2 \\ \vdots \\ Y_n \end{bmatrix} = \begin{bmatrix} 1 & x_{11} & \cdots & x_{1m} \\ 1 & x_{21} & \cdots & x_{2m} \\ \vdots & \vdots & \cdots & \vdots \\ 1 & x_{n1} & \cdots & x_{nm} \end{bmatrix} \begin{bmatrix} \beta_0 \\ \beta_1 \\ \vdots \\ \beta_m \end{bmatrix} + \begin{bmatrix} \varepsilon_1 \\ \varepsilon_2 \\ \vdots \\ \varepsilon_n \end{bmatrix}
$$

or

$$
\mathbf{Y} = X \boldsymbol{\beta} + \boldsymbol{\varepsilon}
$$

The matrix X is often called the *design matrix* of the model. We shall assume that X has the greatest possible column rank, which is $m + 1$. In order to achieve this, we require $n \geq m + 1$, i.e. $n > m$.

In order to make progress, we need to make assumptions about the joint distribution of the random vector $\boldsymbol{\varepsilon}$. In the *normal linear model*, we assume that $\boldsymbol{\varepsilon}$ follows an MVN distribution. Each ε_i, by definition, has expected value 0, so $\mathbb{E}(\boldsymbol{\varepsilon}) = \mathbf{0}$. Sometimes, it is reasonable to assume that the results obtained from each replicate of the experiment are independent but with the same (residual) variance, i.e. $\varepsilon_1, \ldots, \varepsilon_n$ are independent and identically distributed $N(0, \sigma^2)$ random variables. This means that

$$
\boldsymbol{\varepsilon} \sim N_n(\mathbf{0}, \sigma^2 I_n).
$$

Equivalently, assuming that the components of the matrix X are all constants, this means that

$$
\mathbf{Y} \sim N_n(X\boldsymbol{\beta}, \sigma^2 I_n).
$$

So $Y_1, \ldots, Y_n$ are independent, with each $Y_i \sim N(\beta_0 + \beta_1 x_{i1} + \cdots + \beta x_{im}, \sigma^2)$.

Let $\hat{\beta}_0, \hat{\beta}_1, \ldots, \hat{\beta}_m$ denote estimators of $\beta_0, \beta_1, \ldots, \beta_m$. Then the *fitted value* of Y_i is defined to be

$$
\hat{Y}_i = \hat{\beta}_0 + \hat{\beta}_1 x_{i1} + \cdots + \hat{\beta}_m x_{im}.
$$

The *vector of fitted values* is

$$
\hat{\mathbf{Y}} = [\hat{Y}_1 \ldots \hat{Y}_n] = X\hat{\boldsymbol{\beta}}.
$$

The ith *residual* is the discrepancy between the observed value, Y_i, and its fitted value, i.e.

$$
r_i = Y_i - \hat{Y}_i.
$$

When the errors all have the same variance, then it is sensible to obtain estimators of the parameters using the method of (ordinary) least squares. In this approach, $\hat{\beta}_0, \hat{\beta}_1, \ldots, \hat{\beta}_m$ are obtained by minimizing the sum of the squared residuals, $\sum_{i=1}^n r_i^2$.

The *ordinary least-squares estimator* of β can be shown to be

$$\hat{\beta} = (X^T X)^{-1} X^T \mathbf{Y}.$$

This estimator is well defined, since the matrix $(X^T X) \in M_{m+1,m+1}$ has rank

$$r(X^T X) = r(X) = m + 1.$$

This means that $X^T X$ is of full rank, so it is non-singular.

We also wish to show that $X^T X$ is positive definite. In order to do this, note that, if $\mathbf{a}$ is any $m+1$-dimensional vector, then

$$\mathbf{a}^T(X^T X)\mathbf{a} = (X\mathbf{a})^T(X\mathbf{a}) = \mathbf{u}^T\mathbf{u}, \qquad \text{where } \mathbf{u} = X\mathbf{a},$$

so

$$\mathbf{a}^T(X^T X)\mathbf{a} = \sum_{i=1}^{m+1} u_i^2 \geq 0.$$

This means that $X^T X$ is positive semi-definite. Since $X^T X$ is non-singular, as shown above, then it must actually be positive definite. It follows that $(X^T X)^{-1}$ is positive definite as well (see Appendix A4).

Now the vector of fitted values can be written in the form

$$\hat{\mathbf{Y}} = X\hat{\beta} = X(X^T X)^{-1} X^T \mathbf{Y} = H\mathbf{Y}$$

(say). Note that

$$HX = X(X^T X)^{-1} X^T X = X,$$

so

$$[I_n - H]X = 0.$$

Proposition 9.6

The $n \times n$ matrix $H = X(X^T X)^{-1} X^T$ is symmetric idempotent of rank $m + 1$.

Proof. From the definition of H,

$$H^T = \{X(X^T X)^{-1} X^T\}^T = X\{(X^T X)^{-1}\}^T X^T = X\{(X^T X)^T\}^{-1} X^T$$

$$= X\{(X^T X)\}^{-1} X^T = H.$$

This proves that H is symmetric, as stated. Now

$$H^2 = \{X(X^T X)^{-1} X^T\}\{X(X^T X)^{-1} X^T\} = X(X^T X)^{-1}(X^T X)(X^T X)^{-1} X^T$$

$$= X(X^T X)^{-1} X^T = H.$$

This proves that H is idempotent, as stated.

Since H is symmetric idempotent, it follows from Appendix A5 that the rank of H, $r(H)$, is given by

$$r(H) = \mathrm{tr}(H) = \mathrm{tr}\left\{X(X^TX)^{-1}X^T\right\} = \mathrm{tr}\left\{(X^TX)^{-1}X^TX\right\}$$
$$= \mathrm{tr}\{I_{m+1}\} = m+1.$$

∎

It follows from properties of symmetric idempotent matrices that $I_n - H$ is also symmetric idempotent and that it has rank $n - (m+1) = n - m - 1$.

The *vector of residuals* is

$$\mathbf{r} \equiv [r_1 \ldots r_n]^T$$
$$= [Y_1 - \hat{Y}_1 \ldots Y_n - \hat{Y}_n]^T$$
$$= \mathbf{Y} - \hat{\mathbf{Y}}$$
$$= [I_n - H]\mathbf{Y}$$
$$= [I_n - H][X\beta + \varepsilon] = [I_n - H]\varepsilon \quad (\text{since } [I_n - H]X = 0).$$

The *residual sum of squares* (or sum of squared residuals) is

$$\mathrm{rss} = \sum_{i=1}^{n} (Y_i - \hat{Y}_i)^2$$
$$= (\mathbf{Y} - \hat{\mathbf{Y}})^T(\mathbf{Y} - \hat{\mathbf{Y}})$$
$$= \mathbf{r}^T\mathbf{r}$$
$$= \varepsilon^T(I_n - H)^T(I_n - H)\varepsilon$$
$$= \varepsilon^T(I_n - H)\varepsilon.$$

Generally, estimators of σ^2 are based on rss. For example, the ordinary least-squares estimator is $\hat{\sigma}^2 = (1/(n - m - 1))\mathrm{rss}$. The following proposition proves some important properties of the estimators of β and σ^2.

Proposition 9.7

(a) $\hat{\beta} \sim N_{m+1}\left(\beta, \sigma^2(X^TX)^{-1}\right)$

(b) $\mathbf{r} \sim N_n(\mathbf{0}, \sigma^2(I_n - H))$

(c) $\hat{\beta}$ and $\mathbf{r}$ are independent.

Proof. Consider the random vector

$$\mathbf{Z} = \begin{bmatrix} \hat{\beta} \\ \mathbf{r} \end{bmatrix}.$$

Using the above results,

$$\mathbf{Z} = \begin{bmatrix} (X^T X)^{-1} X^T \boldsymbol{\varepsilon} + \boldsymbol{\beta} \\ (I_n - H)\boldsymbol{\varepsilon} \end{bmatrix} = \begin{bmatrix} (X^T X)^{-1} X^T \\ I_n - H \end{bmatrix} \boldsymbol{\varepsilon} + \begin{bmatrix} \boldsymbol{\beta} \\ \mathbf{0} \end{bmatrix}.$$

By Proposition 8.6, since $\boldsymbol{\varepsilon} \sim N_n(\mathbf{0}, \sigma^2 I_n)$, it follows that $\mathbf{Z}$ also has an MVN distribution with

$$\mathbb{E}(\mathbf{Z}) = \begin{bmatrix} (X^T X)^{-1} X^T \\ I_n - H \end{bmatrix} \cdot \mathbf{0} + \begin{bmatrix} \boldsymbol{\beta} \\ \mathbf{0} \end{bmatrix} = \begin{bmatrix} \boldsymbol{\beta} \\ \mathbf{0} \end{bmatrix}$$

and

$$\begin{aligned}
\mathrm{cov}(\mathbf{Z}) &= \begin{bmatrix} (X^T X)^{-1} X^T \\ I_n - H \end{bmatrix} (\sigma^2 I_n) \left[\{(X^T X)^{-1} X^T\}^T \quad (I_n - H)^T \right] \\
&= \sigma^2 \begin{bmatrix} (X^T X)^{-1} X^T X (X^T X)^{-1} & (X^T X)^{-1} X^T (I_n - H)^T \\ (I_n - H) X (X^T X)^{-1} & (I_n - H)(I_n - H)^T \end{bmatrix} \\
&= \sigma^2 \begin{bmatrix} (X^T X)^{-1} & 0 \\ 0 & I_n - H \end{bmatrix}.
\end{aligned}$$

$$\leftarrow m+1 \rightarrow \quad \leftarrow n \rightarrow$$

By the reproductive property of MVN distributions, then,

$$\hat{\boldsymbol{\beta}}(\equiv \mathbf{Z}^{(1)}) \sim N_{m+1} \left(\boldsymbol{\beta}, \sigma^2 (X^T X)^{-1} \right),$$
$$\mathbf{r}(\equiv \mathbf{Z}^{(2)}) \sim N_n \left(\mathbf{0}, \sigma^2 (I_n - H) \right).$$

Since $\Sigma_{12} = 0$, $\hat{\boldsymbol{\beta}}$ and $\mathbf{r}$ are independent. ∎

Proposition 9.8

$$\frac{1}{\sigma^2} \mathrm{rss} \sim \chi^2_{n-(m+1)}.$$

Proof. We have

$$\mathrm{rss} = \mathbf{r}^T \mathbf{r} = \boldsymbol{\varepsilon}^T [I_n - H] \boldsymbol{\varepsilon},$$

where $\boldsymbol{\varepsilon} \sim N_n(\mathbf{0}, \sigma^2 I_n)$ and $I_n - H$ has $m+1$ zero eigenvalues and $n - m - 1$ unit eigenvalues. Using the spectral decomposition theorem (Result A3.4), there is an orthogonal matrix Q such that

$$I_n - H = Q \operatorname{diag}(\underbrace{1 \ldots 1}_{n-(m+1)} \underbrace{0 \ldots 0}_{m+1}) Q^{\mathsf{T}} = Q \cdot \begin{bmatrix} I_{n-(m+1)} & 0 \\ 0 & 0 \end{bmatrix} \cdot Q^{\mathsf{T}}$$

$$\therefore \frac{1}{\sigma^2} \operatorname{rss} = \left\{ \frac{1}{\sigma} \boldsymbol{\varepsilon}^{\mathsf{T}} Q \right\} \begin{bmatrix} I_{n-(m+1)} & 0 \\ 0 & 0 \end{bmatrix} \left\{ \frac{1}{\sigma} Q^{\mathsf{T}} \boldsymbol{\varepsilon} \right\}$$

$$= \boldsymbol{\eta}^{\mathsf{T}} \begin{bmatrix} I_{n-(m+1)} & 0 \\ 0 & 0 \end{bmatrix} \boldsymbol{\eta} \quad \text{where } \boldsymbol{\eta} = \frac{1}{\sigma} Q^{\mathsf{T}} \boldsymbol{\varepsilon}$$

$$= \sum_{i=1}^{n-(m+1)} \eta_i^2.$$

But

$$\boldsymbol{\eta} = \frac{1}{\sigma} Q^{\mathsf{T}} \boldsymbol{\varepsilon} \sim \mathrm{N}_n(\mathbf{0}, I_n)$$

since

$$\left\{ \frac{1}{\sigma} Q^{\mathsf{T}} \right\} \cdot \operatorname{cov}(\boldsymbol{\varepsilon}) \cdot \left\{ \frac{1}{\sigma} Q^{\mathsf{T}} \right\}^{\mathsf{T}} = \frac{1}{\sigma^2} Q^{\mathsf{T}} (\sigma^2 I_n) Q = Q^{\mathsf{T}} Q = I_n.$$

This means that $\eta_1, \ldots, \eta_{n-(m+1)}$ are independent and identically distributed $\mathrm{N}(0, 1)$ random variables and so

$$\frac{1}{\sigma^2} \operatorname{rss} = \sum_{i=1}^{n-(m+1)} \eta_i^2 \sim \chi_{n-(m+1)}^2.$$ ∎

Proposition 9.9

(a) $\hat{\boldsymbol{\beta}} \sim \mathrm{N}_{m+1}\left(\boldsymbol{\beta}, \sigma^2 (X^{\mathsf{T}} X)^{-1}\right)$ is independent of rss $\sim \sigma^2 \chi_{n-(m+1)}^2$.

(b) $(\boldsymbol{\beta} - \hat{\boldsymbol{\beta}})^{\mathsf{T}} (X^{\mathsf{T}} X)(\boldsymbol{\beta} - \hat{\boldsymbol{\beta}}) \sim \sigma^2 \chi_{m+1}^2$ is independent of rss $\sim \sigma^2 \chi_{n-(m+1)}^2$.

(c) $\dfrac{(\boldsymbol{\beta} - \hat{\boldsymbol{\beta}})^{\mathsf{T}} (X^{\mathsf{T}} X)(\boldsymbol{\beta} - \hat{\boldsymbol{\beta}})/(m+1)}{\operatorname{rss}/(n-m-1)} \sim F_{m+1,\, n-(m+1)}$.

Proof. (a) Proposition 9.7(c) tells us that $\hat{\boldsymbol{\beta}}$ is independent of **r**. Therefore, any function of $\hat{\boldsymbol{\beta}}$ is independent of any function of **r**. In particular, $\hat{\boldsymbol{\beta}}$ is independent of rss.

(b) Independence is established by a similar argument to (a) above. $\hat{\boldsymbol{\beta}} \sim \mathrm{N}_{m+1}(\boldsymbol{\beta}, \sigma^2 (X^{\mathsf{T}} X)^{-1})$, where $(X^{\mathsf{T}} X)^{-1}$ is positive definite. Since $\sigma^2 > 0$, it follows that $\sigma^2 (X^{\mathsf{T}} X)^{-1}$ is also positive definite, so $\sigma^2 (X^{\mathsf{T}} X)^{-1}$ is non-singular. By Proposition 8.13, then,

$$(\hat{\boldsymbol{\beta}} - \boldsymbol{\beta})^{\mathsf{T}} \left\{ \sigma^2 (X^{\mathsf{T}} X)^{-1} \right\}^{-1} (\hat{\boldsymbol{\beta}} - \boldsymbol{\beta}) \sim \chi_{m+1}^2$$

that is,

$$(\beta - \hat{\beta})^{\mathrm{T}}(X^{\mathrm{T}}X)(\beta - \hat{\beta}) \sim \sigma^2 \chi^2_{m+1}$$

(c) Using a result proved in Chapter 7, this result follows immediately from (b) since the left-hand side is the ratio of two independent χ^2 random variables, each divided by its degrees of freedom. ∎

Proposition 9.10

Let $\mathbf{b}$ ($\neq \mathbf{0}$) be any vector of constants in $\mathbb{R}^p$. Then

(a) $\mathbf{b}^{\mathrm{T}}\hat{\beta} \sim N(\mathbf{b}^{\mathrm{T}}\beta, \sigma^2 \mathbf{b}^{\mathrm{T}}(X^{\mathrm{T}}X)^{-1}\mathbf{b})$ is independent of rss $\sim \sigma^2 \chi^2_{n-(m+1)}$.

(b) $\dfrac{\mathbf{b}^{\mathrm{T}}(\hat{\beta} - \beta)}{\sqrt{\dfrac{\text{rss}}{n-(m+1)}\mathbf{b}^{\mathrm{T}}(X^{\mathrm{T}}X)^{-1}\mathbf{b}}} \sim t_{n-(m+1)}.$

Proof. (a) Again, the independence follows from the independence of $\hat{\beta}$ and rss. The distribution of $\mathbf{b}^{\mathrm{T}}\hat{\beta}$ is obtained from the usual reproductive property of the MVN distribution.

(b) Since $\mathbf{b} \neq \mathbf{0}, \sigma^2 > 0$ and $(X^{\mathrm{T}}X)^{-1}$ is positive definite, it follows that

$$\sigma^2 \mathbf{b}^{\mathrm{T}}(X^{\mathrm{T}}X)^{-1}\mathbf{b} > 0.$$

So $\mathbf{b}^{\mathrm{T}}\hat{\beta}$ has a non-singular normal distribution. It follows that

$$\frac{\mathbf{b}^{\mathrm{T}}(\hat{\beta} - \beta)}{\sqrt{\sigma^2 \mathbf{b}^{\mathrm{T}}(X^{\mathrm{T}}X)^{-1}\mathbf{b}}} \sim N(0, 1) \text{ independently of } \frac{\text{rss}}{\sigma^2} \sim \chi^2_{n-(m+1)}.$$

So

$$\frac{\mathbf{b}^{\mathrm{T}}(\hat{\beta} - \beta)}{\sqrt{\sigma^2 \mathbf{b}^{\mathrm{T}}(X^{\mathrm{T}}X)^{-1}\mathbf{b}}} \div \frac{\text{rss}/\sigma^2}{n-(m+1)} \sim t_{n-(m+1)}$$

that is,

$$\frac{\mathbf{b}^{\mathrm{T}}\left(\hat{\beta} - \beta\right)}{\sqrt{\dfrac{\text{rss}}{n-(m+1)}\mathbf{b}^{\mathrm{T}}(X^{\mathrm{T}}X)^{-1}\mathbf{b}}} \sim t_{n-(m+1)}. \quad ∎$$

Proposition 9.11

Let q be a positive integer such that $q \leq m+1$. Let $B \in M_{m+1,q}$ be a matrix of constants such that B is of full rank q. Then

(a) $B^T\hat{\beta} \sim N_q(B^T\beta, \sigma^2[B^T(X^TX)^{-1}B])$ independently of $r \sim \sigma^2 \chi^2_{n-(m+1)}$;

(b) $(\beta - \hat{\beta})^T B[B^T(X^TX)^{-1}B]^{-1}B^T(\beta - \hat{\beta}) \sim \sigma^2 \chi^2_q$ independently of $r \sim \sigma^2 \chi^2_{n-(m+1)}$;

(c) $\dfrac{[n-(m+1)](\beta - \hat{\beta})^T B[B^T(X^TX)^{-1}B]^{-1}B^T(\beta - \hat{\beta})}{q \cdot \text{rss}} \sim F_{q,n-(m+1)}.$

Proof. The proof of this proposition is similar to the last two proofs, so the full details are left as an exercise. The most difficult part of the proof is to show that the matrix $\sigma^2 [B^T(X^TX)^{-1}B]$ is non-singular. Since this is a square matrix of dimension $q \times q$, it is enough to show that it has rank equal to q.

Now $\sigma^2(X^TX)^{-1}$ is positive definite, so there is a non-singular matrix $C \in M_{m+1,m+1}$ such that

$$\sigma^2(X^TX)^{-1} = CC^T.$$

This means that the matrix $\sigma^2[B^T(X^TX)^{-1}B]$ has rank

$$r\{\sigma^2[B^T(X^TX)^{-1}B]\} = r\{B^TCC^TB\} = r\{(B^TC)(B^TC)^T\}$$
$$= r(B^TC) = r(B^T) = q. \qquad \blacksquare$$

The normal linear model can be extended to the situation where p response variables are observed during each replicate of the experiment. This means that, on the ith replicate, a random vector of responses is obtained:

$$\mathbf{Y}_i = [Y_{i1} \ Y_{i2} \cdots Y_{ip}]^T.$$

These responses can be written in an $n \times p$ random matrix, Y, and the model specifies that

$$\begin{bmatrix} Y_{11} & Y_{12} & \cdots & Y_{1p} \\ Y_{21} & Y_{22} & \cdots & Y_{2p} \\ \vdots & \vdots & & \vdots \\ Y_{11} & Y_{n2} & \cdots & Y_{np} \end{bmatrix} = \begin{bmatrix} 1 & x_{11} & \cdots & x_{1m} \\ 1 & x_{21} & \cdots & x_{2m} \\ \vdots & \vdots & \cdots & \vdots \\ 1 & x_{n1} & \cdots & x_{nm} \end{bmatrix} \begin{bmatrix} \beta_{01} & \beta_{02} & \cdots & \beta_{0p} \\ \beta_{11} & \beta_{12} & \cdots & \beta_{1p} \\ \vdots & \vdots & & \vdots \\ \beta_{m1} & \beta_{m2} & \cdots & \beta_{mp} \end{bmatrix} + \begin{bmatrix} \varepsilon_{11} & \varepsilon_{12} & \cdots & \varepsilon_{1p} \\ \varepsilon_{21} & \varepsilon_{22} & \cdots & \varepsilon_{2p} \\ \vdots & \vdots & & \vdots \\ \varepsilon_{n1} & \varepsilon_{n2} & \cdots & \varepsilon_{np} \end{bmatrix}$$

$$Y \qquad = \qquad X \qquad\qquad B \qquad + \qquad E.$$

This model is actually a collection of p individual linear models, one for each response variable. Note that the jth column of Y consists of all n observations on the jth response variable while the jth columns of the matrices B and E correspond to the model for the jth response.

The biggest difference between considering the multivariate regression model and p separate linear models for the individual responses is that, in the multivariate model, the response vector $\mathbf{Y}$ has a covariance structure. For a normal model, it is assumed that $\mathbf{Y} \sim N_p(B^T\mathbf{x}, \Sigma)$.

The statistical problem, then, is to estimate B and Σ; the usual estimators are

$$\hat{B} = (X^TX)^{-1}X^TY,$$

$$\hat{\Sigma} = \frac{1}{n}Y^T\left\{I_n - X(X^TX)^{-1}X^T\right\}Y.$$

It can be shown that these estimators are independent random matrices.

Exercises

1 $Y_1, Y_2, \ldots, Y_n$ are independent random variables with each $Y_i \sim N(\beta x_i, \sigma^2)$. Here, β and σ^2 are unknown parameters, to be estimated, and $x_1, \ldots, x_n$ are known constants not all equal to 0.

(a) Write this information in the form of a normal linear model for $\mathbf{Y}$.

(b) Show that a sensible estimator of β is

$$\hat{\beta} = \frac{\sum x_i Y_i}{\sum x_i^2}.$$

(c) Show that the residual sum of squares is

$$\mathrm{rss} = \sum Y_i^2 - \frac{\left\{\sum x_i Y_i\right\}^2}{\sum x_i^2}.$$

(d) Write down the joint distribution of $\hat{\beta}$ and rss. Deduce that

$$\frac{\sqrt{\sum x_i^2}\left(\hat{\beta} - \beta\right)}{\sqrt{\mathrm{rss}/(n-1)}} \sim t_{n-1}.$$

Summary

This chapter has looked at some very important applications of multivariate probability in the area of statistical inference. They have been linked by the idea of sampling. Sampling distributions related to the normal and MVN distributions include the t, F, χ^2, T^2, Wishart and Wilks distributions. These have all been introduced and some of their main properties proved and discussed.

Appendix A
Useful mathematical results

A1 Some important series

The probability mass functions of most of the special discrete probability distributions (e.g. binomial, geometric, hypergeometric and Poisson) are related to standard series. Finding the expected values, variances and moment-generating functions of these distributions requires the summation of these series.

The *arithmetic series* with starting value a and constant difference θ between terms is defined by

$$a + (a + \theta) + (a + 2\theta) + \cdots.$$

Result A1.1

The sum of the first n terms of an arithmetic series is

$$S_n = na + \tfrac{1}{2}n(n-1)\theta.$$

Proof.

$$S_n = a + (a + \theta) + (a + 2\theta) + \cdots + (a + (n-1)\theta)$$
$$= (a + (n-1)\theta) + (a + (n-2)\theta) + \cdots + a$$

So

$$2S_n = \{a + (a + (n-1)\theta)\} + \{(a + \theta) + (a + (n-2)\theta)\} + \cdots$$
$$+ \{(a + (n-1)\theta) + a\}$$
$$= \{2a + (n-1)\theta\} + \{2a + (n-1)\theta\} + \cdots + \{2a + (n-1)\theta\}$$
$$= 2an + n(n-1)\theta$$

and

$$S_n = an + \tfrac{1}{2}n(n-1)\theta. \qquad \blacksquare$$

Result A1.2

$$1 + 2 + \cdots + n = \tfrac{1}{2}n(n+1).$$

Proof. This follows directly from Result A1.1 with $a = \theta = 1$. ■

The *geometric series* with starting value a and constant ratio θ between terms is defined by

$$a + a\theta + a\theta^2 + \cdots .$$

Result A1.3

The sum of the first n terms of a geometric series is

$$S_n = a\frac{1 - \theta^n}{1 - \theta} \qquad (\theta \neq 1).$$

Proof.

$$S_n = a + a\theta + \cdots + a\theta^{n-1}$$

so

$$\theta S_n = a\theta + a\theta^2 + \cdots + a\theta^n$$

and

$$S_n - \theta S_n = a - a\theta^n = a(1 - \theta^n),$$

that is,

$$(1 - \theta)S_n = a(1 - \theta^n)$$

and we have

$$S_n = a\frac{1 - \theta^n}{1 - \theta}.$$

■

Result A1.4

Whenever $-1 < \theta < 1$, the sum to infinity of a geometric series converges to the limiting value

$$S_\infty = \frac{a}{1 - \theta}.$$

Proof. When $-1 < \theta < 1, \theta^n \to 0$ as $n \to \infty$. The required result then follows from Result A1.3. ■

When n is a positive integer, the notation $n!$ (which is read 'n factorial') denotes the product of the first n positive integers: $n! = n(n-1)(n-2)\ldots1$. By convention, this notation is extended to include $0! = 1$.

When n and m are non-negative integers, and $m \leq n$, then the number of different ways of choosing m objects from a set of n distinct objects is known as the number of *combinations* of m from n distinct objects. This is denoted $\binom{n}{m}$, which is often pronounced 'n choose m'. It can be shown that

$$\binom{n}{m} = \frac{n!}{m!(n-m)!} \qquad (n = 0, 1, \ldots, \quad m = 0, 1, \ldots, n).$$

Result A1.5

(a) $\binom{n}{0} = \binom{n}{n} = 1, \binom{n}{1} = \binom{n}{n-1} = n \qquad (n = 0, 1, \ldots).$

(b) $\binom{n}{m} = \binom{n}{n-m} \qquad (n = 0, 1, \ldots, \quad m = 0, 1, \ldots, n).$

Proof. (a) follows by substituting $m = 0, n$ and $1, n-1$ (respectively) into the definition, and (b) follows directly from the definition. ∎

Result A1.6

$$m\binom{n}{m} = n\binom{n-1}{m-1} \qquad (n = 1, 2, \ldots, \quad m = 1, 2, \ldots, n).$$

Proof.

$$\begin{aligned}
m\binom{n}{m} &= m\frac{n!}{m!(n-m)!} = \frac{n!}{(m-1)!(n-m)!} \\
&= n\frac{(n-1)!}{(m-1)!\{(n-1)-(m-1)\}!} = n\binom{n-1}{m-1}.
\end{aligned}$$
∎

Consider now the special case of a geometric series where $a = 1$:

$$1 + \theta + \theta^2 + \cdots = \sum_{n=0}^{\infty} \theta^n.$$

When $-1 < \theta < 1$, Result A1.4 shows that $S_\infty = 1/(1-\theta)$. This is a special case of the following general result.

Result A1.7

$$\sum_{n=m}^{\infty} \binom{n}{m} \theta^{n-m} = \frac{1}{(1-\theta)^{m+1}} \qquad (m = 0, 1, \ldots).$$

Proof. The result for $m = 0$ has already been established (Result A1.4):

$$\sum_{n=0}^{\infty} \theta^n = \frac{1}{1 - \theta}.$$

Now differentiate m times with respect to θ:

$$\sum_{n=m}^{\infty} n(n-1)\cdots(n+1-m)\theta^{n-m} = \frac{1 \cdot 2 \cdots m}{(1-\theta)^{m+1}}$$

that is,

$$\sum_{n=m}^{\infty} \frac{n!}{(n-m)!}\theta^{n-m} = \frac{m!}{(1-\theta)^{m+1}}$$

so

$$\sum_{n=m}^{\infty} \binom{n}{m} \theta^{n-m} = \frac{1}{(1-\theta)^{m+1}}. \qquad \blacksquare$$

$\binom{n}{m}$ is also called a *binomial coefficient*, as a result of the part it plays in the so-called *binomial theorem*, which may be stated as follows:

Result A1.8

Let n be a non-negative integer, and let x and y be any real numbers. Then

$$(x + y)^n = \sum_{m=0}^{n} \binom{n}{m} x^m y^{n-m}.$$

The proof of this result can be found in many standard mathematics textbooks.

Result A1.9

$$2^n = \sum_{m=0}^{n} \binom{n}{m} \qquad (n = 0, 1, \ldots)$$

Proof. This follows immediately from Result A1.8, setting $x = y = 1$. $\qquad \blacksquare$

Result A1.10

For any non-negative integer, n, and any real value θ such that $0 < \theta < 1$,

$$\sum_{m=0}^{n} \binom{n}{m} \theta^m (1 - \theta)^{n-m}.$$

Proof. This follows immediately from Result A1.8 setting $x = \theta$ and $y = 1 - \theta$.

∎

The given definition of $\binom{n}{m}$ was restricted to non-negative integers m and n, with $m \leq n$. It should be noted that, by convention, $\binom{n}{m} = 0$ whenever n is a non-negative integer and either $m < 0$ or $m > n$. This is justified on the common-sense grounds that, in such circumstances, there are 0 ways in which to choose m out of n distinct objects.

Result A1.11. – The hypergeometric identity

Let N, M and n be non-negative integers, with $M \leq N$ and $n \leq N$. Then

$$\sum_{m=0}^{n} \binom{M}{m}\binom{N-M}{n-m} = \binom{N}{n}.$$

Proof. Apply the binomial theorem (Result A1.8) to obtain

$$(x+y)^N = \sum_{m=0}^{N} \binom{N}{m} x^m y^{N-m}.$$

Apply the same theorem twice more to obtain $(x+y)^M$ and $(x+y)^{N-M}$. Then

$$(x+y)^M (x+y)^{N-M} = \sum_{m=0}^{M} \binom{M}{m} x^m y^{M-m} \sum_{k=0}^{N-M} \binom{N-M}{k} x^k y^{N-M-k}.$$

The required result now follows by equating the coefficients of $(x+y)^n$ in these two series.

∎

A2 Gamma and beta functions

The probability density functions of many important continuous distributions (e.g. gamma, beta, normal and Weibull) are related to two special mathematical functions, known as the gamma and beta functions.

The *gamma function*, $\Gamma(\alpha)$, which is a generalization of the factorial function, is defined by

$$\Gamma(\alpha) = \int_0^\infty x^{\alpha-1} e^{-x} \, dx \qquad (\alpha > 0).$$

Result A2.1

For $\alpha > 1$, $\Gamma(\alpha) = (\alpha - 1)\Gamma(\alpha - 1)$.

Proof. Using integration by parts,

$$\Gamma(\alpha) = \int_0^\infty x^{\alpha-1} e^{-x} \, dx$$

$$= [-x^{\alpha-1} e^{-x}]_0^\infty + (\alpha - 1) \int_0^\infty x^{\alpha-2} e^{-x} \, dx$$

$$= 0 + (\alpha - 1)\Gamma(\alpha - 1).$$

∎

Result A2.2

$$\Gamma(n) = (n - 1)! \qquad (n = 1, 2, \dots).$$

Proof.

$$\Gamma(1) = \int_0^\infty e^{-x} \, dx = [-e^{-x}]_0^\infty = 1.$$

The general result now follows from repeated use of Result A2.1. ∎

Result A2.3

$$\Gamma(\tfrac{1}{2}) = \sqrt{\pi}.$$

Proof. The trick required to evaluate this integral is first to find its square by double integration:

$$\{\Gamma(\tfrac{1}{2})\}^2 = \int_0^\infty \frac{1}{\sqrt{x}} e^{-x} \, dx \int_0^\infty \frac{1}{\sqrt{y}} e^{-y} \, dy = \int_0^\infty \int_0^\infty \frac{1}{\sqrt{xy}} e^{-(x+y)} \, dx \, dy.$$

Now make the following change of variables (to polar coordinates):

$$x = R^2 \cos^2 \theta, \qquad y = R^2 \sin^2 \theta.$$

This has Jacobian $4R^3 \cos\theta \sin\theta$, so that the integral becomes

$$\int_0^{\pi/2} \int_0^\infty 4R \exp(-R^2) \, dR \, d\theta = 2\pi \int_0^\infty R \exp(-R^2) \, dR.$$

Make the further change of variable $u = R^2$, so that $du = 2R\,dR$, to obtain

$$2\pi \int_0^\infty \tfrac{1}{2} \exp(-u) \, du = \pi.$$

Finally, taking the square root, it follows that $\Gamma(\tfrac{1}{2}) = \sqrt{\pi}$. ∎

The beta function is defined by

$$\mathbb{B}(\alpha, \beta) = \int_0^1 x^{\alpha-1}(1 - x)^{\beta-1} \, dx \qquad (\alpha, \beta > 0).$$

Result A2.4

$\mathbb{B}(\beta, \alpha) = \mathbb{B}(\alpha, \beta)$, for any $\alpha, \beta > 0$.

Proof. This result follows immediately using the change of variable, $u = 1 - x$. ∎

Result A2.5

For any $\alpha, \beta > 0$,

$$\mathbb{B}(\alpha, \beta) = \frac{\Gamma(\alpha)\Gamma(\beta)}{\Gamma(\alpha + \beta)}$$

Proof. We have

$$\Gamma(\alpha)\Gamma(\beta) = \int_0^\infty \int_0^\infty x^{\alpha-1} y^{\beta-1} e^{-(x+y)} \, dx \, dy.$$

Make the change of variables

$$u = x + y, \qquad v = x.$$

This transformation has Jacobian 1. Noting the restriction $u > v > 0$, the double integral becomes

$$\int_0^\infty \int_0^u v^{\alpha-1}(u - v)^{\beta-1} e^{-u} \, dv \, du$$

$$= \int_0^\infty \left\{ \int_0^u \left(\frac{v}{u}\right)^{\alpha-1} \left(1 - \frac{v}{u}\right)^{\beta-1} dv \right\} u^{\alpha+\beta-2} e^{-u} \, du$$

$$= \int_0^\infty \left\{ \frac{1}{u} \mathbb{B}(\alpha, \beta) \right\} u^{\alpha+\beta-2} e^{-u} \, du$$

$$= \mathbb{B}(\alpha, \beta) \Gamma(\alpha + \beta).$$

The required result now follows by dividing through by $\Gamma(\alpha + \beta)$. ∎

Result A2.6

$$\mathbb{B}(m, n) = \frac{(m - 1)!(n - 1)!}{(m + n - 1)!} \qquad (m, n = 1, 2, \ldots).$$

Proof. This follows by substituting $\Gamma(m) = (m - 1)!$, $\Gamma(n) = (n - 1)!$ and $\Gamma(m + n) = (m + n - 1)!$ (Result A2.2) into Result A2.5. ∎

A3 Eigenvalues and eigenvectors of matrices

The remaining sections of Appendix A are about matrices whose elements are all real numbers. The set of real matrices with n rows and m columns is denoted M_{nm}. The set of real vectors of length n is denoted either M_{n1} or, more usually, $\mathbb{R}^n$.

Let A be a square matrix with n rows and n columns, i.e. $A \in M_{nn}$. The determinant of $A \in M_{nn}$ is denoted $|A|$ and the inverse of A (if it exists) is denoted A^{-1}. This means that $AA^{-1} = A^{-1}A = I_n$.

The scalar λ is an *eigenvalue* of the square matrix $A \in M_{nn}$ if there is a vector $\mathbf{q} \in \mathbb{R}^n$ such that $\mathbf{q} \neq \mathbf{0}$ and $A\mathbf{q} = \lambda\mathbf{q}$. The vector $\mathbf{q}$ is called an *eigenvector* of A corresponding to the eigenvalue λ. If $\mathbf{q}$ is an eigenvector of A, then so too is $k\mathbf{q}$, for any non-zero k, since $A(k\mathbf{q}) = k(A\mathbf{q}) = k(\lambda\mathbf{q}) = \lambda(k\mathbf{q})$. For this reason we often work with the *normalized eigenvector*, which is that eigenvector $\mathbf{q}$ corresponding to the eigenvalue λ which has the property that $\mathbf{q}^T.\mathbf{q} = 1$.

Result A3.1

The eigenvalues of $A \in M_{nn}$ are the roots of the equation $|A - \lambda I_n| = 0$.

Proof.

$$\lambda \text{ is an eigenvalue of } A \Leftrightarrow \exists \mathbf{q} \neq \mathbf{0} \text{ such that } A\mathbf{q} = \lambda\mathbf{q}$$
$$\Leftrightarrow \exists \mathbf{q} \neq \mathbf{0} \text{ such that } (A - \lambda I_n)\mathbf{q} = \mathbf{0}$$
$$\Leftrightarrow |A - \lambda I_n| = 0. \qquad \blacksquare$$

$|A - \lambda I_n|$ is a polynomial of degree n in λ, and so has exactly n (possibly repeated) roots, i.e. A has exactly n (possibly repeated) eigenvalues. In general, the eigenvalues of a real matrix may be either real or complex numbers.

The following result is stated without proof.

Result A3.2

Let $A \in M_{nn}$ be a real symmetric matrix (i.e. $A = A^T$). Then all n eigenvalues of A are real numbers.

We now deal with symmetric matrices only, and write $A \in S_n$ to mean A is an $n \times n$ symmetric real matrix. It is usual to order the eigenvalues of a symmetric matrix. We write $\lambda_1 \geq \lambda_2 \geq \cdots \geq \lambda_n$ for the n real eigenvalues of $A \in M_{nn}$.

Two vectors, $\mathbf{q}, \mathbf{r} \in \mathbb{R}^n$, are said to be *orthogonal* if $\mathbf{r}^T \cdot \mathbf{q} = 0$ (or, alternatively, $\mathbf{q}^T \cdot \mathbf{r} = 0$). A square matrix $Q \in M_{nn}$ is said to be *orthogonal* if $Q^T Q = I_n = QQ^T$ (i.e. $Q^T = Q^{-1}$).

Result A3.3

Suppose $A \in S_n$, and let λ and μ be any (possibly repeated) eigenvalues of A. It is possible to find eigenvectors of A corresponding to λ and μ such that these eigenvectors are orthogonal.

Proof. The proof is given here for distinct eigenvalues, but may be extended to the case where λ and μ are repetitions of the same (repeated) eigenvalue.

Let $\mathbf{q}$ and $\mathbf{r}$ be eigenvectors of A corresponding to λ and μ, respectively. Then

$$A\mathbf{q} = \lambda\mathbf{q}, \qquad A\mathbf{r} = \mu\mathbf{r}.$$

From the latter result,

$$\mathbf{r}^{\mathrm{T}} A^{\mathrm{T}} = \mu\mathbf{r}^{\mathrm{T}}$$

that is,

$$\mathbf{r}^{\mathrm{T}} A = \mu\mathbf{r}^{\mathrm{T}} \qquad (\text{since } A = A^{\mathrm{T}})$$

so

$$\mathbf{r}^{\mathrm{T}} A\mathbf{q} = \mu\mathbf{r}^{\mathrm{T}}\mathbf{q}$$

$$\lambda\mathbf{r}^{\mathrm{T}}\mathbf{q} = \mu\mathbf{r}^{\mathrm{T}}\mathbf{q} \qquad (\text{since } A\mathbf{q} = \lambda\mathbf{q})$$

$$(\lambda - \mu)\mathbf{r}^{\mathrm{T}}\mathbf{q} = 0$$

$$\mathbf{r}^{\mathrm{T}}\mathbf{q} = 0 \qquad (\text{since } \lambda \neq \mu). \qquad \blacksquare$$

Result A3.4 – The spectral decomposition theorem

Let $A \in S_n$. Then there is an orthogonal matrix $Q \in M_{nn}$ and a diagonal matrix $\Lambda = \mathrm{diag}\,(\lambda_1, \ldots, \lambda_n) \in M_{nn}$ such that $A = Q\Lambda Q^{\mathrm{T}}$.

Proof. Let $\lambda_1 \geq \cdots \geq \lambda_n$ be the eigenvalues of A. Choose normalized eigenvectors $\mathbf{q}_1, \ldots \mathbf{q}_n$ corresponding to $\lambda_1, \ldots, \lambda_n$ respectively such that

$$\mathbf{q}_i^{\mathrm{T}}\mathbf{q}_i = 1 \qquad (i = 1, \ldots, n),$$

$$\mathbf{q}_i^{\mathrm{T}}\mathbf{q}_j = 0 \qquad (i \neq j).$$

Write $Q = [\mathbf{q}_1 | \mathbf{q}_2 | \cdots | \mathbf{q}_n]$. The (i, j)th element of $Q^{\mathrm{T}} Q$ is $\mathbf{q}_i^{\mathrm{T}} \mathbf{q}_j$ and so $Q^{\mathrm{T}} Q = I_n$. So Q is orthogonal and

$$AQ = A[\mathbf{q}_1|\mathbf{q}_2|\cdots|\mathbf{q}_n]$$
$$= [A\mathbf{q}_1|A\mathbf{q}_2|\cdots|A\mathbf{q}_n]$$
$$= [\lambda_1\mathbf{q}_1|\lambda_2\mathbf{q}_2|\cdots|\lambda_n\mathbf{q}_n]$$
$$= [\mathbf{q}_1|\mathbf{q}_2|\cdots|\mathbf{q}_n]\begin{bmatrix} \lambda_1 & 0 & \cdots & 0 \\ 0 & \lambda_2 & \cdots & 0 \\ \vdots & \vdots & & \vdots \\ 0 & 0 & \cdots & \lambda_n \end{bmatrix}$$

that is,

$$AQ = Q\Lambda$$
$$\therefore A = Q\Lambda Q^{\mathrm{T}} \qquad (\text{since } QQ^{\mathrm{T}} = I_n).\qquad\blacksquare$$

Result A3.5

Suppose $A \in S_n$ has eigenvalues $\lambda_1 \geq \cdots \geq \lambda_n$. Then

$$|A| = \prod_{i=1}^{n} \lambda_i.$$

Proof.

$$|A| = |Q\Lambda Q^{\mathrm{T}}| = |Q||\Lambda||Q^{\mathrm{T}}|$$

and

$$QQ^{\mathrm{T}} = I_n \Rightarrow |Q||Q^{\mathrm{T}}| = |I_n| = 1$$

so

$$|A| = |Q||\Lambda||Q^{\mathrm{T}}| = |\Lambda| = \prod_{i=1}^{n} \lambda_i \qquad (\text{since } \Lambda \text{ is a diagonal matrix}).\qquad\blacksquare$$

Result A3.6

Suppose $A \in S_n$ has eigenvalues $\lambda_1 \geq \cdots \geq \lambda_n$. Then

$$\mathrm{tr}(A) = \sum_{i=1}^{n} \lambda_i.$$

Proof. $\mathrm{tr}(A) = \mathrm{tr}(Q\Lambda Q^{\mathrm{T}}) = \mathrm{tr}(\Lambda Q^{\mathrm{T}}Q) = \mathrm{tr}(\Lambda) = \sum_{i=1}^{n} \lambda_i.$ $\qquad\blacksquare$

Result A3.7

Suppose $A \in S_n$ has eigenvalues $\lambda_1 \geq \cdots \geq \lambda_n$. Then the rank of A, $r(A)$, is the number of non-zero λ_i.

Proof. We have

$$r(A) = r(Q\Lambda Q^T) = r(\Lambda) \qquad (\text{since } QQ^T = I_n \Leftrightarrow Q^{-1} = Q^T)$$

and $r(\Lambda)$ is the number of non-zero λ_i. ∎

Result A3.8

Suppose $A \in S_n$ has eigenvalues $\lambda_1 \geq \cdots \geq \lambda_n$. Then A is non-singular if and only if all λ_i are non-zero.

Proof. A is non-singular if and only if $r(A) = n$ if and only if λ_i is non-zero for all i (using Result A3.7). ∎

Result A3.9

Suppose $A \in S_n$ has eigenvalues $\lambda_1 \geq \cdots \geq \lambda_n$. Then, when it exists, A^{-1} has eigenvalues $\lambda_1^{-1}, \ldots, \lambda_n^{-1}$.

Proof. Suppose A is non-singular. Then (by Result A3.8) all the λ_i are non-zero. So,

$$A^{-1} = (Q\Lambda Q^T)^{-1} = (Q^T)^{-1}\Lambda^{-1}Q^{-1} = Q\,\mathrm{diag}(\lambda_1^{-1}, \ldots, \lambda_n^{-1})Q^T.$$

This is clearly the spectral decomposition of A^{-1}, and so A^{-1} has the eigenvalues $\lambda_1^{-1}, \ldots, \lambda_n^{-1}$. ∎

A4 Positive definite matrices

The symmetric matrix $A \in S_n$ is said to be *positive definite* (p.d.) if

$$\mathbf{x}^T A \mathbf{x} > 0 \qquad \text{for all } \mathbf{x} \neq \mathbf{0} \in \mathbb{R}^n,$$

and *positive semi-definite* (p.s.d.) if

$$\mathbf{x}^T A \mathbf{x} \geq 0 \qquad \text{for all } \mathbf{x} \in \mathbb{R}^n.$$

Clearly, a positive definite matrix is also positive semi-definite (but not vice versa).

Result A4.1

If $A \in S_n$ has spectral decomposition $A = Q \wedge Q^T$, then:

(i) A is p.d. if and only if $\lambda_i > 0$ for all i;

(ii) A is p.s.d. if and only if $\lambda_i \geq 0$ for all i.

Proof.   (i) Suppose A is p.d. Then

$$\mathbf{x}^T A \mathbf{x} > 0 \qquad \text{for all } \mathbf{x} \neq \mathbf{0},$$

that is,

$$\mathbf{x}^T (Q \wedge Q^T) \mathbf{x} \geq 0 \qquad \text{for all } \mathbf{x} \neq \mathbf{0}.$$

Put

$$
\mathbf{x} = Q \cdot \begin{bmatrix} 0 \\ \vdots \\ 0 \\ 1 \\ 0 \\ \vdots \\ 0 \end{bmatrix} \leftarrow i\text{th position} \qquad
\mathbf{y} = Q^T \mathbf{x} = \begin{bmatrix} 0 \\ \vdots \\ 0 \\ 1 \\ 0 \\ \vdots \\ 0 \end{bmatrix}
$$

Then

$$\mathbf{x}^T (Q \wedge Q^T) \mathbf{x} = (Q^T \mathbf{x})^T \wedge (Q^T \mathbf{x}) = \mathbf{y}^T \wedge \mathbf{y} = \lambda_i$$

so $\lambda_i > 0$.

To prove the reverse direction, suppose $\lambda_i > 0$ for all i. Then, for any $\mathbf{x} \neq \mathbf{0}$,

$$\mathbf{x}^T A \mathbf{x} = \mathbf{x}^T (Q \wedge Q^T) \mathbf{x} = \mathbf{y}^T \wedge \mathbf{y},$$

where $\mathbf{y} = Q^T \mathbf{x}$. But $\mathbf{y} = Q^T \mathbf{x} \neq \mathbf{0}$ since $\mathbf{x} \neq \mathbf{0}$, so

$$\mathbf{x}^T A \mathbf{x} = \mathbf{y}^T \wedge \mathbf{y} = \sum_{i=1}^{n} \lambda_i y_i^2 \qquad \text{where } \mathbf{y} = [y_1 \ldots y_n]^T$$

$$> 0 \qquad \text{since at least one } y_i > 0 \text{ and } \lambda_i > 0$$

that is, A is positive definite.

(ii) The proof of this is similar to that of (i), and is left as an exercise.   ■

Result A4.2

If $A \in S_n$, then:

(i) A is p.d. if and only if there exists a p.d. matrix $C \in M_{nn}$ such that $A = CC^T$;

(ii) A is p.s.d. if and only if there exists a p.s.d. matrix $C \in M_{nn}$ such that $A = CC^T$.

Proof. (i) Assume A is p.d. Then each $\lambda_i > 0$, and we can write

$$A = Q \cdot \text{diag}(\lambda_1, \ldots, \lambda_n) \cdot Q^T$$

Put

$$C = Q \cdot \text{diag}(\sqrt{\lambda_1}, \ldots, \sqrt{\lambda_n}).$$

Clearly, C is positive definite with eigenvalues $\sqrt{\lambda_1}, \ldots, \sqrt{\lambda_n}$. So

$$CC^T = Q \cdot \text{diag}(\sqrt{\lambda_1}, \ldots, \sqrt{\lambda_n}) \cdot \text{diag}(\sqrt{\lambda_1}, \ldots, \sqrt{\lambda_n}) \cdot Q^T$$

$$= Q \cdot \text{diag}(\lambda_1, \ldots, \lambda_n) \cdot Q^T = A$$

Now suppose there exists some non-singular matrix $C \in M_{nn}$ such that $A = CC^T$. Then

$$\mathbf{x}^T A \mathbf{x} = \mathbf{x}^T C C^T \mathbf{x} = (C^T \mathbf{x})^T (C^T \mathbf{x}) = \sum_{i=1}^{n} y_i^2 \geq 0,$$

where $\mathbf{y} = [y_1 \ldots y_n]^T = C^T \mathbf{x}$. But $\mathbf{x}^T A \mathbf{x} = 0 \Leftrightarrow \mathbf{y} = C^T \mathbf{x} = \mathbf{0} \Leftrightarrow \mathbf{x} = (C^T)^{-1} \mathbf{0} = (C^{-1})^T \mathbf{0} = \mathbf{0}$ that is, A is p.d.

(ii) This is left as an exercise, as it is similar to (i). ∎

It follows that a p.d. matrix is non-singular (since none of its eigenvalues is 0), but a p.s.d. matrix that is *not* p.d. is singular (since at least one eigenvalue is 0).

Result A4.3

$A \in S_n$ is p.d. if and only if A^{-1} is p.d.

Proof. Suppose that $A \in S_n$ is p.d. Then A has eigenvalues $\lambda_1 \geq \cdots \geq \lambda_n > 0$. This means that A^{-1} exists and has eigenvalues $\lambda_n^{-1} \geq \cdots \geq \lambda_1^{-1} > 0$ (by Result A3.9), so A^{-1} is p.d.

To prove the reverse direction, use the proof for the forward direction with A^{-1} and A interchanged. ∎

A5 Symmetric idempotent matrices

The matrix $P \in S_n$ is symmetric idempotent if

$$P = P^T \text{(symmetric)} \quad \text{and} \quad P = P^2 \text{ (idempotent)}.$$

It follows that $P = P^2 = P^T P$, etc.

Result A5.1

The eigenvalues of a symmetric idempotent matrix, P, are all either 0 or 1.

Proof. Suppose λ is an eigenvalue of P with corresponding eigenvector $\mathbf{q}(\neq \mathbf{0}) \in \mathbb{R}^n$. Then,

$$P\mathbf{q} = \lambda \mathbf{q} \Rightarrow \mathbf{q}^T P\mathbf{q} = \lambda \mathbf{q}^T \mathbf{q}$$
$$\Rightarrow \mathbf{q}^T (P^T P)\mathbf{q} = \lambda \mathbf{q}^T \mathbf{q} \qquad \text{(since } P = P^T P)$$
$$\Rightarrow (P\mathbf{q})^T (P\mathbf{q}) = \lambda \mathbf{q}^T \mathbf{q}$$
$$\Rightarrow \lambda^2 \mathbf{q}^T \mathbf{q} = \lambda \mathbf{q}^T \mathbf{q}$$
$$\Rightarrow \lambda^2 = \lambda \qquad \text{(since } \mathbf{q} \neq \mathbf{0} \Rightarrow \mathbf{q}^T \mathbf{q} \neq 0)$$
$$\Rightarrow \lambda = 0 \text{ or } 1. \qquad \blacksquare$$

Result A5.2

Let $P \in S_n$ be a symmetric idempotent matrix. Then:

(i) $r(P)$ is the number of eigenvalues of P that are equal to 1;

(ii) $r(P) = \text{tr}(P)$.

Proof. $r(P) = r$ if and only if P has r non-zero eigenvalues if and only if P has r eigenvalues equal to 1. But

$$\text{tr}(P) = \sum_{i=1}^{n} \lambda_i$$

is the sum of r ones and $n - r$ zeros, which equals $r = r(P)$. $\qquad \blacksquare$

Result A5.3

If P is symmetric idempotent of rank r, then $I_n - P$ is symmetric idempotent of rank $n - r$.

Proof.

$$(I_n - P)^T = I_n^T - P^T = I_n - P \text{ (symmetric)}$$
$$(I_n - P)^2 = I_n - P - P + P^2 = I_n - P - P + P = I_n - P \text{ (idempotent)}.$$

Now suppose λ is an eigenvalue of P with corresponding eigenvector $\mathbf{q}$. Then,

$$P\mathbf{q} = \lambda \mathbf{q} \Rightarrow \mathbf{q} - P\mathbf{q} = \mathbf{q} - \lambda \mathbf{q}$$

that is,

$$(I_n - P)\mathbf{q} = (1 - \lambda)\mathbf{q},$$

thus

$1 - \lambda$ is an eigenvalue of $I_n - P$, with corresponding eigenvector $\mathbf{q}$. $\qquad \blacksquare$

By Result A5.2, P has r eigenvalues equal to 1 and $n - r$ eigenvalues equal to 0. So, $I_n - P$ has r eigenvalues equal to 0 and $n - r$ eigenvalues equal to 1. So, by Result A5.2 again, $I_n - P$ has rank $n - r$.

A6 Miscellaneous matrix results

Result A6.1

Suppose that the square matrix $A \in M_{pp}$ can be partitioned as:

$$A = \begin{bmatrix} A_{11} & A_{12} \\ A_{21} & A_{22} \end{bmatrix},$$

where A_{11} is a $(p - q) \times (p - q)$ square sub-matrix, and A_{22} is a $q \times q$ square sub-matrix (with $q < p$). Suppose also that A_{22} is non-singular.

Define $A_{11.2} = A_{11} - A_{12} A_{22}^{-1} A_{21}$. Then:

(a) $|A| = |A_{11.2}||A_{22}|$;

(b) $A^{-1} = \begin{bmatrix} A_{11.2}^{-1} & -A_{11.2}^{-1} A_{12} A_{22}^{-1} \\ -A_{22}^{-1} A_{21} A_{11.2}^{-1} & -A_{22}^{-1} + A_{22}^{-1} A_{21} A_{11.2}^{-1} A_{12} A_{22}^{-1} \end{bmatrix}$.

Proof. (a) Let

$$B = \begin{bmatrix} I_{p-q} & -A_{12} A_{22}^{-1} \\ 0 & I_q \end{bmatrix}, \qquad C = \begin{bmatrix} I_{p-q} & 0 \\ -A_{22}^{-1} A_{21} & I_q \end{bmatrix}.$$

B is upper triangular, so $|B|$ is the product of the diagonal elements of B, which is 1. Similarly, $|C| = 1$. Now

$$BAC = \begin{bmatrix} A_{11} - A_{12} A_{22}^{-1} A_{21} & 0 \\ 0 & A_{22} \end{bmatrix} = \begin{bmatrix} A_{11.2} & 0 \\ 0 & A_{22} \end{bmatrix}$$

So,

$$|BAC| = |A_{11.2}||A_{22}|$$

that is,

$$|B||A||C| = |A_{11.2}||A_{22}|$$

and

$$|A| = |A_{11.2}|A_{22}|.$$

(b)

$$A = B^{-1} \begin{bmatrix} A_{11.2} & 0 \\ 0 & A_{22} \end{bmatrix} C^{-1}$$

so

$$A^{-1} = C \begin{bmatrix} A_{11.2}^{-1} & 0 \\ 0 & A_{22}^{-1} \end{bmatrix} B$$

$$= \begin{bmatrix} A_{11.2}^{-1} & -A_{11.2}^{-1} A_{12} A_{22}^{-1} \\ -A_{22}^{-1} A_{21} A_{11.2}^{-1} & -A_{22}^{-1} + -A_{22}^{-1} A_{21} A_{11.2}^{-1} A_{12} A_{22}^{-1} \end{bmatrix},$$

as required. ∎

Writing $A_{22.1} = A_{22} - A_{21} A_{11}^{-1} A_{12}$, then it can also be shown that

$$|A| = |A_{22.1}||A_{11}|, \text{etc.}$$

Result A6.2

Suppose that $A \in M_{pp}$ is a non-singular matrix, and let $\mathbf{b}, \mathbf{c} \in \mathbb{R}^p$. Then:

$$(a) \quad |A + \mathbf{b}\mathbf{c}^T| = |A| \{1 + \mathbf{c}^T A^{-1} \mathbf{b}\};$$

$$(b) \quad (A + \mathbf{b}\mathbf{c}^T)^{-1} = A^{-1} - \frac{A^{-1}\mathbf{b}\mathbf{c}^T A^{-1}}{\{1 + \mathbf{c}^T A^{-1}\mathbf{b}\}}.$$

Proof. (a) Consider the matrix

$$M = \begin{bmatrix} A & \mathbf{b} \\ -\mathbf{c}^T & 1 \end{bmatrix}.$$

This is partitioned as in Result A6.1, with

$$M_{11.2} = A - \mathbf{b}.1.(-\mathbf{c}^T) = A + \mathbf{b}\mathbf{c}^T$$

so that

$$|M| = |A + \mathbf{b}\mathbf{c}^T| |1| = |A + \mathbf{b}\mathbf{c}^T|,$$

and

$$M_{22.1} = 1 - (-\mathbf{c}^T)A^{-1}\mathbf{b} = 1 + \mathbf{c}^T A^{-1}\mathbf{b}$$

so that

$$|M| = |A| \{1 + \mathbf{c}^T A^{-1}\mathbf{b}\},$$

therefore

$$|A + \mathbf{b}\mathbf{c}^T| = |A| \{1 + \mathbf{c}^T A^{-1}\mathbf{b}\}.$$

(b) Multiply the right-hand side by $A + \mathbf{b}\mathbf{c}^T$ to obtain I_p. ∎

Appendix B
Some statistical tables

Table B1   *Random digits*

82092	89853	20648	60629	11185	44294	38557	06308	12903	03029
70946	96296	50420	77533	06859	99655	06749	12692	10532	99949
28455	52588	56050	12275	56116	61938	33008	10544	34690	21284
60610	51219	53652	02142	06775	32699	83184	69304	53919	21937
78929	00253	48933	67964	76074	21871	09877	68782	92389	01819
51481	01823	38402	59776	24796	44407	34354	88177	72198	67630
69010	10848	67086	79390	61731	29729	20074	13438	32389	06330
02777	72892	99140	43850	13230	17429	57275	10511	58488	69732
00816	01877	12991	70151	28233	09656	95894	32546	46642	36701
93130	45757	34900	01884	18442	44811	42478	33271	63208	83957
79242	69386	05764	92193	82768	32935	51443	12274	55193	18577
22951	90841	43406	07705	61787	13025	75575	40164	72948	80041
77364	81311	50136	54462	67556	06993	63042	81869	50441	55636
34534	40503	21982	69971	45864	11873	67291	47531	00905	08911
13379	05111	20014	84346	44580	32059	27225	87160	76184	08044
74926	04681	62483	48669	15703	25711	35443	33256	06015	04997
14777	12203	54396	34248	61523	74536	61235	05197	29505	69247
32754	47323	28673	23638	37956	07476	80333	24391	92179	08161
11342	56938	92973	41907	49926	39067	28379	96243	14202	08574
43745	46267	93669	71671	48234	63898	80765	01494	83175	96369
73979	55518	89194	09147	40950	71843	47643	91451	90347	01925
67648	19830	17349	99504	97505	00733	06848	74872	05459	91269
19402	91188	41538	22043	36259	58747	57598	02193	61452	08048
81530	10886	41029	59269	84900	68435	38115	24466	26546	80466
51111	45997	34006	77182	23223	93093	25181	64750	83065	40183
95187	28876	91314	07561	90168	13948	71312	45364	59864	94136
71412	09979	12925	97711	38595	43138	65785	44586	85616	41339
97785	13617	54387	53163	34873	10965	83062	45082	49746	40864
30595	27205	26094	93966	46117	01487	71924	86004	13288	68132
89656	61513	40404	87428	14962	61733	06710	32079	14730	86695

(continued)

Table B1 *Continued*

18588	35075	84589	26348	29993	42828	34651	51190	91962	59912
92300	78467	82915	60024	27929	03637	10159	86561	78531	09047
54584	27778	31089	34645	05726	12017	45382	03410	46932	69029
62168	52393	10851	07308	76376	20442	76334	66267	96373	64891
10278	05841	12146	83791	39599	34872	80266	89952	87188	66837
95291	73575	98013	10452	57182	88867	69919	98867	20456	46352
72630	40156	83451	51897	19630	42580	90587	22381	22325	62566
70495	01996	93383	71609	44911	91329	18460	83430	58431	74900
05539	42521	82159	51191	67415	51387	96654	91453	00429	45314
20707	36509	25826	82680	26241	05441	90916	97563	79943	04731
79130	07915	36115	22699	88285	01049	78193	28053	93200	45917
16995	73889	28370	08615	27367	40849	30745	75418	73493	73068
87337	22251	26575	00006	64809	46217	47970	28826	99064	86166
93277	53141	10147	66410	26392	74993	52165	08720	52816	60238
31011	06265	83969	60718	05396	70417	61316	05012	93124	49821
11237	45261	92524	54799	90394	00050	56399	44180	80066	87586
18930	63812	97175	68534	14484	28282	40957	51061	03149	66549
38160	74648	12846	92718	07553	41930	26804	71900	28825	32410
76908	84563	99743	15241	13731	05405	99053	83895	46377	54536
15868	91949	36317	91605	54609	32327	54436	70465	36845	78881
05015	21043	76775	09674	81177	48682	22991	53412	32791	41938
75736	75842	21525	26576	00053	38984	77755	93437	62384	44316
38671	30807	76481	74974	21450	57720	66250	33618	49282	09515
98708	44171	26721	50698	61348	65509	97762	68433	92918	83302
51809	60418	63792	54791	08331	80119	15816	19391	29570	72070
81174	47574	00378	78701	55810	34279	28498	39244	57150	54075
05113	99846	20821	29365	39652	02952	27915	06762	56014	39080
99034	52396	09282	16568	33431	20866	26555	40325	23414	79580
56755	12585	19860	74944	13037	82966	25482	41143	08763	18911
48041	23492	69071	79444	45508	25637	46150	24602	18537	27743
54301	07963	47001	89820	07923	50032	61858	99210	26894	63000
62674	55127	40793	34955	97504	63087	36169	38297	37247	26907
68581	60033	67316	93062	97037	72167	36061	71430	70844	68583
37824	67750	34918	20997	22496	79142	94855	32661	33295	34477
36006	65820	31552	59479	44741	36259	72867	69645	30825	31396
05732	04111	79100	36112	53197	77939	37049	96411	12607	88110
08267	15119	99739	17662	98627	11417	22040	70381	16568	85355
63884	52727	16187	67407	18817	12927	92201	02171	00928	85493
24686	36119	76894	24056	34760	27168	29991	72915	35737	93556
65442	62856	96527	80507	62970	58498	67397	65736	37709	14642

Table B2 *The standard normal cumulative distribution function*

When Z is an $N(0,1)$ random variable, this table gives $\Phi(z) = P(Z \le z)$ for values of z from 0.00 to 3.69 in steps of 0.01.

z	0	1	2	3	4	5	6	7	8	9
0.0	0.5000	0.5040	0.5080	0.5120	0.5160	0.5199	0.5239	0.5279	0.5319	0.5359
0.1	0.5398	0.5438	0.5478	0.5517	0.5557	0.5596	0.5636	0.5675	0.5714	0.5753
0.2	0.5793	0.5832	0.5871	0.5910	0.5948	0.5987	0.6026	0.6064	0.6103	0.6141
0.3	0.6179	0.6217	0.6255	0.6293	0.6331	0.6368	0.6406	0.6443	0.6480	0.6517
0.4	0.6554	0.6591	0.6628	0.6664	0.6700	0.6736	0.6772	0.6808	0.6844	0.6879
0.5	0.6915	0.6950	0.6985	0.7019	0.7054	0.7088	0.7123	0.7157	0.7190	0.7224
0.6	0.7257	0.7291	0.7324	0.7357	0.7389	0.7422	0.7454	0.7486	0.7517	0.7549
0.7	0.7580	0.7611	0.7642	0.7673	0.7704	0.7734	0.7764	0.7794	0.7823	0.7852
0.8	0.7881	0.7910	0.7939	0.7967	0.7995	0.8023	0.8051	0.8078	0.8106	0.8133
0.9	0.8159	0.8186	0.8212	0.8238	0.8264	0.8289	0.8315	0.8340	0.8365	0.8389
1.0	0.8413	0.8438	0.8461	0.8485	0.8508	0.8531	0.8554	0.8577	0.8599	0.8621
1.1	0.8643	0.8665	0.8686	0.8708	0.8729	0.8749	0.8770	0.8790	0.8810	0.8830
1.2	0.8849	0.8869	0.8888	0.8907	0.8925	0.8944	0.8962	0.8980	0.8997	0.9015
1.3	0.9032	0.9049	0.9066	0.9082	0.9099	0.9115	0.9131	0.9147	0.9162	0.9177
1.4	0.9192	0.9207	0.9222	0.9236	0.9251	0.9265	0.9279	0.9292	0.9306	0.9319
1.5	0.9332	0.9345	0.9357	0.9370	0.9382	0.9394	0.9406	0.9418	0.9429	0.9441
1.6	0.9452	0.9463	0.9474	0.9484	0.9495	0.9505	0.9515	0.9525	0.9535	0.9545
1.7	0.9554	0.9564	0.9573	0.9582	0.9591	0.9599	0.9608	0.9616	0.9625	0.9633
1.8	0.9641	0.9649	0.9656	0.9664	0.9671	0.9678	0.9686	0.9693	0.9699	0.9706
1.9	0.9713	0.9719	0.9726	0.9732	0.9738	0.9744	0.9750	0.9756	0.9761	0.9767
2.0	0.9772	0.9778	0.9783	0.9788	0.9793	0.9798	0.9803	0.9808	0.9812	0.9817
2.1	0.9821	0.9826	0.9830	0.9834	0.9838	0.9842	0.9846	0.9850	0.9854	0.9857
2.2	0.9861	0.9864	0.9868	0.9871	0.9875	0.9878	0.9881	0.9884	0.9887	0.9890
2.3	0.9893	0.9896	0.9898	0.9901	0.9904	0.9906	0.9909	0.9911	0.9913	0.9916
2.4	0.9918	0.9920	0.9922	0.9925	0.9927	0.9929	0.9931	0.9932	0.9934	0.9936
2.5	0.9938	0.9940	0.9941	0.9943	0.9945	0.9946	0.9948	0.9949	0.9951	0.9952
2.6	0.9953	0.9955	0.9956	0.9957	0.9959	0.9960	0.9961	0.9962	0.9963	0.9964
2.7	0.9965	0.9966	0.9967	0.9968	0.9969	0.9970	0.9971	0.9972	0.9973	0.9974
2.8	0.9974	0.9975	0.9976	0.9977	0.9977	0.9978	0.9979	0.9979	0.9980	0.9981
2.9	0.9981	0.9982	0.9983	0.9983	0.9984	0.9984	0.9985	0.9985	0.9986	0.9986
3.0	0.9987	0.9987	0.9987	0.9988	0.9988	0.9989	0.9989	0.9989	0.9990	0.9990
3.1	0.9990	0.9991	0.9991	0.9991	0.9992	0.9992	0.9992	0.9992	0.9993	0.9993
3.2	0.9993	0.9993	0.9994	0.9994	0.9994	0.9994	0.9994	0.9995	0.9995	0.9995
3.3	0.9995	0.9995	0.9996	0.9996	0.9996	0.9996	0.9996	0.9996	0.9996	0.9997
3.4	0.9997	0.9997	0.9997	0.9997	0.9997	0.9997	0.9997	0.9997	0.9997	0.9998
3.5	0.9998	0.9998	0.9998	0.9998	0.9998	0.9998	0.9998	0.9998	0.9998	0.9998
3.6	0.9998	0.9998	0.9999	0.9999	0.9999	0.9999	0.9999	0.9999	0.9999	0.9999

For $z \ge 3.70$, use $\Phi(z) = 1.0000$. When $z < 0.00$, $\Phi(z)$ can be found from this table using the relationship $\Phi(-z) = 1 - \Phi(z)$.

Hints and solutions for selected exercises

Section 1.1

1 (a) $S_a = \{0, 1, 2, \ldots\}$ – countably infinite – or $S'_a = \{0, 1, \ldots, N\}$ – finite

(b) $S_b = \{0, 1, \ldots, 30\}$ – finite

(c) $S_c = \{x : x \geq 0\}$ – uncountably infinite

(d) $S_d = \{(x_1, x_2, \ldots, x_{10}) : x_1 \geq 0, x_2 \geq 0, \ldots, x_{10} \geq 0\}$ – uncountably infinite

(e) $S_e = \{(x_1, x_2, \ldots, x_{20}) : \text{each } x_i = A, B, C, D\}$ – finite

(f) $S_f = \{(x, y) : x > 0, y > 0\}$ – uncountably infinite

2 (a) $E = \{0, 1, \ldots, 9, 999\}$

(b) $E = \{0, 1, 2, 3, 4, 5, 6\}$

(c) $E = \{x : x > c\}$

(d) $E = \{(x_1, x_2, \ldots, x_{10}) : x_1 \leq c, x_2 \leq c, \ldots, x_{10} \leq c\}$

(e) $E = \{(D, D, D, x_4, \ldots, x_{20}) : \text{each } x_i = A, B, C, D\}$

(f) $E = \{(x, y) : x > 0, y > 20x^2\}$

3 $\emptyset, \{s_1\}, \{s_2\}, \{s_3\}, \{s_4\}, \{s_1, s_2\}, \{s_1, s_3\}, \{s_1, s_4\}, \{s_2, s_3\}, \{s_2, s_4\}, \{s_3, s_4\},$

$\{s_1, s_2, s_3\}, \{s_1, s_2, s_4\}, \{s_1, s_3, s_4\}, \{s_2, s_3, s_4\}, \{s_1, s_2, s_3, s_4\}$

$0, \frac{1}{4}, \frac{1}{4}, \frac{1}{4}, \frac{1}{4}, \frac{1}{2}, \frac{1}{2}, \frac{1}{2}, \frac{1}{2}, \frac{1}{2}, \frac{1}{2}, \frac{3}{4}, \frac{3}{4}, \frac{3}{4}, \frac{3}{4}, 1$

4 Let E be any event. Then s_i $(i = 1, 2, \ldots, k)$ is either in E or not – 2 possibilities for each outcome – so, 2^k possibilities in total.

Section 1.2

3 Note that E is the disjoint union of the m events $\{s_{i_1}\}, \{s_{i_2}\}, \ldots, \{s_{i_m}\}$.

4 Note that F is the disjoint union of E and $E'F$.

6 $P(EF') = P(E) - P(EF)$

$P(EF' \cup E'F) = P(EF') + P(E'F) = P(E) + P(F) - 2P(EF)$

$P(0 \text{ event occurs}) = 1 - P(E \cup F) = 1 - P(E) - P(F) + P(EF),$

$P(1 \text{ event occurs}) = P(E) + P(F) - 2P(EF),$

$P(2 \text{ events occur}) = P(EF)$

10 Note that $E_1 \cup E_2 \cup \cdots \cup E_n$ is the disjoint union of $E_1, E'_1 E_2, E'_1 E'_2 E_3, \ldots,$ $E'_1 E'_2 \ldots E_n$.

11 This follows from Exercise 10 since $E'_1 E_2 \subseteq E_2 \Rightarrow P(E'_1 E_2) \leq P(E_2)$, etc.

Section 1.3

2 Either $P(E) = 0$ or $P(F) = 0$.

3 (a) $P(AS) = P(A) = P(A).1 = P(A)P(S)$, so A and S are independent.

(b) $P(A\emptyset) = P(\emptyset) = 0 = P(A)P(\emptyset)$, so A and $\emptyset$ are independent.

(c) $P(AA) = P(A)P(A) \Rightarrow P(A) = 0$ or 1.

5 (b), (c) A counterexample uses $S = \{1, 2, 3, 4, 5, 6\}, E = \{1, 3, 5\}, F = \{6\}$.

7 Note that $FG \subseteq F$, so $P(F) \geq P(FG) > 0$. Hence, $P(E|F) = P(E)$ by property of independence already discussed. $P(E|G) = P(E)$ similarly. Also,

$$P(E|FG) = \frac{P(EFG)}{P(FG)} = \frac{P(E)P(F)P(G)}{P(F)P(G)} = P(E).$$

8 $P(A) = P(B) = P(C) = \frac{1}{2}, P(AB) = P(AC) = P(BC) = \frac{1}{4}, P(ABC) = 0$

12 Note that $(E_1 \cup E_2 \cup \cdots \cup E_n)' = E'_1 E'_2 \ldots E'_n$.

13 Note the connection with Exercise 12.

14 $1 - (1 - \theta)^2$

15 θ^2

16 (a) $1 - (1 - \theta)^n$

(b) θ^n

For $0 < \theta < 1, (1 - \theta)^n < 1 - \theta < 1 - \theta^n$, so $\theta^n < 1 - (1 - \theta)^n$.

19 $P(\text{red ball}) = \frac{1}{k} \sum_{j=1}^{k} \frac{x_j}{n_j}, P(\text{bag } i \mid \text{red ball}) = \frac{x_i/n_i}{\sum_{j=1}^{k} x_j/n_j}$

20 (a) $\dfrac{n_i}{N}$

(b) $\dfrac{in_i}{Nm}$

(c) $\dfrac{in_i}{Nm} > \dfrac{n_i}{N} \Leftrightarrow \dfrac{i}{m} > 1 \Leftrightarrow i > m$

21 $P(\text{no claim}) = 0.825, P(\text{good risk} \mid \text{no claim}) = 0.230$

Section 2.1

1 $P(D=-1)=\frac{3}{16}, P(D=0)=\frac{10}{16}, P(D=1)=\frac{3}{16}$

3 $P(X=x)$ increases for as long as $(n+1)\theta > x$. When $(n+1)\theta$ is an integer, then $P(X=x)$ takes its maximum value at both $x=(n+1)\theta-1$ and at $x=(n+1)\theta$. Otherwise, its maximum occurs at that x which is the largest integer less than or equal to $(n+1)\theta$.

4 This distribution is positively skewed, but becomes more symmetric as θ increases.

5 $P(X=x)$ increases with x as long as $\theta/x > 1$. When θ is a positive integer, then $P(X=x)$ takes its maximum value at both $x=\theta-1$ and $x=\theta$. Otherwise, its maximum occurs at that x which is the largest integer that is not greater than θ.

6 $f(x) = \begin{cases} 2x/\theta_1\theta_2 & (0 < x \le \theta_1) \\ 2(\theta_2-x)/\theta_2(\theta_2-\theta_1) & (\theta_1 < x < \theta_2) \end{cases}$

$P(X \le \theta_1) = \theta_1/\theta_2$

Section 2.2

1 $\mathbb{E}(X)=\frac{1}{\theta}, \quad \mathbb{E}(X(X-1))=\frac{2(1-\theta)}{\theta^2}, \quad \text{var}(X)=\frac{(1-\theta)}{\theta^2}$

2 $\mathbb{E}(X)=n\theta, \quad \mathbb{E}(X(X-1))=n(n-1)\theta^2, \quad \text{var}(X)=n\theta(1-\theta)$

3 $\mathbb{E}(X)=n\frac{M}{N}, \quad \text{var}(X)=n\frac{M}{N}\left(1-\frac{M}{N}\right)$

4 $\mathbb{E}(X)=\theta, \quad \mathbb{E}(X(X-1))=\theta^2, \quad \text{var}(X)=\theta$

5 $\mathbb{E}(X)=\frac{k}{\theta}, \quad \mathbb{E}(X(X+1))=\frac{k(k+1)}{\theta^2}, \quad \text{var}(X)=\frac{k(1-\theta)}{\theta^2}$

6 Use the fact that $\dfrac{1}{x(x+1)}=\dfrac{1}{x}-\dfrac{1}{(x+1)}$.

8 Integrate separately from $-\infty$ to 0 and from 0 to ∞.

9 $\mathbb{E}(U)=\dfrac{m}{m+n}, \quad \mathbb{E}(U^2)=\dfrac{m(m+1)}{(m+n)(m+n+1)},$

$\text{var}(U)=\dfrac{mn}{(m+n)^2(m+n+1)}$

10 Integrate separately from $-\infty$ to c and from c to ∞.

12 (b) Write the integrand as $F(x)=1\cdot F(x)$ and integrate by parts.

13 $\mu=\mathbb{E}(X)=\dfrac{1}{2}(a+b), \quad \sigma^2=\text{var}(X)=\dfrac{1}{4}(b-a)^2, \quad a=\mu-\sigma, \quad b=\mu+\sigma$

14 $\mu = \mathbb{E}(X) = b + \theta(a - b)$, $\sigma^2 = \text{var}(X) = (b - a)^2\theta(1 - \theta)$

16 (c) Put $\alpha = 1$ for $\text{Ex}(\theta)$ distribution, which has expected value $1/\theta$ and variance $1/\theta^2$.

17 (b) Put $\alpha = 1$ for $\text{Ex}(\theta)$ distribution.

21 $\mathbb{E}(X) = \int_0^\infty xf(x)\,dx \geq \int_k^\infty xf(x)\,dx \geq k\int_k^\infty f(x)\,dx = kP\,(X > k)$

Section 2.3

1 $M_X(t) = \exp\{(e^t - 1)\theta\}$

$M_X'(t) = \exp\{(e^t - 1)\theta\} \cdot e^t\theta \therefore \mathbb{E}(X) = \theta$

$M_X''(t) = \exp\{(e^t - 1)\theta\} \cdot (e^t\theta)^2 + \exp\{(e^t - 1)\theta\} \cdot (e^t\theta)$

$\therefore \mathbb{E}(X^2) = \theta^2 + \theta$, $\text{var}(X) = \theta$

2 $M_Y(t) = M_X(kt) = \dfrac{\theta/k}{\theta/k - t}$ $\therefore Y \sim \text{Ex}(\theta/k)$

3 $X \sim N(50, 100)$, so $P(X \geq 63) = 1 - \Phi(1.3) = 0.0968$

4 Use Bayes' theorem. $P(\text{Type G} \mid \text{diagnosed as Type A}) = 0.107$
$P(\text{correctly diagnosed}) = 0.6\Phi(4c - 20.8) + 0.4\Phi(28.5 - 5c)$
Differentiate this expression with respect to c and discover a local maximum at $c = 5.5$.

6 (a) $\mathbb{E}(Y^k) = \mathbb{E}(e^{Xk}) = M_X(k)$

(b) $\mathbb{E}(Y) = \exp\left\{\mu + \dfrac{1}{2}\sigma^2\right\}$, $\mathbb{E}(Y^2) = \exp\{2\mu + 2\sigma^2\}$

$\text{var}(Y) = \exp\{2\mu + \sigma^2\} \cdot [\exp(\sigma^2) - 1]$

Section 3.1

1 $R_Y = \{-n, -n+3, \ldots, 2n-3, 2n\}$, $p_Y(y) = \dbinom{n}{(y+n)/3}\theta^{(y+n)/3}(1 - \theta)^{(2n-y)/3}$

$Y = 3X - n$, where $X \sim \text{Bi}(n, \theta)$, so $\mathbb{E}(Y) = n(3\theta - 1)$, $\text{var}(Y) = 9n\theta(1 - \theta)$

3 $Y = k + 10X$, where $X \sim \text{Bi}(k, 1 - (1 - \theta)^{10})$

$\mathbb{E}(Y) = k + 10k(1 - (1 - \theta)^{10})$, $\text{var}(Y) = 100k(1 - (1 - \theta)^{10})(1 - \theta)^{10}$

5 $Y = \dfrac{X - a}{b - a} \sim \text{Un}(0, 1)$, $\mathbb{E}(X) = \dfrac{1}{2}(a + b)$, $\text{var}(X) = \dfrac{1}{12}(b - a)^2$

6 $F_D(d) = \sin^{-1}\left(\dfrac{dg}{v^2}\right)$, $f_D(d) = \dfrac{g}{v^2} \cdot \dfrac{1}{\sqrt{1 - d^2g^2/v^4}}$

8 The χ_1^2 distribution has expected value 1 and variance 2.

10 (a) $U \sim \text{We}(\alpha, \theta/k^{\alpha})$

(b) $V \sim \text{We}(\alpha/k, \theta)$, so $X^{\alpha} \sim \text{Ex}(\theta)$

13 Notice that $h(x) = 1/x$ is strictly decreasing on R_X and use Proposition 3.1.

14 $f_Y(y) = \dfrac{1}{2\theta\sqrt{y}}(0 < y < \theta^2)$, $\mathbb{E}(Y) = \dfrac{1}{3}\theta^2$, $\text{var}(Y) = \dfrac{4}{45}\theta^4$

15 $f(x) = \begin{cases} \dfrac{1}{\theta} + \dfrac{1}{\theta^2}x & (-\theta < x < 0) \\[2mm] \dfrac{1}{\theta} - \dfrac{1}{\theta^2}x & (0 \le x < \theta) \end{cases}$

$f_Y(y) = \dfrac{1}{\theta\sqrt{y}} - \dfrac{1}{\theta^2} \qquad (0 \le y < \theta^2)$

16 Use Proposition 3.2.

17 Define $Y = X^2$. Then Proposition 3.2 gives $f_Y(y) = \dfrac{1}{\sqrt{y}}\text{e}^{-\sqrt{y}} \quad (0 < y)$.

Section 3.2

2 $f_U(u) = \dfrac{\theta}{u^{1+\theta}}(1 < u)$, $\mathbb{E}(U) = \dfrac{\theta}{\theta-1}$, $\mathbb{E}(U^2) = \dfrac{\theta}{\theta-2}$,

$\text{var}(U) = \dfrac{\theta}{(\theta-2)(\theta-1)^2}$

Approximate values are $\mathbb{E}(U) \approx \text{e}^{1/\theta}\dfrac{1+2\theta^2}{2\theta^2}$, $\text{var}(U) \approx \text{e}^{2/\theta}\dfrac{1}{\theta^2}$.

3 $f_Y(y) = \dfrac{1}{y} \quad (1 < y < \text{e})$, $\mathbb{E}(Y) = \text{e} - 1$, $\mathbb{E}(Y^2) = \dfrac{1}{2}(\text{e}^2 - 1)$,

$\text{var}(Y) = -\dfrac{1}{2}\text{e}^2 + 2\text{e} - \dfrac{3}{2}$

Approximate values are $\mathbb{E}(Y) \approx \dfrac{25}{24}\text{e}^{1/2}$, $\text{var}(Y) \approx \dfrac{1}{12}\text{e}^1$.

4 $\mathbb{E}(Y) = \dfrac{\theta}{\alpha-1}$, $\text{var}(Y) = \dfrac{\theta^2}{(\alpha-1)^2(\alpha-2)}$

Approximate values are $\mathbb{E}(Y) = \dfrac{\theta}{\alpha^2}(\alpha+1)$, $\text{var}(Y) = \dfrac{1}{\alpha^2}$.

Section 3.3

4 (a) The pseudo-random variates are: $1, 1, 1, 2, 3$.

(b) The pseudo-random variates are: $1, 2, 2, 3, 4$.

(c) The pseudo-random variates are: $1, 1, 1, 1, 2$.

(d) The pseudo-random variates are: $1, 2, 2, 4, 6$.

5 (a) The pseudo-random variates are: $0.0851, 0.4659, 0.6247, 1.5721, 3.1725$.

(b) The pseudo-random variates are: $0.0284, 0.1553, 0.2082, 0.5240, 1.0575$.

(c) The pseudo-random variates are: $0.0018, 0.0543, 0.0976, 0.6179, 2.5162$.

6 If u is a random real number between 0 and 1, a pseudo-random variate from this distribution is $x = \left\{ -\dfrac{1}{\theta} \log (1 - u) \right\}^{1/\alpha}$.

9 $f_Y(y) = \dfrac{y^{n-1} e^{-y^2/4}}{2^{n-1} \Gamma(n/2)} \qquad (0 < y)$

10 One possible method is to simulate a value, x, from the $N(0, 1)$ distribution, using the distribution function method, and then set $y = |x|$.

Section 4.1

1 (a) $P(X = -1) = \frac{1}{2}(1 - \theta), P(X = 0) = \theta, P(X = 1) = \frac{1}{2}(1 - \theta)$

$P(Y = -1) = \frac{1}{2}(1 - \theta), P(Y = 0) = \theta, P(Y = 1) = \frac{1}{2}(1 - \theta)$

(b) $\mathbb{E}(X) = \mathbb{E}(Y) = 0, \mathbb{E}(X^2) = \mathbb{E}(Y^2) = 1 - \theta, \text{var}(X) = \text{var}(Y) = 1 - \theta$

(c) $\mathbb{E}(XY) = 0, \text{cov}(X, Y) = 0$

2 (a) $P(X = 0) = \frac{1}{2}(1 + \theta), P(X = k) = \frac{1}{2}(1 - \theta)$

$P(Y = 0) = \frac{1}{2}(1 + \theta), P(Y = k) = \frac{1}{2}(1 - \theta)$

(b) $\mathbb{E}(X) = \mathbb{E}(Y) = \frac{1}{2}k(1 - \theta), \text{var}(X) = \text{var}(Y) = \frac{1}{4}k^2(1 - \theta)(1 + \theta)$

(c) $\mathbb{E}(XY) = 0, \text{cov}(X, Y) = -\frac{1}{4}k^2(1 - \theta)^2$

3 (a) $\mathbb{E}(X) = 0.9, \mathbb{E}(Y) = 1.8$

(b) $\mathbb{E}(X + Y) = 2.7$

(c) $\mathbb{E}(X - Y) = -0.9$

4 (a) $k = 1$

(b) (i) 0.375, (ii) 0.375, (iii) 0.125

(c) $f_X(x) = x + \frac{1}{2} \quad (0 < x < 1), f_Y(y) = y + \frac{1}{2} \quad (0 < y < 1)$

(d) $\mathbb{E}(X) = \mathbb{E}(Y) = \dfrac{7}{12}, \text{var}(X) = \text{var}(Y) = \dfrac{11}{144}$

(e) $\mathbb{E}(XY) = \dfrac{1}{3}, \text{cov}(X, Y) = -\dfrac{1}{144}$

5 (b) $P(X > 1, Y > 1) = \frac{5}{2}e^{-2}$

(c) $f_X(x) = \frac{1}{4}(2 + x^2)e^{-x}$ $\qquad (x > 0)$

Section 4.2

1 (a)

x	-1	0	1
$p(x \mid Y = -1)$	$\frac{1}{2}$	0	$\frac{1}{2}$
$p(x \mid Y = 0)$	0	1	0
$p(x \mid Y = 1)$	$\frac{1}{2}$	0	$\frac{1}{2}$

(b) $\operatorname{cov}(X, Y) = 0$

(c) $P = Q = \frac{1}{2}\theta(1 - \theta) + \frac{1}{16}(1 - \theta)^2$, so $\gamma = 0$

2 (a)

y	0	k
$p(y \mid X = 0)$	$\dfrac{2\theta}{1 + \theta}$	$\dfrac{1 - \theta}{1 + \theta}$
$p(y \mid X = k)$	1	0

(b) $\operatorname{cov}(X, Y) = -\frac{1}{4}k^2(1 - \theta)(1 + \theta)$

(c) $P = 0, Q = \frac{1}{4}(1 - \theta)^2$, so $\gamma = -1$

4 (a) (i) $\frac{1}{4}$, (ii) $\frac{1}{2}$, (iii) $\frac{1}{4}$

(b) (i) $\frac{3}{4}$, (ii) $\frac{1}{4}$

5 (a) 9/16

(b) The missing probabilities are $\frac{3}{4} - \theta, \frac{3}{4} - \theta, \theta - \frac{1}{2}$

6 Notice that $I_{AB} = I_A I_B$.

7 $OR = 5, P = 0.0090, Q = 0.0018, \gamma = 2/3$

8 $\operatorname{cov}(X, Y) = -1/144, \rho_{XY} = -1/11$

10 $f_X(x) = 6x(1 - x)$ $\quad (0 < x < 1)$, i.e. $X \sim \operatorname{Be}(2, 2)$ so $\mathbb{E}(X) = 0.5, \operatorname{var}(X) = 0.05$

$f_Y(y) = 3y^2$ $\quad (0 < y < 1)$, i.e. $Y \sim \operatorname{Be}(3, 1)$ so $\mathbb{E}(Y) = 0.75, \operatorname{var}(Y) = 0.0375$

$\mathbb{E}(XY) = 0.4, \operatorname{cov}(X, Y) = 0.025, \rho_{XY} = 0.5774$

12 $\rho_{XY} = -0.5$

15 (a) $P(U = u) = (e^{-\theta})^u(1 - e^{-\theta})$, $\qquad u = 0, 1, \ldots$

(b) $f(v|u) = \dfrac{\theta e^{-\theta v}}{1 - e^{-\theta}}$ $(0 \le v < 1)$

20 The correlation is undefined since $\text{var}(X) = 0$.

Section 4.3

1 (a) $p_{XY}(x, y) = \dfrac{\theta^x (1 - \theta)^{y-x} e^{-\lambda} \lambda^y}{(y - x)! x!}$

(b) $X \sim \text{Po}(\lambda\theta)$

3 (a) R_{XY} consists of the following line segments: $(0, 0)$ to $(0, 1)$; $(1, 0)$ to $(1, 1)$; ... $(k, 0)$ to $(k, 1)$.

4 (a) $p_Y(y) = (1 - \phi)^y \phi, \quad y = 0, 1, 2, \ldots$

$$\mathbb{E}(Y) = \frac{1 - \phi}{\phi}, \quad \text{var}(Y) = \frac{1 - \phi}{\phi^2}$$

(b) $p_Y(y) = \left(\dfrac{1}{1 + \theta}\right)^y \dfrac{\theta}{1 + \theta}, \quad y = 0, 1, 2, \ldots$

$$\mathbb{E}(Y) = \frac{1}{\theta}, \quad \text{var}(Y) = \frac{1 + \theta}{\theta^2}$$

6 (a) $(0.97)^{10} = 0.7374$

(b) 0.4557

(c) 0.0883

(d) $10 \times 0.03 = 0.3$

7 (a) 0.2252

(b) 0.3585

(c) 0.0686

(d) $20 \times 0.05 = 1$

8 (a) 0.7061

(b) 0.5445

(c) 0.3778

(d) The conditional distribution is $\text{Bi}(27, 4/94)$. The probability is 0.6938.

9 $P = \dfrac{120}{64^2}, \quad Q = \dfrac{812}{64^2}, \quad \gamma = -\dfrac{692}{932} = -0.7425$

10 (a) $X \sim \text{Hyp}(n, N, M_1), \mathbb{E}(X) = n \cdot \dfrac{M_1}{N}, \text{var}(X) = n \cdot \dfrac{M_1}{N} \cdot \dfrac{N - M_1}{N} \cdot \dfrac{N - n}{N - 1}$

$$Y \sim \text{Hyp}(n, N, M_1), \mathbb{E}(Y) = n.\frac{M_2}{N}, \text{var}(Y) = n.\frac{M_2}{N}.\frac{N-M_2}{N}.\frac{N-n}{N-1}$$

(b) The conditional distribution is $\text{Hyp}(n - y, N - M_2, M_1)$

$$\mathbb{E}(X \mid y) = (n - y).\frac{M_1}{N - M_2}$$

$$\text{var}(X \mid y) = (n - y).\frac{M_1}{N - M_2}.\frac{N - M_1 - M_2}{N - M_2}.\frac{(N - M_2) - (n - y)}{(N - M_2) - 1}$$

12 (a) $p_{XY}(x, y) = \begin{cases} (1 - \theta_1 - \theta_2)^{x-1}\theta_1(1 - \theta_2)^{y-x-1}\theta_2 & (x < y) \\ (1 - \theta_1 - \theta_2)^{y-1}\theta_2(1 - \theta_1)^{x-y-1}\theta_1 & (x > y) \end{cases}$

(b) $X \sim \text{Ge}(\theta_1), Y \sim \text{Ge}(\theta_2)$

Section 4.4

3 (b) $\mathbb{E}(Y) = c + d\alpha, \text{var}(Y) = \sigma^2 + d^2\alpha^2\left(\frac{4}{\pi} - 1\right)$

Using Proposition 4.10(b), $\rho_{XY}^2 = \dfrac{\text{var}(Y) - \sigma^2}{\text{var}(Y)} = \dfrac{d^2\alpha^2(4/\pi - 1)}{\sigma^2 + d^2\alpha^2(4/\pi - 1)}$

4 (a) $X \sim \text{Ex}(1)$ so $\mathbb{E}(X) = 1, \text{var}(X) = 1$

$Y \sim \text{Ga}(2, 1)$ so $\mathbb{E}(Y) = 2, \text{var}(X) = 2$

(b) $f_{Y|X}(y|x) = e^{-(y-x)}, y > x$, so $\mathbb{E}(Y \mid x) = 1 + x$

(c) $f_{X|Y}(x|y) = \dfrac{1}{y}, 0 < x < y$, so $\mathbb{E}(X \mid y) = \frac{1}{2}y$

(d) $\rho_{XY} = \dfrac{1}{\sqrt{2}}$

(e) $\alpha = 1, \beta = 1, \dfrac{\rho_{XY}\sqrt{\text{var}(Y)}}{\sqrt{\text{var}(X)}} = 1 = \beta, \mathbb{E}(Y) - \beta\mathbb{E}(X) = 1 = \alpha$

6 (a) $f_X(x) = 2e^{-x}(1 - e^{-x}), 0 < x$, so $\mathbb{E}(X) = \frac{3}{2}, \text{var}(X) = \frac{5}{4}$

$f_Y(y) = 2e^{-2y}, 0 < y$, so $\mathbb{E}(Y) = \frac{1}{2}, \text{var}(Y) = \frac{1}{4}$

(b) $\mathbb{E}(XY) = 1, \text{cov}(X, Y) = \dfrac{1}{4}, \rho_{XY} = \dfrac{1}{\sqrt{5}}$

7 $X \sim \text{Be}(4, 1), \mathbb{E}(X) = \dfrac{4}{5}, \text{var}(X) = \dfrac{4}{150}$

$Y \sim \text{Be}(3, 2), \mathbb{E}(Y) = \dfrac{3}{5}, \text{var}(X) = \dfrac{6}{150}$

$E(X \mid y) = \frac{1}{2}(1+y)$, so using Proposition 4.10(a), $\dfrac{\rho_{XY}\sqrt{\text{var}(X)}}{\sqrt{\text{var}(Y)}} = \dfrac{1}{2}$

$E(Y \mid x) = \frac{3}{4}x$, so using Proposition 4.10(a), $\dfrac{\rho_{XY}\sqrt{\text{var}(Y)}}{\sqrt{\text{var}(X)}} = \dfrac{3}{4}$

Taking these two results together, $\rho_{XY}^2 = \dfrac{1}{2} \times \dfrac{3}{4} = \dfrac{3}{8}$, i.e. $\rho_{XY} = \sqrt{\dfrac{3}{8}}$

8 $E(X \mid y) = \frac{1}{2}(1-y)$, so using Proposition 4.10(a), $\dfrac{\rho_{XY}\sqrt{\text{var}(X)}}{\sqrt{\text{var}(Y)}} = -\dfrac{1}{2}$

$E(Y \mid x) = \frac{1}{2}(1-x)$, so using Proposition 4.10(a), $\dfrac{\rho_{XY}\sqrt{\text{var}(Y)}}{\sqrt{\text{var}(X)}} = -\dfrac{1}{2}$

Taking these two results together, $\rho_{XY} = -\dfrac{1}{2}$.

Section 5.1

1 (b) $[X_1 \ X_2]^T \sim \text{Tri}\left(4, \frac{1}{4}, \frac{1}{4}\right)$, $\text{cov}(X_1, X_2) = -\frac{1}{4}$

(c) $X_1 \sim \text{Bi}\left(4, \frac{1}{4}\right)$, $E(X_1) = 1$, $\text{var}(X_1) = \frac{3}{4}$

(d) $E(\mathbf{X}) = \begin{bmatrix} 1 \\ 1 \\ 1 \end{bmatrix}$, $\text{cov}(\mathbf{X}) = \begin{bmatrix} \frac{3}{4} & -\frac{1}{4} & -\frac{1}{4} \\ -\frac{1}{4} & \frac{3}{4} & -\frac{1}{4} \\ -\frac{1}{4} & -\frac{1}{4} & \frac{3}{4} \end{bmatrix}$

2 (a) $\dfrac{100!}{(20!)^5}\left(\dfrac{1}{5}\right)^{100}$

(b) $\dfrac{100!}{(20!)^3 40!}\left(\dfrac{1}{5}\right)^{60}\left(\dfrac{2}{5}\right)^{40}$

3 (a) $k = \dfrac{1}{3}$, $f_{\mathbf{X}}(\mathbf{x}) = \dfrac{2}{3}(x_1 + x_2 + x_3)$

(b) $f_{12}(x_1, x_2) = \dfrac{2}{3}\left(x_1 + x_2 + \dfrac{1}{2}\right)$, $f_1(x_1) = \dfrac{2}{3}(x_1 + 1)$

(c) $\dfrac{1}{6}, \dfrac{1}{2}, \dfrac{7}{12}$

4 (a) $k = 48$

(b) $f_{12}(x_1, x_2) = 24x_1 x_2(1 - x_2^2)$, $f_1(x_1) = 6x_1(1 - x_1^2)$

Section 5.2

2 Given $X_3 = x_3, [X_1 \ X_2]^\mathsf{T} \sim \text{Tri}\left(3 - x_3, \frac{1}{3}, \frac{1}{3}\right), \text{cov}(X_1, X_2) = -\dfrac{(3 - x_3)}{9}$,

$\rho_{12|3} = -\frac{1}{2}\gamma = -\dfrac{53}{83}$ $(x_3 = 0)$, $-\dfrac{11}{15}$ $(x_3 = 1)$, -1 $(x_3 = 2)$

3 (a) $f_{12}(x_1, x_2) = 24x_1x_2^3$, $f_{13}(x_1, x_3) = 24x_1x_3(x_1^2 - x_3^2)$, $f_{23}(x_2, x_3) = 24x_2x_3$

$(1 - x_2^2)f_1(x_1) = 6x_1^5$, $f_2(x_2) = 12x_2^3(1 - x_2^2)$, $f_3(x_3) = 6x_3(1 - x_1^2)^2$

(b) $f_{1|2}(x_1 \mid x_2) = \dfrac{2x_1}{(1 - x_2^2)}$, $f_{3|2}(x_3 \mid x_2) = \dfrac{2x_3}{x_2^2}$, $f_{13|2}(x_1, x_3 \mid x_2) = \dfrac{4x_1x_3}{x_2^2(1 - x_2^2)}$

4 $f_{3|12}(x_3 \mid x_1, x_2) = \dfrac{x_1 + x_2 + x_3}{x_1 + x_2 + \frac{1}{2}}$, $\mathbb{E}(X_3 \mid x_1, x_2) = \dfrac{3x_1 + 3x_2 + 2}{6x_1 + 6x_2 + 3}$

5 $f_{123}(x_1, x_2, x_3) = f_{12}(x_1, x_2)f_3(x_3)$ $\Rightarrow$ $f_{13}(x_1, x_3) = f_3(x_3)\int f_{12}(x_1, x_2)\, dx_2 = f_3(x_3)f_1(x_1)$

6 $f_{123}(x_1, x_2, x_3) = f_{12|3}(x_1, x_2 \mid x_3)f_3(x_3) = g(x_1, x_2)f_3(x_3)$

$\therefore f_{12}(x_1, x_2) = g(x_1, x_2)\int f_3(x_3)\, dx_3 = g(x_1, x_2)$

$\therefore f_{123}(x_1, x_2, x_3) = f_{12}(x_1, x_2)f_3(x_3)$

7 $f_{12|3}(x_1, x_2 \mid x_3) = f_{1|3}(x_1 \mid x_3)f_{2|3}(x_2 \mid x_3)$

$\therefore f_{1|23}(x_1 \mid x_2, x_3) = \dfrac{f_{123}(x_1, x_2, x_3)}{f_{23}(x_2, x_3)} = \dfrac{f_{12|3}(x_1, x_2 \mid x_3)f_3(x_3)}{f_{23}(x_2, x_3)}$

$\qquad = \dfrac{f_{1|3}(x_1 \mid x_3)f_{2|3}(x_2 \mid x_3)f_3(x_3)}{f_{23}(x_2, x_3)} = f_{1|3}(x_1 \mid x_3)$

Section 5.3

3 Exercise 5 on Section 5.2 showed that X_1 is independent of X_3 and that X_2 is independent of X_3. Therefore, $\rho_{13} = \rho_{23} = 0$ and $\rho_{12 \cdot 3} = \rho_{12}$. Proving that $f_{12|3}(x_1, x_2 \mid x_3) = f_{12}(x_1, x_2)$ shows that $\rho_{12|3} = \rho_{12}$.

Section 5.4

1 $M_Y(t) = \left(\dfrac{1}{1 - t}\right)^m$, $Y \sim \text{Ga}(m, 1)$

2 (b) $M_X(t) = \dfrac{1}{(1 - t_3)(1 - t_2 - t_3)(1 - t_1 - t_2 - t_3)}$

$\qquad M_1(t) = \dfrac{1}{(1 - t)}$, $X_1 \sim \text{Ex}(1)$

$$M_2(t) = \frac{1}{(1-t)^2}, \quad X_2 \sim \text{Ga}(2,1)$$

$$M_3(t) = \frac{1}{(1-t)^3}, \quad X_3 \sim \text{Ga}(3,1)$$

(c) $M_Y(t) = \dfrac{1}{(1-t_3)(1-t_2)(1-t_1)}$

$$\mathbb{E}(\mathbf{X}) = \begin{bmatrix} 1 \\ 2 \\ 3 \end{bmatrix}, \quad \text{cov}(\mathbf{X}) = \begin{bmatrix} 1 & 1 & 1 \\ 1 & 2 & 2 \\ 1 & 2 & 3 \end{bmatrix}$$

Section 6.1

1 (a) θ, $\dfrac{\theta}{n^2} \sum_{i=1}^{n} \dfrac{1}{t_i}$

(b) θ, $\dfrac{\theta}{\sum_{i=1}^{n} t_i}$

2 $S = $ number of trials until kth success $\sim \text{NeBi}(k, \theta)$.

3 Use Proposition 4.10.

4 $P(X_1 = x_1, X_2 = x_2, \cdots, X_n = x_n | S =) =$

$$\frac{P(X_1 = x_1, X_2 = x_2, \cdots, X_n = s - x_1 - \cdots - x_{n-1})}{P(S = s)}$$

Use this to show that the conditional distribution is the multinomial distribution with s trials and jth category probability $\theta_j / (\theta_1 + \theta_2 + \cdots \theta_n)$.

5 $\dfrac{\lambda}{\theta}$, $\dfrac{(2-\theta)\lambda}{\theta^2}$

6 The exact values are $P\{|X| \ge k\} = \begin{cases} 1, & 0 < k \le 1 \\ 0, & 1 < k \end{cases}$

Chebyshev's inequality gives $P\{|X| \ge k\} \le \dfrac{1}{k^2}$

8 (a) 0.0340

(b) $\dfrac{4}{1 - 0.0340} = 4.1408$

10 (b) $\text{Ga}(n, \frac{1}{2})$ or χ^2_{2n}

11 $\sum_{i=1}^{n} X_i \sim \text{Ga}(\alpha_1 + \cdots + \alpha_n, \theta)$, $\bar{X} \sim \text{Ga}(\alpha_1 + \cdots + \alpha_n, n\theta)$

12 $M_S(t) = \left((1-\theta)e^{-t} + \theta e^t\right)^n = e^{-nt}\left((1-\theta) + \theta e^{2t}\right)^n$

Section 6.2

1 The limiting distribution is $N(0, 1)$.

2 (a) 0.0336

  (b) 0.4286

  (c) 0.0668

  (d) 0.0146

3 $100 + 2.33\sqrt{1000} = 173.68$ hours

4 $n \geq 660.49$, so the smallest integer answer is $n = 661$.

6 $n \leq 214.7$, so the largest integer answer is $n = 214$.

7 S can be decomposed as the sum of k independent $\text{Ge}(\theta)$ random variables. The central limit theorem gives $S \approx N(k/\theta, k(1 - \theta)/\theta^2)$.

9 (a) $\chi_n^2 \approx N(n, 2n)$

  (b) $Z_i^2 \sim \chi_1^2$ $\therefore \sum_{i=1}^{n} Z_i^2 \sim \chi_n^2 \approx N(n, 2n)$

Section 7.1

3 (a) $p_{12}(x_1, x_2) = \dfrac{1}{100}$, $x_1 = 0, 1, \ldots, 9$; $x_2 = 0, 1, \ldots, 9$

  (b) $P(S = s) = \begin{cases} \dfrac{s+1}{100}, & s = 0, 1, \ldots, 9 \\ \dfrac{19 - s}{100}, & s = 10, 11, \ldots, 18 \end{cases}$

  (d) $\mathbb{E}(S) = 9$, $\text{var}(S) = \dfrac{99}{6}$, $\mathbb{E}(M) = \dfrac{9}{2}$, $\text{var}(M) = \dfrac{99}{24}$

Section 7.2

1 $R_Y = (0, \theta)$, $F_Y(y) = 1 - \left(1 - \dfrac{y}{\theta}\right)^2$, $f_Y(y) = \dfrac{2}{\theta}\left(1 - \dfrac{y}{\theta}\right)$,

  $\mathbb{E}(Y) = \dfrac{\theta}{3}$, $\text{var}(Y) = \dfrac{\theta^2}{18}$

2 $T \sim \text{Ga}(2, \theta)$

3 (a) $R_Y = (0, \infty)$, $F_Y(y) = (1 - e^{-\theta y})^2$, $f_Y(y) = 2\theta e^{-\theta y}(1 - e^{-\theta y})$

  (b) $f_Y(y) = \dfrac{\theta}{2}e^{-\theta|y|}$

7 Could use $Y_1 = \dfrac{\theta X_1}{\phi X_2}, Y_2 = \phi X_2$

Notice that $\theta X_1, \phi X_2 \sim \mathrm{Ex}(1) \approx \chi_2^2$.

8 $Y_1 \sim \mathrm{Ga}(\alpha_1 + \alpha_2 + \alpha_3, \theta), \quad Y_2 \sim \mathrm{Be}(\alpha_2 + \alpha_3, \alpha_1), \quad Y_3 \sim \mathrm{Be}(\alpha_3, \alpha_2)$

9 (b) $Y \sim \mathrm{We}(\alpha, n\theta)$

Section 7.3

1 $f_U(u) = \begin{cases} 2^{m-1} m u^{m-1}, & 0 < u \le 0.5 \\ 2^{m-1} m (1-u)^{m-1}, & 0.5 < u < 1 \end{cases}$

2 (a) $g_\mathbf{Y}(\mathbf{y}) = m!, \quad 0 < y_1 < \cdots < y_m < 1$

(b) $Y_j \sim \mathrm{Be}(j, m-j+1), \quad \mathbb{E}(Y_j) = \dfrac{j}{m+1}, \quad \mathrm{var}(Y_j) = \dfrac{j(m-j+1)}{(m+1)^2(m+2)}$

(d) $R \sim \mathrm{Be}(m-1, 2), \quad \mathbb{E}(R) = \dfrac{m-1}{m+1} \to 1, \quad \mathrm{var}(R) = \dfrac{2(m-1)}{(m+1)^2(m+2)} \to 0$

3 $P(Y_m < y_m) = \{F(y_m)\}^m$

$P(y_1 < Y_1, Y_m < y_m) = \{F(y_m) - F(y_1)\}^m$

$G_{1m}(y_1, y_m) = \{F(y_m)\}^m - \{F(y_m) - F(y_1)\}^m$

$g_{1m}(y_1, y_m) = m(m-1)\{F(y_m) - F(y_1)\}^{m-2} f(y_1) f(y_m)$

4 $\mathbb{E}(Y_j) = \dfrac{1}{\theta m} + \dfrac{1}{\theta(m-1)} + \cdots + \dfrac{1}{\theta(m+1-j)}$

5 (a) $g_2(y_2) = m(m-1)\theta\, e^{-(m-1)\theta y_2}(1 - e^{-\theta y_2})$

$G_2(y_2) = (m-1)\, e^{-m\theta y_2} - m\, e^{-(m-1)\theta y_2} + 1$

(b) $R_{\mathrm{TMR}}(t) = m\, e^{-(m-1)\theta t} - (m-1)\, e^{-m\theta t}, \quad t > 0$

$R_{\mathrm{component}}(t) = e^{-\theta t}, \quad t > 0$

Section 7.4

1 $a = 0, b = 1, c = \frac{3}{2}$, so $P(\text{rejection}) = \frac{1}{3}$

Section 8.1

2 (b) Part (a) showed that $[Y_1 \ Y_2 \ \cdots \ Y_{n-1}]^\mathrm{T}$ is independent of Y_n.

So, any function of $[Y_1 \ Y_2 \ \cdots \ Y_{n-1}]^\mathrm{T}$ is independent of any function of Y_n.

In particular, then, S^2 is independent of $\bar{X}$.

3 (b) The rows of the covariance matrix all add to the same value, so the matrix is not of full column rank. The reason is that $\mathbf{Y}$ is of higher dimension than $\mathbf{X}$ itself.

Section 8.2

1 (a) $N(4 + 3x_1 + 6x_2, 4.8)$, $R^2_{3(12)} = 1 - \dfrac{4.8}{9} = 0.467$

(b) $N\left(\begin{bmatrix} 6 + 6x_1 \\ 4 + 3x \end{bmatrix}, \begin{bmatrix} 5 & 4 \\ 4 & 8 \end{bmatrix}\right)$, $\rho_{23 \cdot 1} = \dfrac{4}{\sqrt{40}}$

3 (a) $\mathbb{E}(X_3 | x_1, x_2) = \mu_3 + \rho_{13}\sigma_1\sigma_3(x_1 - \mu_1) + \rho_{23}\sigma_2\sigma_3(x_2 - \mu_2)$

$\text{var}(X_3 | x_1, x_2) = \sigma_3^2(1 - \rho_{13}^2 - \rho_{23}^2)$

(b) $\mathbb{E}(X_3 | x_1) = \mu_3 + \rho_{13}\sigma_1\sigma_3(x_1 - \mu_1)$

$\text{var}(X_3 | x_1) = \sigma_3^2(1 - \rho_{13}^2)$

4 $\mathbb{E}(X_3 | x_1, x_2) = \mu_3 + \rho_{13}\sigma_1\sigma_3(x_1 - \mu_1)$

$\text{var}(X_3 | x_1, x_2) = \sigma_3^2 \left(\dfrac{1 - \rho_{13}^2 - \rho_{23}^2}{1 - \rho_{12}^2} \right)$

Section 9.1

1 (a) $\mathbb{E}(X_i) = \dfrac{M}{N}$, $\mathbb{E}(X_i^2) = \dfrac{M}{N}$, $\text{var}(X_i) = \dfrac{M}{N}\left(1 - \dfrac{M}{N}\right)$

$\mathbb{E}(X_i X_j) = \dfrac{M(M-1)}{N(N-1)}$, $\text{cov}(X_i, X_j) = -\dfrac{M(N-M)}{N^2(N-1)}$

(b) $S \sim \text{Hyp}(n, N, M)$

2 (c) $\dfrac{(\overline{X} - \overline{Y}) - (\mu_1 - \mu_2)}{\sqrt{S_p^2(1/n + 1/m)}} \sim t_{n+m-2}$

Section 9.2

1 Use the fact that $\sqrt{\dfrac{n}{n+1}}(\mathbf{X}_i - \overline{\mathbf{X}}) \sim N_p(0, \Sigma)$ independently of $S \sim W_p(n-1, \ \Sigma)$.

2 Use the fact that $\sqrt{\dfrac{n-1}{n}}(\mathbf{X}_i - \overline{\mathbf{X}}_{-i}) \sim N_p(0, \Sigma)$ independently of $S_{-i} \sim W_p(n-2, \Sigma)$.

References

Copas, J. B. and Heydari, F. (1997) Estimating the risk of re-offending by using exponential mixture models, *Journal of the Royal Statistical Society, Series A*, **160**, 237–52.

Deadman, D. and MacDonald, Z. (2004) Offenders as victims of crime? An investigation into the relationship between criminal behaviour and victimization, *Journal of the Royal Statistical Society, Series A*, **167**, 53–67.

Feller, W. (1968) *An Introduction to Probability Theory and Its Applications, Volume 1*, third edition, Wiley, New York.

Goodman, L. A. and Kruskal, W. H. (1954) Measures of association for cross classifications. *Journal of the American Statistical Association*, **49**, 732–64.

Holmes, P. (1994) Computer generated thinking, in D. Green, *Teaching Statistics at Its Best*, Teaching Statistics Trust, University of Sheffield.

Kaminsky, F. C. and Kirchhoff, R. H. (1988) Bivariate probability models for the description of average wind speed at two heights, *Solar Energy*, **40**, 49–56.

McColl, J. H. (1995) *Probability*, Butterworth Heinemann, London.

Matthews, R. (1996) Why is weather forecasting still under a cloud?, *Mathematics Today*, November/December, 168–70.

Shiryaev, A. N. (1995) *Probability*, second edition, Springer, New York.

Strachan, D. P., Butland, B. K. and Anderson, H. R. (1996) Incidence and prognosis of asthma and wheezing illness from early childhood to age 33 in a national British cohort, *British Medical Journal*, **312**, 1195–99.

Stuart, A. and Ord, K. (1994) *Kendall's Advanced Theory of Statistics, Volume 1, Distribution Theory*, sixth edition, Edward Arnold, London.

Stuart, A. and Ord, K. (1996) *Kendall's Advanced Theory of Statistics, Volume 2A, Classical Inference and Relationship*, sixth edition, Edward Arnold, London.

Index